Hybrid Micromachining and Microfabrication Technologies

Scrivener Publishing
100 Cummings Center, Suite 541J
Beverly, MA 01915-6106

Innovations in Materials and Manufacturing

Series Editor: Prasenjit Chatterjee

Scope: "Innovations in Materials and Manufacturing" series addresses recent developments and research issues related to the productive use of materials and manufacturing processes, which necessitates the development of newer materials and manufacturing processes in order to produce high-quality products at lower prices in less time. This series aims to provide a scientific platform for researchers, practitioners, professionals, and academics to discuss the most recent technological developments in metals, polymers, ceramics, composites, biomaterials, nanomaterials, special materials, metals, microforming, powder metallurgy, ceramics processing, non-traditional machining, high speed machining, micro and nanomachining, and laser processing. Tribological analysis, friction behavior, modeling, and optimization techniques in materials, machining, and manufacturing are also covered in this series.

Submission to the series:
Dr. Prasenjit Chatterjee, Department of Mechanical Engineering, MCKV Institute of Engineering, Howrah - 711204, West Bengal, India
E-Mail: dr.prasenjitchatterjee6@gmail.com

Publishers at Scrivener
Martin Scrivener (martin@scrivenerpublishing.com)
Phillip Carmical (pcarmical@scrivenerpublishing.com)

Hybrid Micromachining and Microfabrication Technologies

Principles, Varieties and Applications

Edited by

Sandip Kunar

Department of Mechanical Engineering, Aditya Engineering College, A.P., India

Golam Kibria

Department of Mechanical Engineering, Aliah University, Kolkata, India

Prasenjit Chatterjee

Department of Mechanical Engineering, MCKV Institute of Engineering, Howrah, India

and

Asma Perveen

Department of Mechanical & Aerospace Engineering, School of Engineering & Digital Sciences, Nazarbayev University, Republic of Kazakhstan

Scrivener Publishing

WILEY

This edition first published 2023 by John Wiley & Sons, Inc., 111 River Street, Hoboken, NJ 07030, USA and Scrivener Publishing LLC, 100 Cummings Center, Suite 541J, Beverly, MA 01915, USA

For more information about Scrivener publications please visit www.scrivenerpublishing.com.

Wiley Global Headquarters
111 River Street, Hoboken, NJ 07030, USA

For details of our global editorial offices, customer services, and more information about Wiley products visit us at www.wiley.com.

Library of Congress Cataloging-in-Publication Data

ISBN 978-1-394-17447-8

Cover image: Pixabay.Com
Cover design by Russell Richardson

Set in size of 11pt and Minion Pro by Manila Typesetting Company, Makati, Philippines

Printed in the USA

10 9 8 7 6 5 4 3 2 1

The Editors would like to dedicate this book to their parents, life partners, children, students, scholars, friends and colleagues.

Contents

Preface

The advancement and subsequent widespread usage of hybrid micromachining and microfabrication technologies have been propelled by the significant need for superior quality in manufactured goods, particularly those of micro-level size and constructed from advanced materials. Hybrid micromachining and microfabrication technologies are variations of advanced machining processes, which combine two or more advanced manufacturing processes with a traditional metalworking procedure (often some types of machining process) to produce outputs that are not attainable with the constituent processes acting alone.

To improve the machining rate and/or surface quality, hybrid micromachining, and microfabrication technologies may also use a fluid, abrasive medium, and an additional external energy source. Due to better machining and surface quality, hybrid micromachining and microfabrication technologies are gradually making their way into the majority of production across the globe. Hybrid micromachining and microfabrication technologies have proven to have a considerable advantage over traditional applications. The main benefits are their capacity for machining exceptionally advanced materials, as well as the production of increased surface integrity characteristics.

This book primarily aims to offer a broader perspective on some of the most significant hybrid micromachining and microfabrication technologies utilized for those goals, and to act as a source for understanding the fundamental principles of various methods. As an opening stage into the world of hybrid micromachining and microfabrication technologies, this book intends to satisfy the requirements of academicians, professionals, and researchers in the domains of mechanistic and manufacturing engineering.

Chapter 1 introduces unique categorization and analyses of the previous and current exploration and functions of the hybrid micromachining and microfabrication procedures, and it emphasizes the influence on performance characteristics.

Chapter 2 demonstrates a brief review of the work done on ECDM in various fields, including history, design, workpiece, electrolyte, tool electrode, and input and output parameters used in ECDM.

Chapter 3 provides an overview of laser-assisted micro milling and hard-to-machine materials such as steel, Ti alloy, Ni alloy, cementite carbides, and other ceramics that are used in laser-assisted micro milling.

Chapter 4 describes how various processing properties are affected by machining factors including electrode feed rate, operating voltage, and ultrasonic vibration amplitude, and the outcomes of experimentation demonstrates that the periodic pressure difference for the electrolyte is produced by the vibrating electrode array supported by ultrasonics.

Chapter 5 discusses the reduction of tool wear and overcut without the addition of any external force, and how the concentration of NaCl electrolyte played a vital role in improving the performance features of micro-electrochemical machining of SS 204.

Chapter 6 explains the laser-aided ECDM for glass micro grooving, which unites ECDM and laser micromachining to address these issues. Single and hybrid processing techniques are examined for their morphological characteristics, and the outcomes demonstrate that ECDM produced tubular protrusions at the microgrooves' base.

Chapter 7 explains various laser-based hybridized shaping processes with their applicability for processing and shaping materials at the micron/submicron level.

Chapter 8 discusses the mechanisms for material removal and process energy distribution in hybrid laser-assisted jet electrochemical micromachining.

Chapter 9 demonstrates the fabrication of microstructure with a high aspect ratio on brittle materials using ultrasonic vibration-assisted micro wire WEDM.

Chapter 10 reveals the influence of input factors, i.e., the concentration of electrolyte, voltage, the rotation speed on average Material Removal Rate (MRR), and average hole diameter, as well as the average machined depth on drilling on soda-lime glass.

Chapter 11 introduces a miniature electromagnetic induction type generator by combining the Micro Electromechanical Systems (MEMS) process, a microfabrication technology, and the multilayer ceramic technology for miniatured electronic components fabrication.

Chapter 12 provides a detailed analytical overview of various hybrid-micromachining process bibliometric surveys, classification and machining mechanisms, and the effective utilization of process parameters with particular emphasis on the hybrid-micromachining (nano, macro and micro-level) domain.

Chapter 13 describes the basic ideas behind the Spark Assisted Chemical Engraving technique. Many processes, including electrochemical discharge-aided melting and vaporization, etching at high temperatures, differential thermal expansion of components, random thermal stress cycles, and thermo-mechanical shocks caused by expanding gases and electrolyte movement, which contribute to this micromachining mode, are discussed.

A fundamental overview of the processes in issue, equipment specifics, operating principles, relevant process parameters, and significant applications are all included in various chapters' material. Appropriate references to ongoing and past research as it relates to the described methods are provided throughout. This book will give a thorough understanding of numerous advanced hybrid micromachining and microfabrication techniques, as well as their future directions, providing researchers and engineers who work in hybrid micromachining with a necessary orientation. It intends to be a research-focused reference book used by scholars of hybrid microfabrication and micromachining procedures and aims to aid engineers, scientists, and academics in the field. Among the other crucial elements of the laboratories and institutes, the systematic importance stands out and should stimulate and inspire other researchers to pursue new avenues of exploration into hybrid microfabrication and micromachining technologies. The editors thank Scrivener Publishing for this prospect and their great assistance.

The Editors
March 2023

Acknowledgement

The starting and completing this book is unquestionably owed to the support of a number of colleagues and authors for their consistent direction, helpful suggestions, ongoing association, support, encouragement, and invaluable advice at every stage of this book's development, from the embryonic to the end stages. This book would not have advanced as smoothly toward completion without their helpful suggestions at the right moment. They are in fact owed by the editors for the important time they gave to this publication.

Without the significant intellectual contributions of scholars from all across the world, this work would not have been conceivable.

All of the individuals who helped with the book's editing and evaluation are sincerely appreciated by the editors.

Words alone cannot adequately express the editors' sincere gratitude to Martin Scrivener and the entire Scrivener Publishing team for maintaining their commitment and illuminating the proper course of action in order to complete this high level book.

The editors acknowledge the unending encouragement and support of their families and friends.

The editors would like to use this opportunity to express their gratitude to all of the readers and the hope that this book will continue to inspire and direct them in their future endeavours.

The Editors

1

Overview of Hybrid Micromachining and Microfabrication Techniques

Sandip Kunar[1,2*], Akhilesh Kumar Singh[3], Devarapalli Raviteja[1,2], Golam Kibria[4], Prasenjit Chatterjee[5], Asma Perveen[6] and Norfazillah Talib[7]

[1]*Department of Mechanical Engineering, Aditya Engineering College, Surampalem, India*
[2]*Jawaharlal Nehru Technological University Kakinada, Kakinada, East Godavari, India*
[3]*Department of Mechanical Engineering, Aditya College of Engineering, Surampalem, Andhra Pradesh, India*
[4]*Department of Mechanical Engineering, Aliah University, Kolkata, India*
[5]*Department of Mechanical Engineering, MCKV Institute of Engineering, Howrah, India*
[6]*Mechanical & Aerospace Engineering Department, School of Engineering & Digital Sciences, Nazarbayev University, Republic of Kazakhstan*
[7]*Department of Manufacturing Engineering, Faculty of Mechanical and Manufacturing Engineering, Universiti Tun Hussein Onn Malaysia, Batu Pahat, Johor, Malaysia*

Abstract

Hybrid micromachining and microfabrication techniques utilize concurrent deed of two or more micromachining procedures with assistance of some vitality in removal of material to augment the advantages and diminish the prospective difficulties observed in specific material ejection methods. There are different instances, like compound processes, energy aided micromachining methods, thermally aided micromachining, pulse-aided micromachining, and combined hybrid micromachining processes. This study introduces a unique categorization and analyses of the previous and current exploration and functions of the hybrid micromachining and microfabrication procedures and emphasizing its influences on performance characteristics. Even though it is an enthusiastic research field

**Corresponding author*: sandip.sandip.kunar@gmail.com

Sandip Kunar, Golam Kibria, Prasenjit Chatterjee and Asma Perveen (eds.) Hybrid Micromachining and Microfabrication Technologies: Principles, Varieties and Applications, (1–16) © 2023 Scrivener Publishing LLC

in unconventional and significant micromachining and microfabrication methods subsequently, inadequate acquaintance about proficiencies in shape, sizes, and controlling mechanisms is still the most important restraint in the progress of these procedures. In the adjacent future, there is plenty of opportunity for hybrid micromachining and microfabrication processes in studying of material removal at microlevel, consequence of residue strains, and working environment.

Keywords: Hybrid, micromachining, microfabrication

1.1 Introduction

From the last decades, there has been an enhanced benefit in microfabrication and micromachining techniques that have gained the creativity of investigators and industrial engineers from manufacturing sectors, mainly aviation, biomedical, and automobile. Emerging micromachining and microfabrication techniques are evidently continuing advances in microlevel industries, metrology, and machines to accomplish the requirements associated with the characteristics of microfeature [1]. Hybrid techniques state the conception of high-quality characteristics with shapes and sizes varying in micro level for different materials. The prerequisite for microproduct reduction continues to create several methodological restraints on discrete precision machining procedures, which are not constantly achievable to independently create a microcomponent meeting prerequisite, i.e., accuracy, surface quality, and shape intricacy. Also, the micro level measurements influence the specific method on functional characteristics and limit the application of self-regulating machining procedure. For instance, production techniques such as lithography are the utmost prevalent micromachining and microfabrication approaches that are accomplished to create microfeatures. Though, the limited choice of substrate materials, higher investment, and inability is unable to produce complex structures and inevitable clean room surroundings limit the usage of MEMS methods in micromachining and microfabrication fields [2].

Due to challenging in microfabrication and micromachining area and restrictions of discrete machining methods, researchers are concentrated to emerging hybrid micromachining and microfabrication methods, in which two or more techniques are composed together for micromachining to improve the benefits of fundamental procedures, while at that time diminishing their adversative problems when they are functional independently. The performing qualities of a hybrid procedure are significantly distinct from those methods in terms of efficiency, machining quality, and precision [3].

Hybrid micromachining is described as the combination of two or more micromachining methods to eradicate material [4]. The functional characteristics of hybrid micromachining and microfabrication approaches are significantly diverse from those that are distinctive for the constituent procedures when accomplished distinctly [5]. The hybrid machining method is to signify the amalgamation of different machining techniques with dissimilar capability of material exclusion [6]. These techniques are the approaches in which material eradication is owing to instantaneous action of two or more micromachining or microfabrication techniques or perform with the assistance of some energy in which the performance of the method enhanced pointedly, or else which are incredible when the procedures are utilized independently.

This chapter efforts to categorize recent hybrid micromachining and microfabrication techniques into pertinent classification and elucidate several methods utilized by researchers. The aim of this chapter is to deliver an appraisal on the different investigation works stated in the field of hybrid micromachining and microfabrication techniques to realize the machining capability of hybrid methods. This chapter efforts to deal with different hybrid methods in micro domain.

1.2 Classification of Hybrid Micromachining and Microfabrication Techniques

The category of hybrid micromachining and microfabrication techniques is classified with the utilization of procedural capability and/or energy/tools resources. Hybridization of microfabrication and micromachining methods is carried out with the following conditions:

(i) Hybridization of different methods into a compound procedure in which material removal arises because of concurrent action of different methods.
(ii) With aiding of various energy resources in material elimination to enhance performance of primary micromachining method.
(iii) Utilization of specifically devised hybrid tool that can perform the hybrid micromachining operation at a same time.

In the first type, compound process comprises techniques in which two or more techniques are employed for concurrent action of material removal such that substantial modifications in process operation can be accomplished. For instance, in EDG method, effective material removal occurs because of electro-discharge and grinding action compared to electro-discharge machining method. The second type, energy aided methods, the material removal is occurred owing to preliminary machining action, only certain amount of energy is utilized to enhance machining performances. For instance, in laser aided milling, the laser initially warms the material for removal, improving the material elimination capability of the milling approach. By the laser source for preheating substrate, the advanced materials are machined effortlessly. The third type utilizes a specific hybrid tool that performs single-time machining for different surfaces.

1.2.1 Compound Processes

Microfabrication and micromachining of microparts can be accomplished with distinct advantages with the assistance of chemical and electro-physical procedures as there is no explicit contact between the tool and substrate between these procedures. Combining these methods to create hybrid technique, further advances in unique benefits over specific method because of their less advantages [2]. In compound methods, there is instantaneous action of two or more machining techniques that completely entail in material removal action. Significant research outcomes are stated in these methods combining various techniques through various methods; some approaches are described.

In electro discharge micromachining (EDM), discharge energy considerably influences the machining quality. The surface finish reduces with greater voltage and current. The heating effect creates microdefects, residual stresses, etc. in EDM. To avoid these problems, innumerable studies are conveyed in hybrid electro-discharge micromachining since last decades. In compound hybrid micromachining processes, EDM is hybridized with other techniques, i.e., abrasive electro-discharge grinding (AEDG), electro discharge grinding (EDG), etc.

EDG confiscates from advanced materials by quick sparking phenomena between substrate and revolving tool that are divided by the dielectric fluid. This process utilizes a revolving tool, resulting in improved flushing proficiency of the method. Consequently, the material is efficiently removed from disparity and the unwanted materials are not stored in the machining space, whereas the accumulation of debris is a key difficulty with undesirable impact on functioning of the method in EDM. Several investigators are reported to find the significant parameters that influence the performances of the EDG method. Material removal and surface integrity can be enhanced by suitable input factors. To acquire greater material removal, greater machining current and higher duty cycle with positive polarity of electrode are utilized [7]. Researchers are concentrated for machining of delicate materials using EDG because no mechanical forces are applied through performance time, and it provides better machining than EDM because of the rotary wheel [8].

AEDG is a procedure in which the combined impact of grinding method and EDM is applied to improve the machining characteristics [9]. In this process, the grinding wheel is used instead of metallic or graphite electrode. Therefore, mechanical abrasion and discharge erosion take place for material removal. It is also known as electro-discharge abrasive grinding (EDAG) in various research papers. This procedure is suitable for machining engineering ceramics, and metal composites. Minimum surface roughness values are compared using three kinds of machining, i.e., electro-erosion grinding, hybrid grinding (AEDG), and traditional grinding [10]. The surface textures are associated with traditional grinding, and it is observed that irregular peaks are more in micro level in AEDG than traditional grinding method.

ECDM is a hybrid micromachining technique, where material ejection takes place through electrical discharges and electrolysis. This method is mostly used for microfabrication and micromachining of fragile materials such as quartz, refractory bricks, etc. [11]. The material removal is as much as 4 to 45 times than EDM and ECM. The combined form of this method is utilized for hole drilling, which is known as 3D micro structuring [12–15]. This method is specifically efficient for machining of HSTR alloys. This method is applied as microgrinding to improve the quality of coarse surface on glass material [16]. The procedure enhances the surface quality of ECDM structures and decreases the total structuring time of hybrid procedure compared to other traditional grinding procedure.

ECDG is a hybrid procedure of EDG and ECM, which merges the electrolysis, electro-discharge erosion and mechanical abrasion of the grinding procedure. Excess material is eliminated from the substrate by the electrolysis, the mechanical action of abrasives, and the erosion because of the spark phenomena. It is applied for conductive material's machining. This method can grind hard and brittle materials for greater harder materials. The surface quality is achieved up to 0.13 to 0.75 μm, and the dimensional precision is obtained up to 0.0013 mm [17]. The grinding aided ECDM is used for finishing of metal matrix composites and achieved the surface roughness up to 10 times lower compared to that of the workpiece machined exclusive of grinding assistance [18].

1.2.2 Methods Aided by Various Energy Sources

(i) Thermally aided machining

The advanced materials have extensive uses in different areas, i.e., biomedical, aviation, etc. and their necessity is expanding; however, they are very tricky to process because of their properties like higher strength and lower thermal conductivity. Generation of précised 3D structures is carried out by micromilling and micromachining. The machining force at the microlevel is much greater compared to the macro-level, which creates micromachining of advanced materials [19]. The reduced surface finish of substrate and failure of the tool are occurred due to higher cutting force. Thermally aided micromachining assists for generating the high-aspect-ratio and intricate 3D microfeatures in different advanced materials. This assisted method uses exterior thermal resources to improve the temperature of the machining area for assisting the machining of material as decreases the residual stress of materials with higher heat. This method lowers the powered processing energy on the tool. The heat resources are pertinent for thermal assisted machining, which have the subsequent features: (i) higher thermal energy for quick heating of the material; (ii) easy controlling of the heated areas; and (iii) satisfactory cost of traditional machines. Until now, plasma [20, 21], induction coils [22, 23] and laser [24–36] are employed. However, the characteristics, i.e., higher heating intensity and controlling of heat resource become laser more favourite for thermally assisted machining.

Laser-aided machining is a prominent procedure that affects the rapid heating ability of laser with concentrated beam, challenging to machining by mechanical machining method. It is extremely concerned as stated by the progress of elevated power laser and improved in its pertinency. Laser is utilized for aiding grinding, electrochemical machining, and electro discharge machining processes to enhance their functioning. The material removal and laser heating occur concurrently in laser-assisted turning. This method is applied in various types of numerous diverse materials. It has demonstrated the effective way in lowering tool wear and forces when utilized to machine the advanced materials, i.e., titanium alloys, silicon nitride, Inconel, etc. Most of these findings are performed by Nd:YAG laser and CO_2 lasers. Most of these research explore that laser assisted turning procedure helps for reducing the cutting forces, boosting the tool life and surface quality. In laser-aided milling method, the combination of milling machine and laser is occurred in which the tool is revolving.

This occurrence of rotating tool is a problematic task. This method utilizes lower laser power to heat up the material, which is eliminated by the tool instantly behind it. This method is carried out utilizing microball end-milling tool on tool steel and demonstrated higher machining precision, and lower surface roughness. The surface finish improves as the cutting speed rises owing to the presence of laser heating. Improvement of thermal assisted milling method is safer, reliable, and accessible for engineering applications.

(ii) Media-aided machining

In this method, mechanical and heating characteristics of higher pressurized jet of emulsion and water are focused into the machining zone to enhance traditional machining of difficult materials for aerospace and defence applications. The pressurized liquid can offer advantages such as effective chip instability and decrease in machining forces, particularly in advanced materials. It can also enhance the lubricating characteristics and decrease heat responses on the cutting tools [37]. The turning assisted method is carried out with pressurized coolant for machining operation and surface finishing investigation is conducted on steel, Inconel 718, etc. [38]. The utilization of coolant with this method is responsible for hydraulic pressure between the chip and the rake face of tool, lowering the cutting forces and chip size and reducing tool erosion. More understandings into the effect of contact length, and surface finish in jet-aided turning of Inconel 718 are introduced. The machining capability of Inconel 718 is experimentally investigated in traditional and alternative higher pressurized cooling conditions [39]. The experimental outcomes demonstrate that the tool erosion and cutting forces significantly reduce with the supply of pressurized coolant to the machining zone. Cryogenic machining is utilized to affect the performance and is applied to reduce the machining temperature and improve the chemical constancy of the cutting tool and workpiece and is anticipated to enhance the performance capability of advanced materials [40–43].

(iii) Vibration-aided machining

This method is employed to different techniques from facing to grinding. These combined method performs precision machining with tool vibration and small amplitude. Here, the tool tip is moved with slight reciprocating motion. The tool can intermittently drop contact with the chip for proper sequences of frequency and cutting velocity. As a result, machining forces are lowered, and tiny chips are produced. This advances to enhance the surface quality, accuracy and burr free compared to traditional machining [44]. The constant flow of debris particles assists the stable performance in micro-EDM because of the application of tubular type substrate. However, a revolving spindle is not utilized in the creation of quadrilateral holes [45]. Many researchers have concentrated on ultrasonically aided micro-EDM to explore the influences of frequency on accuracy, machining rate, and so on [46]. The utilization of quivering unit is mostly used to improve dielectric movement, subsequent in efficient elimination of metal from craters [47–49], succeeding in lowest micro-racks and modifications of microstructure on substrate at greater material removal [50]. The vibration assisted workpiece has a substantial outcome on the machined characteristics of micro-EDM technique [51]. This method is more prominent for higher depth drilling on tungsten carbide [52]. The vibration assisted workpiece suggestively improves material removal, geometric accuracy, surface quality, and lower tool wear. Ultrasonic vibration-aided grinding is also known as rotary ultrasonic machining (RUM), is applied to investigate the machining capability of titanium [53], steels [54], etc. and most findings are stated that enhanced surface finish and better tool life are achieved at greater pulsation frequencies and smaller feed rates.

(iv) Pulse-assisted machining

Pulse-assisted electrochemical machining [55–57] offers an inexpensive and efficient technique for machining of advanced materials to generate intricate profiles, i.e., micro cavities, die, molds, etc. PECM improves the precision and machined quality because of a pulsed voltage with greater current density compared to electrochemical machining.

(v) Electromagnetic-aided machining

Many investigators are concentrated their attempts on magnetic force applications to support the production method presently and explored the advantageous consequences of magnetic force support in material removal. The viability and consistency of the magnetic media for machining are studied. This media is employed in different areas of surface polishing. The magnetic abrasive polishing method is applied to smooth the machined surface of steel [58]. Enhanced characteristic of EDM-machined surface is followed because of magnetic abrasive polishing technique [59]. The magnetic field around the workpiece is applied to enhance the machining capability in the abrasive flow machining (AFM) method [60]. The use of magnetic field is responsible for enhancing debris flow in EDM [61]. The greater material removal is achieved due to improved debris exclusion from gap in aid of magnetic fields [62, 63].

1.2.3 Processing Using a Hybrid Tool

A hybrid tool is used for machining of two or more planner surfaces. The tools of inconsistent type, multifunctional tools and tools of consistent type are distinguished for hybrid micromachining. These hybrid tools are more intricate and costly than traditional tools and may only be suggested for heavy production [64]. These tools of incompatible type comprise different comprehensive tools. They have excellent machining capability since they significantly improve the efficiency. Consistent tools are created by sequence of different cutting tools of several types. They are generally compound tools, but one-piece production is feasible. These hybrid methods accomplish two distinct types of cutting, deformation, and machining in different classifications.

1.3 Challenges in Hybrid Micromachining

Requirements for improvement of better instruments for ability of machining of précised intricate parts:

- Hybrid micromachining techniques are evidently expanding for production of complicated 3D microparts. The development of precision instrument is very significant to gather requirements.
- The microtool with précised movement with slight vibration will generate the precise microproducts.
- Précised controlling of machining factors is important since little variation will affect the machining performance.

Improvement of accurate actuators and positioning mechanism

- Actuators with high precision are incredibly significant. Efficient evaluation, design techniques and modelling for positioning and accurate high-speed actuators should be studied.
- The capability of Piezoelectric actuators for précised positioning and their design is studied. Moreover, study of thermal influences on accuracy is important.
- Advancement of receptive controller to meet higher functioning is crucial. Precision operation is reliable for producing precise surfaces.

Advancement of multi-functional tool for meeting micromachining requirements and improving research resources

- While generating intricate parts, consumers incessantly require higher accuracy with enhanced feature. To gather these requirements, multi-operational tooling at decreased investment is necessary.
- Controlling of microtools and microcomponents are complicated and required resources are supplied on machine tool.

On-machine production of electrode for hybrid electro discharge machining processes

- In this method, managing of micro tools is a challenging task because of their possibility of twisting during assignments.
- Moreover, precision of performance characteristics is influenced owing to higher tool wear in EDM.

Advancement of multi-operational CAD tool supporting computer-aided hybrid micromachining processes

- CAD and CAM software enabling production through hybrid micromachining and microfabrication processes are significant.
- For production of intricate parts, a particular CAD/CAM method is required for creating tool path and associated machining activity.

1.4 Conclusions

Hybrid micromachining and microfabrication is one of the dynamic research areas in production of microcomponents. This manuscript demonstrates a unique categorization of hybrid methods and evaluations of literature review, specifically, hybrid approaches, energy-aided methods, and procedures utilizing a hybrid tool. In micro level, hybrid methods are developing as crucial techniques. This methods demonstrate open new path of research for improving method abilities, diminishing their limitations, and spreading functional areas. The main limitations in the progress of hybrid schemes are inadequate acquaintance about compatibilities in sizes, and employing strategies associated to manufacturing products.

Additionally, the essential acquaintance will affect the product superiority of different materials is still inadequate, and there is sufficient possibility in studying concerns in material removal at microlevel, and consequence of residual strains in the future.

1.5 Future Research Opportunities

Hybrid micromachining and microfabrication techniques require several types of energy, concurrently which are affected at the same fabrication area. Fundamental issues of material elimination capability and the mechanism related with hybrid system are one of the upcoming research purposes. Additional forthcoming research opportunity is simulation and modeling of hybrid micromachining and microfabrication procedures. Simulation of such procedures is explored with dissimilar approaches, i.e., molecular simulation, multiscale modeling, etc. Simulation and modeling support improved fundamental aspects of the hybrid procedures. There is prerequisite of ultraprecision machines that keeps high stiffness, control mechanisms, accurate feed drives, and are furnished to recompense for dynamic and static aligning faults. The accuracy of these techniques is improved further with diminished holding and rearrangement of tool faults.

References

1. Ehmann, K.F., Bourell, D., Culpepper, M.L. *et al.*, *Micromanufacturing*, pp. 1–362, Springer, Dordrecht, 2007.
2. Rajurkar, K.P., Levy, G., Malshe, A. *et al.*, Micro and nano machining by electro-physical and chemical processes. *CIRP Ann. Manuf. Technol.*, 55, 643, 2006.
3. Leuven, K.U., Surface integrity in hybrid machining processes. *Proc. Eng.*, 19, 241, 2012.
4. Rajurkar, K.P., Zhu, D., McGeough, G.A. *et al.*, New developments in electro-chemical machining. *CIRP Ann. Manuf. Technol.*, 48, 567, 1999.
5. Kozak, J. and Rajurkar, K.P., Hybrid machining process evaluation and development, in: *Proceedings of 2nd International Conference on Machining and Measurements of Sculptured Surfaces*, The Institute of Metal Cutting (IOS, Krakow, pp. 501–536, September 20–22, 2000.
6. Menzies, I. and Koshy, P., Assessment of abrasion-assisted material removal in wire EDM. *CIRP Ann. Manuf. Technol.*, 57, 195, 2008.
7. Shih, H.R. and Shu, K.M., A study of electrical discharge grinding using a rotary disk electrode. *Int. J. Adv. Manuf. Technol.*, 38, 59, 2008.

8. Yadav, R.N. and Yadava, V., Electric discharge grinding: A review. *Proceedings of the National Conference on Trends and Advances in Mechanical Engineering*, YMCA University of Science & Technology Faridabad, Haryana, pp. 590–597, October 19th-20th, 2012.
9. Kozak, J., Abrasive electro discharge grinding (AEDG) of advanced materials. *Arch. Civil Mech. Eng.*, 2, 83, 2002.
10. Dabrowski, L. and Marciniak, M., Investigation into hybrid abrasive and electrodischarge machining. *Arch. Civ. Mech. Eng.*, 2, 5, 2005.
11. Jawalkar, C.S., Sharma, A.K., Kumar, P., Micromachining with ECDM: research potentials and experimental investigations. *IJMAE*, 6, 7, 2012.
12. Sarkar, B.R., Doloi, B., Bhattacharyya, B., Parametric analysis on electrochemical discharge machining of silicon nitride ceramics. *Int. J. Adv. Manuf. Technol.*, 28, 873, 2005.
13. Wei, C. and Ni, J., Electrochemical discharge machining using micro-drilling tools. *Trans. NAMRI/SME*, 38, 105, 2010.
14. Wei, C., Xu, K., Ni, J. *et al.*, A finite element-based model for electrochemical discharge machining in discharge regime. *Int. J. Adv. Manuf. Technol.*, 54, 987, 2011.
15. Zheng, Z.P., Wu, K.L., Hsu, Y.S. *et al.*, Feasibility of 3D surface machining on pyrex glass by electrochemical discharge machining (ECDM), in: *Proceedings of AEMS07*, Nagoya, Japan, pp. 98–103, November 28–30, 2007.
16. Cao, X.D., Kim, B.H., Chu, C.N., Hybrid micromachining of glass using ECDM and micro grinding. *Int. J. Precis. Eng. Manuf.*, 5, 5, 2013.
17. Rajgopal, S.P., Ganesh, V., Lanjewar, A.V. *et al.*, Past and current status of hybrid electric discharge machining (H-EDM) processes. *IJAMMC*, 3, 111, 2013.
18. Liu, J.W., Yue, T.M., Guo, Z.N., Grinding-aided electrochemical discharge machining of particulate reinforced metal matrix composites. *Int. J. Adv. Manuf. Technol.*, 68, 2349, 2013.
19. Sun, S., Brandt, M., Dargusch, M.S., Thermally enhanced machining of hard-to-machine materials—A review. *Int. J. Mach. Tools Manuf.*, 50, 663, 2010.
20. Leshock, C.E., Kim, J.N., Shin, Y.C., Plasma enhanced machining of Inconel 718: Modeling of workpiece temperature with plasma heating and experimental results. *Int. J. Mach. Tools Manuf.*, 41, 877, 2001.
21. Shin, Y.C. and Kim, J.N., Plasma enhanced machining of Inconel 718. *ASME Int. Mech. Eng. Congress Expo.*, 4, 243, 1996.
22. Amin, A.K.M.N., Dolah, S.B., Mahmud, M.B. *et al.*, Effects of workpiece preheating on surface roughness, chatter and tool performance during end milling of hardened steel D2. *J. Mater. Process. Technol.*, 201, 466, 2008.
23. Lajis, M.A., Amin, A.K.M.N., Karim, A.N.M. *et al.*, Hot machining of hardened steels with coated carbide inserts. *Am. J. Eng. Appl. Sci.*, 2, 421, 2009.
24. Kim, K.S., Kim, J.H., Choi, J.Y. *et al.*, A review on research and development of laser assisted turning. *Int. J. Precis. Eng. Manuf.*, 12, 753, 2011.

25. Raghavan, S., Melkote, S., Hashimoto, F., Laser tempering based turning process for efficient machining of hardened AISI 52100 steel. *J. Manuf. Process.*, 15, 318, 2013.
26. Masood, S.H., Armitage, K., Brandt, M., An experimental study of laser-assisted machining of hard-to-wear white cast iron. *Int. J. Mach. Tools Manuf.*, 51, 450, 2011.
27. Navas, V.G., Arriola, I., Gonzalo, O. *et al.*, Mechanisms involved in the improvement of Inconel 718 machinability by laser assisted machining (LAM). *Int. J. Mach. Tools Manuf.*, 74, 19, 2013.
28. Lee, J., Lim, S., Shin, D. *et al.*, Laser assisted machining process of HIPed silicon nitride. *J. Laser Micro/Nanoeng.*, 4, 207, 2009.
29. Dandekar, C.R., Shin, Y.C., Barnes, J., Machinability improvement of titanium alloy (Ti–6Al–4V) via LAM and hybrid machining. *Int. J. Mach. Tools Manuf.*, 50, 174, 2010.
30. Rahman Rashid, R.A., Sun, S., Wang, G. *et al.*, An investigation of cutting forces and cutting temperatures during laser-assisted machining of the Ti–6Cr–5Mo–5V–4Al beta titanium alloy. *Int. J. Mach. Tools Manuf.*, 63, 58, 2012.
31. Wang, Y., Yang, L.J., Wang, N.J., An investigation of laser-assisted machining of Al2O3 particle reinforced aluminum matrix composite. *J. Mater. Process. Technol.*, 129, 268–271, 2002.
32. Melkote, S.N., Kumar, M., Hashimoto, F. *et al.*, Laser assisted micro milling of hard-to-machine materials. *CIRP Ann. Manuf. Technol.*, 58, 45, 2009.
33. Kumar, M. and Melkote, S.N., Process capability study of laser assisted micro milling of a hard-to-machine material. *J. Manuf. Process.*, 14, 41, 2012.
34. Shelton, J.A. and Shin, Y.C., Experimental evaluation of laser-assisted micro-milling in a slotting configuration. *J. Manuf. Sci. Eng. Trans. ASME*, 132, 1, 2010.
35. Yang, J., Sun, S., Brandt, M. *et al.*, Experimental investigation and 3D finite element prediction of the heat affected zone during laser assisted machining of Ti6Al4V alloy. *J. Mater. Process. Technol.*, 210, 2215, 2010.
36. Ding, H., Shen, N., Shin, Y.C., Thermal and mechanical modeling analysis of laser-assisted micro-milling of difficult-to-machine alloys. *J. Mater. Process. Technol.*, 212, 601–613, 2012.
37. Courbon, C., Kramar, D., Krajnik, P. *et al.*, Investigation of machining performance in high-pressure jet assisted turning of Inconel 718: an experimental study. *Int. J. Mach. Tools Manuf.*, 49, 1114, 2009.
38. Sanz, C., Fuentes, E., Gonzalo, O., Turning performance optimisation of aeronautical materials by using high pressure cooling technology. *Int. J. Mach. Mach. Mater.*, 2, 270, 2007.
39. Cxolak, O., Investigation on machining performance of Inconel 718 under high pressure cooling conditions. *J. Mech. Eng.*, 58, 683, 2012.
40. Kenda, J., Pusavec, F., Kopac, J., Analysis of residual stresses in sustainable cryogenic machining of nickel based alloy—Inconel 718. *J. Manuf. Sci. Eng. Trans. ASME*, 133, 041009, 2011.

41. Dilip Jerold, B. and Kumar, P., The influence of cryogenic coolants in machining of Ti–6Al–4V. *J. Manuf. Sci. Eng. Trans. ASME*, 135, 031005, 2013.
42. Barletta, M. and Tagliaferri, V., Development of an abrasive jet machining system assisted by two fluidized beds for internal polishing of circular tubes. *Int. J. Mach. Tools Manuf.*, 46, 271, 2006.
43. Barletta, M., Ceccarelli, D., Guarino, S. *et al.*, Fluidized bed assisted abrasive jet machining (FB-AJM): precision internal finishing of Inconel 718 components. *J. Manuf. Sci. Eng. Trans. ASME*, 129, 1, 2007.
44. Brehl, D.E. and Dow, T.A., Review of vibration-assisted machining. *Precis. Eng.*, 32, 153, 2008.
45. Endo, T., Tsujimoto, T., Mitsui, K., Study of vibration assisted micro-EDM—The effect of vibration on machining time and stability of discharge. *Precis. Eng.*, 32, 269, 2008.
46. Cao, G. and Liu, Z., A study of ultrasonically aided micro electrical-discharge machining by the application of work piece vibration. *J. Mater. Process. Technol.*, 139, 226, 2003.
47. Zhang, Q.H., Du, R., Zhang, J.H. *et al.*, An investigation of ultrasonic-assisted electrical discharge machining in gas. *Int. J. Mach. Tools Manuf.*, 46, 1582, 2006.
48. Singh, J., Walia, R.S., Satsangi, P.S. *et al.*, Parametric optimization of hybrid electric discharge machining process with continuous and discontinuous ultrasonic vibrations on workpiece. *Int. J. Mater. Sci. Eng.*, 2, 1, 2011.
49. Shabgard, M.R., Sadizadeh, B., Kakoulvand, H., The effect of ultrasonic vibration of workpiece in electrical discharge machining of AISIH13 tool steel. *IJMAE*, 4, 161, 2010.
50. Kremer, D., Effects of ultrasonic vibrations on the performances in EDM. *CIRP Ann. Manuf. Technol.*, 38, 199, 1989.
51. Cao, G. and Liu, Z., A study of ultrasonically aided micro electrical-discharge machining by the application of work piece vibration. *J. Mater. Process. Technol.*, 139, 226, 2003.
52. Jahan, M.P., Rahman, M., Wong, Y.S. *et al.*, On-machine fabrication of high-aspect-ratio micro-electrodes and application in vibration-assisted micro-electro discharge drilling of tungsten carbide. *Proc. Inst. Mech. Eng. B J. Eng. Manuf.*, 224, 795, 2010.
53. Tawakoli, T., Azarhoushang, B., Rabiey, M., Ultrasonic assisted grinding of soft steel. *Ind. Diam. Quart.*, 1, 40, 2009.
54. Qin, N., Pei, Z., Treadwell, C. *et al.*, Physics-based predictive cutting force model in ultrasonic-vibration-assisted grinding for titanium drilling. *J. Manuf. Sci. Eng. Trans. ASME*, 131, 1, 2009.
55. Kock, M., Kirchner, V., Schuster, R., Electrochemical micromachining with ultrashort voltage pulses—A versatile method with lithographical precision. *Electrochem. Acta*, 48, 3213, 2003.
56. Kozak, J., Gulbinowicz, D., Gulbinowicz, Z., Investigations of Micro electrochemical machining with ultrashort pulses, in: *Proceedings of the 5th*

International Conference of the European Society for Precision Engineering and Nanotechnology, Montpellier, May 8–11, 2005.

57. Burkert, S., Schulze, H., Gmelin, T. *et al.*, The pulse electrochemical micromachining (PECMM)—Specifications of the pulse units. *J. Mater. Process. Technol.*, 2, 645, 2009.
58. Khairy, A.B., Aspects of surface and edge finish by magnetoabrasive particles. *J. Mater. Process. Technol.*, 116, 77, 2001.
59. Lin, Y.C. and Lee, H.S., Machining characteristics of magnetic force assisted EDM. *Int. J. Mach. Tools Manuf.*, 48, 1179, 2008.
60. Singh, S. and Shan, H.S., Development of magneto abrasive flow machining process. *Int. J. Mach. Tools Manuf.*, 42, 953, 2002.
61. De Bruijn, H.E., Delft, T.H., Pekelharing, A.J., Effect of a magnetic field on the gap cleaning in EDM. *CIRP Ann. Manuf. Technol.*, 27, 93, 1978.
62. Lin, Y.C. and Lee, H.S., Optimization of machining parameters using magnetic-force-assisted EDM based on gray relational analysis. *Int. J. Adv. Manuf. Technol.*, 42, 1052, 2009.
63. Lin, Y., Chen, Y.F., Wang, D.A. *et al.*, Optimization of machining parameters in magnetic force assisted EDM based on Taguchi method. *J. Mater. Process. Technol.*, 209, 3374, 2009.
64. Ponomarev, A.V., Yakukhin, V.G., Maksimov, A.D., Hybrid machining technologies. *Russ. Eng. Res.*, 31, 717, 2011.

2

A Review on Experimental Studies in Electrochemical Discharge Machining

Pravin Pawar[1]*, Amaresh Kumar[2] and Raj Ballav[2]

[1]*Department of Mechanical Engineering, National Institute of Technology Goa, Farmagudi, Ponda, Goa, India*
[2]*Department of Production & Industrial Engineering, National Institute of Technology Jamshedpur, Jharkhand, India*

Abstract
The electrochemical discharge machining (ECDM) is a hybrid nontraditional process, which precisely ideal for machining on conductive and nonconductive materials. Due to advanced technologies, miniaturized products have specific demand in most of the fields, which can easily be fulfilled by ECDM. The present study is a brief review of work done on ECDM in various fields, which comprises history, design, workpiece, electrolyte, tool electrode, input and output parameters used in ECDM. The conclusion gives a precise idea for enhancement and efficiently working with ECDM for the future research study.

Keywords: ECDM, electrolyte, applied voltage, tool electrode, material removal rate, glass, ceramic

2.1 Introduction

The machining has a special position in manufacturing industries due to its abilities, i.e., high accuracy, better surface finish, and stable particular property of bulk work material. The complex shapes can be manufactured by different machining processes [4]. The various nonconventional machining processes like abrasive jet process, laser beam process, and ultrasonic process are extensively used for machining of advanced materials [5].

**Corresponding author*: pravinpawar@nitgoa.ac.in

Sandip Kunar, Golam Kibria, Prasenjit Chatterjee and Asma Perveen (eds.) Hybrid Micromachining and Microfabrication Technologies: Principles, Varieties and Applications, (17–100) © 2023 Scrivener Publishing LLC

Electrochemical machining (ECM), as well as electric discharge machining (EDM) processes are very effective when they are used for micromachining with good precision, but they are only applicable to electrically conductive and nonconductive materials. So, to solve this problem, integration of ECM and EDM material removal processes has been invented, which is called electrochemical discharge machining (ECDM). This nontraditional technology is specially used to fabricate all types of materials at a microscale. The electrochemical discharge machining (ECDM) process is also termed as electrochemical arc machining (ECAM), spark-assisted chemical engraving (SACE), and electrochemical spark machining (ECSM) [1–3, 6]. It shows combined ECM action, which is aided by the thermal erosive influence of the electrical discharge process in the machining gap occupied with electrolyte [7]. This process has various machining characteristics, such as depth of machined hole and microchannel, surface topography, machined surface accuracy, tool wear, material removal rate, heat affected zone, etc. The ECDM process has advantages like productivity, quality, and ultimately, cost efficiency. Hence, this process has a demand in the field of precision parts manufacturing for automotive and biomedical components, which will lead to the advanced manufacturing industry utilizing the micros-ECDM process more effectively [8]. In this process, the applied voltage goes more than the critical voltage to an electrochemical container, the discharge creates among one electrode and the nearby electrolyte, which is named electrochemical discharge [9]. The material removal occurs because of the combined process of localized sparks and electrolysis in an electrolytic cell [10].

2.2 Historical Background

The spark discharge machining was applied for the erosion of materials, which was previously noticed by Priestley in 1762. This process was first utilized for the dissolution of metals and to produce small particles achieved by Svedberg (1906) who worked on oscillatory spark circuits to generate colloidal solutions using the electric pulverization of the spark gap tool electrode, which was dipped in chemical [11]. Sir Humphry Davy (1778–1829) first detected and recorded the occurrence of discharging the light. The electrochemical discharge phenomenon indicates a major effect in the fabricating of nonconducting materials. This phenomenon was prominent in the work of Fizeau and Foucault in 1844. Fizeau and Foucault first worked on the intensity of light discharged by the carbon in the experiment of Davy in the year 1844, and they recorded the experimental spectral results of light sources. They relate the spectra initiated by electrical discharges among two carbon electrodes and those from the sun. They also proposed a new approach by considering the phenomenon as a stochastic process. In this model, they considered the process as a phase transition [12, 13]. The ECDM was first used by Kurafuji *et al.* in 1968 for producing microholes drilling in glass material [14]. Cook *et al.* examined the influence of the electrolyte during the process on nonconductive materials [15]. Tsuchiya *et al.* developed a new process called wire electrochemical discharge machining in which a wire is used as a tool-electrode, and it is used to cut the different ceramic, glass, and composite materials [16]. Esashi *et al.* first made a MEMS device by utilizing the electrochemical discharge drilling process [17]. Basak *et al.* made a model to calculate the physical characteristics of the MRR for changing the various factors [9]. Skrabalak *et al.* scrutinized a model for investigation of the current of electro-discharge machining and electrochemical dissolution of the ECDM. They had explored the fuzzy-logic controller for the ECDM process [18]. The first review paper was published in the year 2005 based on the ECDM process. In this report, they emphasized the electrochemical characteristics and showed some limiting causes and their results [19]. Yang *et al.* observed improvement in surface roughness and overcut when working with SiC abrasive particles, which were added into the chemical in the wire ECDM process [20]. Didar *et al.* investigated the modeling and characterization of 2D glass micromachining through constant velocity by utilizing spark-assisted chemical engraving [21]. Cao *et al.* formed 3D microstructures on a glass sample test piece by applying the ECDM process [22]. Yang *et al.* found a method to solve problems observed during the drilling of the microhole using a spherical end tool, which helped to enhance efficiency and accuracy [23]. Liu *et al.* designed a new process, i.e., grinding aided ECDM to raise the enactment of the conventional ECDM, which was used to machine particulate reinforced metal matrix composites material [24]. Jiang *et al.* experimented with a stochastic model for a spark energy assessment in which the FEM model was developed to study a correlation between geometry and spark energy of eroded material [25]. Ziki *et al.* had clarified the machining temperature, according to them the temperature can be dependent on the machining structure geometry and drilling strategy [26]. Goud *et al.* developed a 3D finite element model for the virtual simulation analysis of material removal in ECDM drilling [27]. Hof *et al.* produced micropatterned glass templates with the help of SACE to produce micropatterned glass templates. They invented a new technique that has the potential of microfabrication, and it could be beneficial for MEMS product developers [28]. Hof *et al.* have fabricated a low-cost SACE machine. This machine was successfully used to fabricate glass material, which is also applicable for the fabrication of optic telecommunications, smartphones, green energy devices, and advanced medical devices such as Lab-on-Chip [29]. Goud *et al.* deeply reviewed past to present studies of the ECDM process [30]. Table 2.1 shows developments in the ECDM process from past to present.

2.3 Principle of Electrochemical Discharge Machining Process

In this process, the high voltage is provided to the cathode and anode, which was placed in conductive chemical. The electrical discharges in an electrolyte chemical raised due to confined areas of grown gas and electrolyte vapor [31]. The fine hole drilling achieved through ECAM, and the inter-electrode gap width enhanced due to applied voltage [32]. In ECDM, several intermediate processes like electrochemical reactions after that nucleate pool boiling, and then the breakdown of hydrogen bubbles, producing the electrons and these electrons drifting near the workpiece due to which the material removal get initiated [33]. In this process, the quick creation of bubbles takes place because of the electrochemical reaction between cathode and anode. Later on, these bubbles get burst because of a collision that generates sparking. This process continuously happened because of contact between a cathode and electrolyte [34]. The process of electrolysis in an aqueous solution having a high current density is called the aqueous anode effect. When the anode temperature reaches up to the boiling of electrolyte at that time the transition period gets started. Later on, bubbles wall gets vaporized due to which electrical resistance gets increased at the anode and voltage is also raised [35]. The dense and steady gas film can be achieved at an extreme voltage known as transition voltage as a result of electrochemical discharge action occurs [36]. Figure 2.1 shows the ECDM experimental setup.

2.4 Basic Mechanism of Electrochemical Discharge Machining Process

In this process, the machining gap between the workpiece and cathode tool point is retained for a steady electrochemical reaction and discharge mechanism inside the nonconducting container. The anode and cathode

Table 2.1 Developments in the ECDM process from past to present.

Year	Progressive developments in the ECDM process from past to present
1762	Spark discharge machining process detected by Priestley [11]
1778	Detected occurrence of emitting light for instance when electrical arcs between two carbon electrodes by Sir Humphry Davy [13]
1844	Electrochemical discharge phenomenon identified by Fizeau and Foucault [12, 13]
1906	Dissolution of metals into small particles obtained by Svedberg [11]
1968	The first article conveyed based on electrical discharge drilling on glass by Kurafuji and Suda [14]
1973	First characterizations by Cook *et al.* [15]
1985	Invented new modified traveling wire-ECDM by Tsuchiya *et al.* [16]
1990	First MEMS devices were formed by Esashi *et al.* [17]
1997	Invented first theoretical model by Basak *et al.* [9]
2004	Investigated fuzzy-logic controller for ECDM by Skrabalak *et al.* [18]
2005	A first review paper published by Wuthrich *et al.* [19]
2006	Added abrasive mixed into electrolyte by Yang *et al.* 2006 [20]
2008	2D-glass micromachining using SACE by Didar *et al.* 2008 [21]
2009	3D microstructures on glass surface by using ECDM by Cao *et al.* [22]
2011	Made spherical tool electrode by Yang *et al.* 2011 [23]
2013	New process Grinding aided ECDM by Liu *et al.*, 2013 [24]
2014	Stochastic model for spark energy estimation by Jiang *et al.* [25]
2015	Clarification about the machining temperature by Abou Ziki *et al.* [26]
2016	3D FEM model for simulation of material erosion in ECDM by Goud *et al.* [27]
2017	Made micropatterned glass templates by Hof *et al.* 2017 [28]
2018	Invented novel SACE machine suitable for Industry 4.0., by Hof. *et al.* 2018 [29]

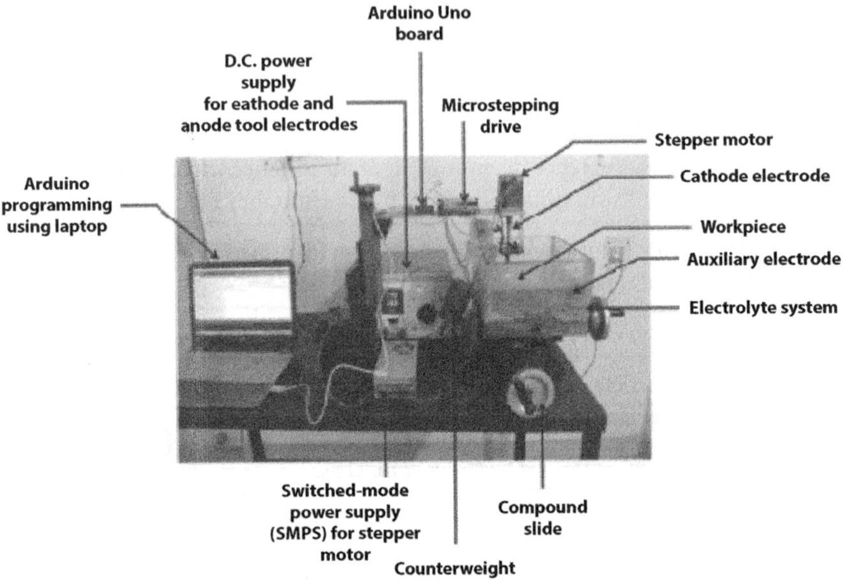

Figure 2.1 ECDM experimental setup.

electrodes are immersed in an electrolyte container and DC voltage is given among them. The spark at the tooltip generates when voltage is greater than critical voltage. The disintegration of metal occurs in the anode electrode, so it is larger than the cathode. After that, the electrolysis initiates hydrogen bubbles formed nearby the cathode and oxygen bubbles formed nearby the area of anode electrode. The bubble layer produces nearby the cathode and anode tools, which integrate bubbles into a vapor layer. The light release could be observed in electrical discharge and therefore named electrochemical discharges. This discharge produces between the tool and electrolyte boundary as a result of the maximum gradient established across the hydrogen bubbles. The electrons transferred near the workpiece when the cathode moved towards the workpiece. The workpiece material melts, and the removal of material occur in the form of small particles [19, 37, 38]. The vital characteristics of this process are the local Joule heating, variation in wettability of the tool, hydrodynamic inconsistency, hydrodynamic effects, and a combination of wettability. The limitations of the ECDM are quality of machining, efficiency, and precision [19]. The machining depends upon electrolyte properties like concentration, electrical conductivity, viscosity, as well as the temperature of the electrolyte. Electrolytes are generally specified into three types, i.e., alkaline, neutral, and acidic. The different electrolytes like H_2SO_4, NaCl,

NaOH, KCl, $NaClO_3$ $NaNO_3$, and KOH are mostly utilized. The majority of the researchers used alkaline-based electrolytes because of high ions mobility and improved superiority of etching. The maximum electrolyte concentration triggers the formation of the dense gas film. The ECDM process causes health hazards because of fumes generated during machining conditions [39]. In this process, the discharge activity across the vapor bubble films produced on the workpiece and therefore erosion of material takes place [40].

2.5 Application of ECDM Process

The different processes can be done by ECDM process, i.e., dressing of microgrinding tools, machining of cylindrical parts, microdrilling, microdies, blind holes, machining of complex and intricate microprofiles [1]. It offers enhanced accuracy and material removal in microscale and mesoscale machining of hard materials. Nowadays, the industries show less interest in ECDM due to low repeatability and controllability [41]. This process is applied to make several channel surfaces texture on nonconducting ceramic materials in addition to composite materials and microtexturing on the glass surface etc. [42–45]. This process can be used to machine very hard materials, like silicon carbide, diamond materials, etc. [46–48, 145].

2.6 Literature Review on ECDM

The work of literature from past to present, related to the ECDM machining process is deeply reviewed, and based on it this literature, review is categorized into the following areas: past work was done based on theoretical modeling in ECDM, internal behavioral studies of ECDM, machine design of ECDM, materials used into the machine by using ECDM, tooling material and its design of tooling in ECDM, electrolyte chemical used in ECDM, and optimization techniques used in ECDM process.

2.6.1 Literature Review on Theoretical Modeling

McGeough *et al.* [7] developed a theoretical model and observed the effect on the metal removal rate influenced by varying the machining voltage. They found that the sparking element plays a key role in increasing the metal removal rate when the higher the voltage and higher the feed rate.

Khairy *et al.* [49] developed a theoretical model for the specific MRR, power utilization factor. The specific MRR was directly calculated from the ratio between the removed volume of workpiece material and the amount of Coulomb by using Coulomb\meter during machining. Equation 2.1 shows a theoretical model for the specific material removal rate.

$$smrr = 23.55 d_a^2 / c \tag{2.1}$$

where smrr is mm^3/KC, KC is specific energy or cutting pressure, c is the total amount of Coulomb on meter.

The power utilization factor (puf) model was to find out rate of the volume of metal eroded from the die-cavity mm^3 to the total amount of power (q) counted by a power meter. The equation 2.2 shows for power utilization factor.

$$puf = 23.55 d_a^2 / q \tag{2.2}$$

Basak [50] studied the mechanism of electrochemical discharge, He developed a theoretical model of electrochemical discharge (ECD) and developed a quantitative model of material removal in ECDM. The model of ECD is based on the switching phenomenon. The switching action among the electrolyte solution and tool electrode occurred because of the blanketing of the tool electrode by a layer of gas. The water vapor is produced because of the interface heating and the hydrogen gas generated from the electrochemical reactions, which assists in blanking the electrode surface. The material removal rate increases considerably when the total circuit inductance is increased. Basak and Ghosh [51] proposed a theoretical model used for detecting voltage and current for investigating the discharge formed between electrolyte and electrodes. A theoretical discharge phenomenon resulted in the tooltip. The developed model is presented in equation 2.3.

$$V_C = J_c A_c [R_{1C} + R_{2C} + R_{3C}] \tag{2.3}$$

where Vc sparking start voltage, Jc-current density, Ac-surface area of the electrode, R_{1C}, R_{2C}, R_{3C} are the critical values of R_1, R_2, R_3, which are interfaced resistance among the tool and the electrolyte.

Basak and Ghosh [9] developed a simplified model to calculate the characteristics of the material erosion rate for altering input factors and to check the possibility of improving the capability of the process. The theoretical model (equation 2.4) indicates an enhancement in MRR, which can be succeeded due to additional inductance.

$$M = \frac{0.\dot{2}5L^{0.5}(V_o - V_c)^3(R_2 - 0.5R_{3c})}{R_c^3 In \dfrac{2R_c}{R_{3c}}} mg.min^{-1} \tag{2.4}$$

where M is the material erosion rate, V_0 is applied voltage, V_C is critical voltage, L is inductance of circuit in Henry, R_2 is resistance of the larger electrode-electrolyte interface, R_{3c} critical value of resistance, and $R_c = R_2 + R_{3c}$.

Jain *et al.* [37] have identified the electrochemical discharge model, which was related to the arc discharge valves theory. This model is used for the analysis of the material erosion rate by modeling the 3D unsteady state heat conduction. Mediliyegedara *et al.* [52] developed a model of a cathode electrode location control system. The PID controller and ARX model were designed and developed. Mediliyegedara *et al.* [53] invented the pulse classification of the electrochemical discharge process by using Fuzzy Logic. The ECDM was developed by integrating altering the control system and an electrolyte system. The different ECDM input process factors are utilized for developing the pulse classification rule base. Simulation results indicated that the suggested pulse classification system can be efficiently utilized in the pulse classification of the ECDM process. Fascio *et al.* [54] investigated two models for SACE. The first model was developed by using percolation theory, which identified current and voltage. The second model was to investigate the spark's features, i.e., duration and amplitude. Bhondwe *et al.* [55] have investigated the FEM model for the evaluation of material erosion rate. The mathematical formulation for mass MRR was also investigated, which is shown in equation 2.5.

$$MRR_m = MRR_v \times \rho \tag{2.5}$$

where MRRm = mass MRR, MRRv = voltage into a number of sparks per unit time, ρ = density of the workpiece material.

Allagui and Wuthrich [56] developed a two-step algorithm, i.e., Meyer wavelet and center-clipping method was given for the current signal. Liu *et al.* [57] developed a new method in which they applied electric field acts on a hydrogen bubble. They also developed a model shown in equation 2.6.

$$E_0 = \frac{161}{240} MaxE_{in} \tag{2.6}$$

where *E0* is original electric field strength, Max Ein is the dielectric strength of hydrogen.

El-Haddad and Wuthrich [58] developed a model for the prediction of current-voltage characteristics of two electrodes in which the dynamics of gas film form during the ECD phenomenon. In this phenomenon, the mean-field version model denotes a good qualitative covenant but overrates the hysteresis outcome and predicts enormous current densities for the cell operation as soon as the gas film was made. Wei *et al.* [59] computed a FEM model for identified depth of machined area in the electrochemical discharge machining process, which is shown in equation 2.7.

$$Z_\infty \left\{ 1 - \exp\left[\frac{2tr_0^2 z_0 A_1 u^2 \exp\left({}^{-B_1}\!/_{u} \right)}{z_\infty (4dl + d^2)} \right] \right\} \tag{2.7}$$

where *u* is the voltage, *d* is the cathode diameter, *Z∞* is the machinable depth, l is the tool immersion depth, *t* is the time, *ro* and *zo* is the intercepts of the isotherm and coordinate axes.

Panda and Yadava [60] developed two models of die-sink ECDM process by using the artificial neural network and FEM. Krotz *et al.* [61] developed a model and simulated the heat transfer into a workpiece in single discharges of ECDM. Kamaraj *et al.* [62] proposed a mathematical model shown in equation 8 to estimate the overcut of a machined surface having less variation and the tool diameter indicates a minor influence on overcut.

$$\frac{1}{\frac{\{T(r,t) - t_0\} \times 4\pi K}{(V \times I_m - I_m^2 \times R)} + \frac{1}{\pi}\sqrt{\frac{pc}{KC_f t}}} \tag{2.8}$$

where *t* is total machining time, *V* is voltage, I_m is the mean machining current, *K* is thermal conductivity, *R* is interelectrode resistance of the electrolyte, C_f is the ratio of mean machining current to peak current, *p* is density, *c* is heat capacity, *r* is overcut.

Paul and Korah [63] developed a model to investigate the MRR with pulsed DC and direct current. They detected that the pulse DC was more significant than direct current with lesser temperature and improved MRR. Goud and Sharma [27] proposed a 3D FEM simulation model in which the variance of temperature distribution on the workpiece, as well as the variance of MRR by input factors like electrolyte concentration and voltage. The soda-lime glass and alumina materials were taken for the given simulation study. They observed that the experimental results agreed with the shape of material removal. Ladeesh and Manu [64] proposed a model for calculating the edge chipping thickness. The high-energy DC supply can be utilized to enhance the MRR. Hajian *et al.* [65] assessed equivalent temperature in ECD milling by utilizing a thermo-physical method, which was based on the FEM method. In this investigation, they considered two temperatures, i.e., 600°C and 850°C, as the temperature criterion for the material erosion temperature in the FEM modeling.

2.6.2 Literature Review on Internal Behavioral Studies

Crichton *et al.* [66] correlated electrochemical and electrodischarge machining to ECAM. The effect of metal removal occurred because of electrical discharge in electrolytes. They reviewed investigations of off-sparks in dielectric and electrolytes. De silva [67] has analyzed the gap phenomena in electrochemical arc machining. They used radiofrequency emission from the gap to distinguish between abnormal and normal discharges. The portable electrochemical arc machining setup was designed and built. This machine was used to drill holes in a nickel alloy, cobalt alloy, chrome steel, low alloy steel and titanium materials. Raghuram *et al.* [68] have applied a separate circuit to electrolytes, current and voltage intended for drilling experiments in quartz and analyzing the internal behavior process of capacitance and inductance in electrolytes. Kulkarni *et al.* [69] found the temperature and synchronized time-varying current in the ECDM process. They investigated the machine, which creates the discharge and removes the material. The pyrometers were used to measure the transient temperatures. The discharge was a discrete occurrence. It occurs in bursts. Therefore, it enhances the temperature of the workpiece in response to these discharge bursts in a localized zone causing the material removal. The advantage of pyrometers is minimized uncertainties in measurements with a higher sensing temperature range. Kulkarni and Karnik [70] executed temperature measurements in the machining region of ECDM with different temperature sensing schemes. The temperature at various radii of the workpiece was investigated. Mediliyegedara *et al.* [71] developed the algorithm in a software system that can be advantageous to find out ideal factors. Wuthrich *et al.* [72] examined the discharge regime and hydrodynamic regime during the SACE process. They found that the results of needle-shaped tools showed marginally higher machining speeds. Wuthrich and Hof [73] stated that the gas film made nearby cathode in SACE. The gas film was not steady and frequently changing, which results in nonreproducible machining with SACE. Wuthrich *et al.* [74] stated that the current signal with high bandwidth data on sparks develops, which permits monitoring of the machining process. They observed that there was no direct relation in MRR and actual spark activity. Kim *et al.* [75] observed the machining of Pyrex glass by using rectangular voltage pulses, which cause decreasing heat affecting zones. The MRR reduces and the microdrilled surface becomes smoother as the voltage pulse frequency enhances. West and Jadhav [76] developed the ECDM setup, which was rapid, and low cost to produce complete holes in thin and thick glass material. The K-type thermocouple thermometers were operated for monitoring and to regulate the temperature of the electrolyte. They create unique spherical cavity microstructures using this setup. Kulkarni [44] studied the transient and synchronized measurements done in the ECDM process. The synchronized study showed the increase in discharge temperature, which is due to the bombardment of electrons in ECDM. Maillard *et al.* [77] obtained various types of microholes in the SACE process, i.e., cylindrical holes having a fine surface, jagged outline holes, holes with thermal cracks and holes having a heat-affected zone. They found out the relative roundness error of the hole enhanced with drilling depth from 10% to 20%, which was influenced by the applied voltage. Chak [78] found that the pulsed DC has improved the performance of the process and decreased the tendency of cracking at the higher voltage as a result of better control of spark steadiness. Sandison *et al.* [79] produced glass apertures that were appropriate for the making of artificial bilayer lipid membranes by using SACE. The glass apertures were meltdown hydrophobic with a silanization technique followed by a combined microfluidic device. They achieved glass apertures having smooth and variability of diameters. This process offers rapidity and simplicity during the creation of glass apertures. Didar *et al.* [21] developed 2D machining on the glass such as well-defined linear microchannel edges and jagged outline contours.

Cao *et al.* [22] produced a stable gas film by using a load cell and the slight dipping depth of electrode, which helps to decrease the essential voltage. They observed the decreased hole size as well as surface quality achieved by using pulse voltage. A feed rate that was too slow creates and rougher surface. The microstructures less than 100 μm in sizes like 3D microstructure 10μm-thin-wall and Ø 60 μm microholes were developed. Jalali *et al.* [80] stated that material erosion happened because of a hybrid mechanism coalescing chemical etching and local heating. Tastekin *et al.* [81] supplied low-loss electrical energy to EDM process, as well as ECM process in addition to all these process combinations as ECDM provides supply without the necessity for any circuit modifications. The mode of a process was described for EDM, ECM, and ECDM, which indicates that the suggested electrical supply was open-circuit and short-proof. Nandi *et al.* [82] determined the behavior of bubbles created nearby cathode and anode in the ECDM process. They found out the voltage-current characteristics produce the discharge, as well as create bubbles during this process. Sankar *et al.* [83] enhanced the performance of the ECDM process in which the ECDM process was coupled with an electroplating gold layer. Abou Ziki [43] found that the electrolyte viscosity was the greatest dominant parameter influencing the microtexture on the glass surface. Mochimaru *et al.* [84] developed a two-step method that was advantageous to achieve the precise hole. They achieved microholes up to 12 μm on borosilicate glass material. De Souza [85] found that the machining time, tool-electrode wear and geometry of microholes were strongly influenced by input factors such as electrolyte conductivity, the chemical composition of the electrolyte, rotational speed of cathode. Abou Ziki and Wuthrich [86] observed that the tool-glass bond was developed because of a chemical reaction during the microdrilling process. Abou Ziki and Wuthrich [87] determined the tool-workpiece gap in constant velocity feed microdrilling of glass material. In the case of tool thermal expansion, they observed the machining gap smaller than 10 μm and its size, which were influenced by the flushing and local temperature. Abou Ziki *et al.* [26] observed that the local temperature was influenced by the machined surface geometry and the machining configuration, which has impact on the local flushing. Jiang *et al.* [88] investigated the modeling of a gas film in the ECDM process, which contains departure on the electrode, gas film progression bubble growth, and electrolysis characteristics transient performance. They found that the gas film behavior was influenced by voltage, current, as well as physical properties of the electrolyte. Behroozfar and Razfar [89] inspected the material removal and characteristics of the plasma channel in ECDM of glass material. They detected the spark can be focused in the desired position by a tapered tip electrode tool. Gupta *et al.* [90] examined the influence of pulse duration on the aspect ratio of machined glass by the ECDM process. The aspect ratio increases with increase in the duty factor. They noticed higher etching action attained for NaOH electrolyte as related to NaCl. Gupta *et al.* [91] found that the duty ratio was the most remarkable factor, which increased in heat-affected zone, depth of penetration, material removal rate, and surface damage for drilling on the ceramic material.

2.6.3 Literature Review on Design of ECDM

Wuthrich *et al.* [92] developed the ECDM machine for micromachined glass substrates in which the tool holder permits scanning the substrate surface and machining it in a closed loop. Yang *et al.* [93] developed an innovative machine that integrated with microelectrical discharge process and electrochemical discharge process to drill holes in the borosilicate glass material. They observed that the machine improved surface roughness and MRR. The diameter expansion of the microhole increases when increasing the machining time. Fascio *et al.* [94] examined current/voltage measurements integrated with snaps of the cathode tool under various regimes during the SACE process. Wuthrich *et al.* [95] manufactured gravity feed assisted spark-assisted chemical engraving setup has a facility of observing the movement of a tool electrode. The special actuator was used for the making of cathode tool vibrations, which enhance the mean material removal. Chak and Rao [96] developed the ECDM, which gives precise vertical feed movement to the cathode tool. The trepanning technique was utilized to eradicate microcracks when drilling occurred, it also helps to enhance the machining performance. Chak and Rao [97] made a setup applied for producing deep holes in aluminum oxide ceramics by using a setup with pulsed DC connected with the abrasive rotary tool, which improved the quality of the holes. Morrison *et al.* [98] applied a feedback controller in the SACE micromachining process. The used controller decreases the variability of the process considerably. Due to the decrease in variability, it improved the quality of machined holes. Lal *et al.* [99] developed a process for producing metallic nanoparticles by applying electrochemical discharges. Coteata *et al.* [100] invented a slotter ram mechanism that helps to provide in-line motion to cathode tool for making smaller diameter holes on steel material. Coteata *et al.* [101] modified a universal milling machine in which the rotary motion of the shaft was transformed into linear alternate motion by a crank mechanism. They experimentally observed that the higher MRR and improved quality of holes rotary when supply voltage increases by rotary abrasive tool electrode. Han *et al.* [102] invented ultrasonic vibration aided ECDM has offered an accurate electrolyte flow due to which it enhanced the surface quality of the hole. Liu [103] invented a unique type of grinding-assisted electrochemical discharge machining (G-ECDM). This machining process has been successfully used for machining of MMCs with improved surface finish and the smaller number of defects. Cheng *et al.* [104] observed that the magnetic field rises the flow of electrolyte offers speed of higher machining accuracy and higher efficiency. Wei *et al.* [105] developed a gravity feed aided ECDM machine in which tool rotation was controlled through a stepper motor. They observed with microdrill tools give maximum MRR and significantly increased machining depth at the same time radial overcut decreases. Wei *et al.* [106] investigated a new process, i.e., the electrochemical discharge dressing process in which the micro end grinding bit was applied. The output results showed that surface roughness and normal grinding force were decreased by half after dressing. Coteata *et al.* [107] modified the milling machine and transformed it into ECDM machine setup to cut steel material. Kulkarni *et al.* [108] made innovative automated five degrees of freedom ECSM machine, which can be beneficial for micromachining and surface modification. Doan [109] developed a method for glass micromachining named as spark-assisted micromachining, which was an integrated process of mechanical squeezing and electrochemical discharge machining. The 2D and 3D structures were successfully done on glass material with high surface quality due to the mechanical squeezing process. Lu *et al.* [110] integrated micromilling technology with electrochemical discharge machining technology. The results represent the best surface quality and machining efficiency could be achieved by using the ECDM micromilling process on nonconductive materials. This setup required simple equipment in the processing and hence, the cost was very low. Rusli and Furutani [111] developed the ultrasonic vibration aided ECDM setup, which increases the electrolyte circulation and fluctuates the discharge behavior resulting in reduced MRR and enhanced surface quality. Abou Ziki and Wuthrich [112] used a PID controller to provide continuous feed force to stainless steel cathode electrode between 0 and 5 N with dimensions of 500 μm and 250 μm in diameter for machining of glass material through the SACE process. Liu *et al.* [24] have worked on conventional ECDM to G-ECDM technology performance for particulate reinforced metal matrix composite material by utilizing a coated hard reinforcement phase of a diamond particle cathode tool electrode. In G-ECDM, the grinding action can efficiently eliminate recast material that was deposited on the machine surface. The G-ECDM provides improved surface quality surface and greater machining efficiency than conventional ECDM. Jiang *et al.* [113] developed vibration-assisted ECDM, which gives maximum machinable depth compared to conventional ECDM. Paul and Hiremath [114] made a prototype of ECDM setup with the attachment of microlinear actuators for cathode tool movement and X-Y scanning mechanism used to machine hole in a silicon wafer workpiece material by applying stainless steel wire. The material removal rate rises with increasing voltage. Paul *et al.* [115] developed a setup in which cathode tool feed was controlled through stepper motor and stepper motor rotation was regulated by Arduino controller, which was interfaced to the computer. Paul and Hiremath [116] developed ECDM setup, which has automated control X-Y axis of the workpiece, as well as tool feeding motion, is also controlled by a motor. Gao *et al.* [117] fabricated the gravity feeding mechanism ECDM in which the workpiece was fixed on a fixture and dipped in electrolyte cell, which gives a high aspect ratio. Krotz and Wegener [118] investigated a new technique named as spark-assisted electrochemical machining (SAEM), which integrates electrochemical arc machining with contact arcs and is used for the micro-drilled hole and finishing of a drilled hole. This machine improves the machining efficiency. Kulkarni and Jain [119] fabricated and designed a modular electrochemical spark micromachining setup. The motion control card on the LabVIEW platform was used for the automation of machine axis. In this machine setup, they also provide an online data recording of the process gap voltage and current. The machine was versatile and can be used for machining, deposition, and surface modification purposes. Furutani and Kojima [120] invented a new type of ECDM, i.e., electrochemical discharge machining with the lathe-type machine. In this machine, a tool electrode was located with a two-axis rigid support and provisions of a force sensor. This flexible force sensor conceives the friction among the workpiece and cathode tool electrode deformed in its lateral direction. Zhang *et al.* [121] developed the gravity feed mechanism ECDM setup in which the workpiece movement is controlled by the gravity feed mechanism. The cathode tool downward, upward motion, and rotation speeds are controlled by a computer control system. Hajian *et al.* [122] investigated the impact of voltage, electrolyte concentrations and magnetic field orientation on ECDM enactment of glass material. They occurred the effect of magnetic field on magneto hydrodynamic convection by which bubbles motion get increased. The direction of bubbles movement was influenced by the magnetic field orientation. Madhavi and Hiremath [123] developed a new μ-ECDM setup, which was efficiently used to make channels and holes on borosilicate, as well as soda-lime glass. The maximum MRR was achieved for input parameters of voltage 60 V, duty factor 70% and concentration 25 wt%. Ladeesh and Manu [124] investigated the effect of machining factors during the process of borosilicate glass by using G-ECDD, which gives accurate holes with better repeatability. They observed the maximum MRR of 0.05896 g/min, which was attained by using optimum parameters, i.e., higher voltage 110 V, lower pulse on-time 0.0002 s and higher concentration 4 M. Ladeesh and Manu [125] found the most important factor was duty cycle then voltage, concentration, and cycle time for G-ECDD. The maximum tool wear was detected for a high voltage of 110 V and a higher frequency beyond 5 kHz [126]. Similarly, the feed rate, voltage, duty factor, and frequency were the utmost important factors for surface roughness in G-ECDM process. Jiang [127] invented a hybrid machining technique, which was the integration of electrochemical discharge as well as mechanical cutting named as electrochemical discharge assisted cutting. The mechanical cutting can be advantageous as the material was softened due to the high heat creation in electrochemical discharging, which enhances surface quality the material removal rate was enhanced noticeably using the hybrid process. Singh and Dvivedi [128] fabricated the fixture for ECDM in which a pressurized feeding mechanism was done through compression springs. The pressurized feeding was utilized to control the constant working gap in the workpiece and cathode tool. This machine has advantages to enhancement in machining depth parameters. Elhami and Razfar [129] developed an ultrasonic-assisted electrochemical discharge machine to make ultrasonic vibration, which causes an increase in the number of discharges and cooling mechanism on the tool. The ultrasonic vibration decreased the wear of tool in between 3% and 14% and improved the material removal up to 35%. Xu *et al.* [130] investigated a new technique called counter-resistant microhole drilling. The reducing contact force was applied at the tooltip by using the new setup with flexure beams, which enriches the consistency and speed. In this setup, the electrolyte can be certainly refilled to the machining point to improve the electrolyte flow in the hydrodynamic regime. This setup assists in enhanced repeatability of machining with the new fixture by reducing the deformation and clearance, which was better than the conventional gravity-feed method. Hof and Wuthrich [29] investigated new manufacturing methods fitting Industry 4.0, which can be used for inexpensive rapid prototyping technology and machining setup. This approach can be useful for the fabrication of tool making, reducing lead times and costs as related to the conventional SACE machining process.

2.6.4 Literature Review on Workpiece Materials Used in ECDM

Cook *et al.* [15] studied discharge machining of glass in which they observed the erosion of glass that takes place due to discharges between a tool and an electrolyte. They achieved drilling rates up to 0.1 in/min. They also observed bubbles that were formed during electrolysis seem to be strongly charged. Tokura *et al.* [131] studied electrical discharge hybrid processing on alumina and silicon nitride ceramic materials using a needle electrode. The lower electrode wear and high removal rate were found when the needle tool electrode was negative. When the high voltage was applied during the machining process the greater volume gets removed.

Tandon *et al.* [132] the cutting and drilling operation were done by using ECDM to produce holes in composite materials. Singh *et al.* [133] the traveling wire electrochemical spark machining was utilized for machined carbon fiber epoxy composites and piezoelectric ceramic materials. Gautam and Jain [134] produced holes on quartz and borosilicate glass materials with the help of a rotational tool, which enhanced the process performance. The maximum depth of the machining hole can be achieved by an orbital rotational tool. Doloi *et al.* [135] determined the drilling operation on zirconium oxide ceramic workpiece by the ECDM process. The interelectrode gap, voltage, and concentration of electrolyte were the three main factors that considerably influence the overcut and MRR. Fascio *et al.* [136] produced the three-dimensional micropattering structure. This machine can be useful to make 3D microstructuring on glass material and surface modification. Jain *et al.* [137] used an abrasive cutting tool in the ECSM drilling process on alumina and borosilicate glass materials. This process enhances the material removal and increases machined depth. Lee *et al.* [138] fabricated the reversible interconnections for a glass-based microfluidic device with the application of the ECDM process as well as it also produced the three-dimensional vias in glass specimen. Peng and Liao [139] investigated the traveling wire electrochemical discharge machine to cut the quartz bars and optical glass materials. Wuthrich *et al.* [42] stated that the SACE process was used to obtain rapid prototyping of fused silica material having a microcrack free surface and high aspect ratio. The conical hole drilling of 450 μm depth and 300 μm diameter was acquired in nearly 30 s. This process does not require a clean environment or heavy machinery. Sarkar *et al.* [140] machined silicon nitride ceramic material by using the ECDM process. They observed that the voltage was the utmost dominant parameter for radial overcut, heat-affected zone and material removal rate. Zheng *et al.* [141] achieved complex three-dimensional microstructures on Pyrex glass material by ECDM process. They found that the optimum combination factors of tool rotational rate and pulse voltage were improved machining accuracy. Zheng *et al.* [142] the pulse voltage was persuaded to enhance the spark steadiness to improve the accuracy. The machining accuracy noticeably enhances as the rotational rate was increased. Jain and Adhikary [143] observed that in electrochemical spark machining by using inverse polarity slices the quartz material at a higher rate. However, results obtained by using reverse polarity, i.e., tool wear, surface roughness and overcut were greater as related to the direct polarity. Furutani and Maeda [144] verified the machining performance of a rotating glass rod using ECDM. The surface roughness, as well as the width and depth of the machined grooves of their bottom, were improved by increasing the voltage. Kudla [145] obtained microholes partly in diamond crystals as well as in borosilicate glass through the ECDM process and these produced microholes by using different shapes of the tool electrode tip. Didar *et al.* [146] obtained the two-dimensional micromachining on the soda-lime glass material by SACE. The density and hardness of the machined zones get decreased during this process. Ozhikandathil *et al.* [147] machined material of silica layer on a silicon substrate by using SACE. Manna and Narang [148] machined the e-glass-fiber-epoxy composite by newly designed different shapes of cathode tool electrodes. Jawalkar *et al.* [149] produced microchanneling on soda-lime glass by ECDM. Paul and Hiremath [150] machined borosilicate glass to attain a blind hole by using a developed ECDM setup. Jain and Priyadarshini [6] produced microchannel on quartz by using ECDM and achieved a high aspect ratio. Bhuyan and Yadava [151] fabricated a tabletop traveling wire ECSM setup for cutting of borosilicate glass. Coteata and Cretu [152] carried out drilling of spring steel using ECDM. For MRR and TWR the density of electrolyte, voltage and capacitance of the capacitors were the most influencing parameters. Pawar *et al.* [153] concluded that the various researchers have taken workpiece materials as a variety of nonconducting materials, i.e., glass, ceramic, and composites, as well as some conducting materials. Sarkar *et al.* [154] investigated electrochemical discharge micromachining on silicon carbide material by using stainless steel tool electrode. They found that radial overcut and MRR increased with increasing machining voltage and decreased with the rise in the inter-electrode gap. He *et al.* [155] analyzed processed products by applying SACE process on ZrO_2 ceramics. As a result, they observed black crystalline debris as physical removal and white amorphous debris as chemical removal. Wang *et al.* [156] integrated electrochemical discharge with a diamond wire sawing machine have been used to improve the surface roughness and MRR of alumina material.

Figure 2.2 shows the percentage of materials used as workpieces during the present review study. There are near about 22 types of materials that have been used for the ECDM study including different types of glass and metals, which have conductive and nonconductive properties in it. So, according to observations most of the glass materials is used as workpiece followed by metals as a workpiece. Same observations have been noticed by Goud *et al.* [30]. Due to specific properties such as chemical resistance and transparency as compared to metals glass gives faster results when machining with ECDM. It is very easy to obtain a precise result on glass material in less time as compared to other workpiece materials. On the other hand, glass material has huge demands in various fields so most the researcher prefers glass material followed by other metals workpiece.

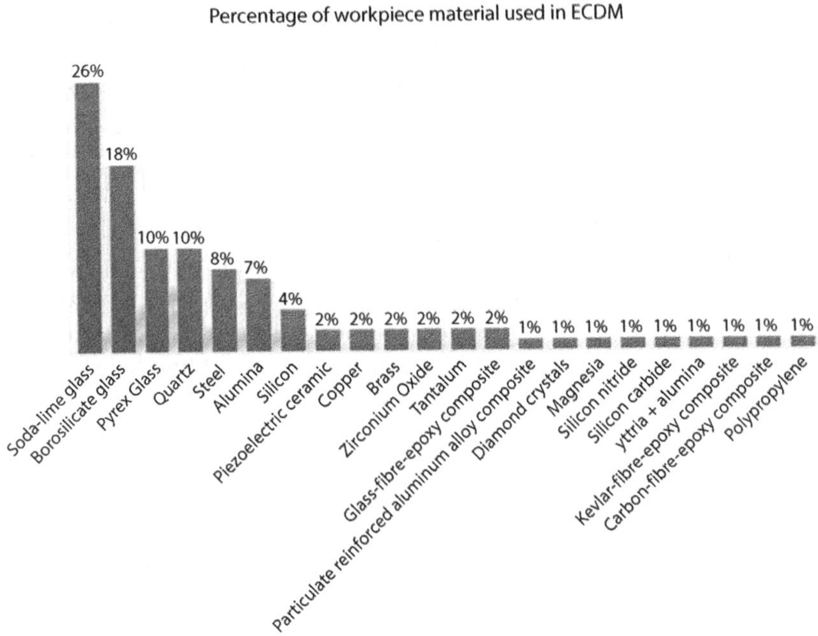

Figure 2.2 Percentage of materials used as workpiece in ECDM.

2.6.5 Literature Review on Tooling Materials and Its Design in ECDM

Schopf *et al.* [157] used the ECDM process for trueing and dressing metal bonded diamond grinding wheel tools, which offers better roundness and surface quality. Lim *et al.* [158] combined two machining systems, i.e., first a tungsten tool made using an electrochemical process (ECP), further, this tool is used for machining ceramic material in the ECDM process. The blind holes and a through a hole on the ceramic material were machined by using the tool made up of the ECP process. Wuthrich *et al.* [159] developed a setup based on atomic force microscopy, which provides extremely sharp tool electrodes and gives the resolution features in glass around 25 μm machining by using SACE. Yan *et al.* [160] improved the ECDM quality of microhole by applying the electrophoretic deposition grinding (EPDG) technique. In this technique first ECDM was utilized to drill a microhole and then EPDG was used to polish the hole. Zheng *et al.* [161] the flat sidewall-flat front tool electrode was fabricated, and this tool was used to decrease taper phenomena as a result of the sidewall discharge, which improves the quality of the microhole as well as process performance of ECDM. Han *et al.* [162] made a side-insulated tool that controlled a continuous contact surface area with the electrolyte, this electrode was utilized in the ECDM. This electrode forms steadier spark discharges as compared to conventional tool electrodes. The surface roughness of the microchannel was increased by a side-insulated electrode. Han *et al.* [163] Similarly, the side insulation tool was utilized to produce microgroove on the borosilicate glass with the increase in machining resolution. Mousa *et al.* [164] studied the process performance of stainless steel, tungsten, high carbon steel and copper tool electrodes. They observed that the electrodes had properties of low heat capacity and high thermal conductivity, which were given quicker drilling in the gravity-feed SACE drilling process. Yang *et al.* [165] fabricated electrodes of different materials i.e., Tungsten, 304 Stainless steel as well as Tungsten carbide by using wire electrical discharge grinding then these electrodes were used in the ECDM process. Abou Ziki *et al.* [166] studied the effect of forces applied to the cathode tool during steady velocity feed. They observed that the thermal expansion of the cathode and the workpiece happened because of the pressure of gas film formation. Yang *et al.* [23] produced through holes in quartz material by using a spherical end tool electrode, which improved the shape accuracy of a hole. Han *et al.* [167] formed surface textures on the brass wire tool by using the EDM process. Then this surface texture tool was used for cutting soda-lime glass, which gives the smallest surface roughness. Abou Ziki and Wuthrich [168] measured the tool wear of stainless steel, tungsten, and steel electrodes in the SACE process. They found that the tungsten tool electrode has the maximum tool wear compared to stainless steel and steel materials. Jui *et al.* [169] invented the high aspect ratio microtools and these tools were used for obtaining the deep microhole on the glass by ECDM process. Cao *et al.* [170] achieved 3D microstructures with high surface quality on glass material by a polycrystalline diamond tool in the ECDM process. Huang *et al.* [171] developed the high-speed tool rotation ECDM machine in which cathode tool electrode rotate at 42,000 rpm. They drilled the hole on stainless steel by using a rotational tool, which enhances the circularity and reduces the tool wear. Abou Ziki [172] developed a methodology for measuring the local machining zone factors by considering the force utilized on the tool. The thermal expansion, tool bending, and wear were considered during measuring and analyzing the machining forces. Jiang *et al.* [25] reported that the tapered cathode tool can improve the steadiness of spark generation, which gives a maximum depth of machine holes compared to conventional tool electrode.

Furutani *et al.* [173] machined holes and grooves on glass materials by using electrochemical discharge machining with multiple electrodes were carried out. They found that many cracks were frequently nearby a machined hole because of the accumulation of discharge by switching the electrodes to decrease the number of cracks. The performance of ECDM was related to those of several electrodes connected electrically in parallel and of a single electrode. The machined roundness of holes was better electrically in parallel than those with a single electrode. Razfar *et al.* [174] used a tungsten carbide tool electrode to produce holes on soda-lime glass material through a gravity feed ECD drilling process. They observed that vibration was not influenced greatly when a rotating tool electrode was used. Jiang *et al.* [41] fabricated a conic tool electrode to provide steadier discharging. They found sparks were steady and concentrated with conic tools therefore the temperature of the tip-point increases quickly and could wear out the tool. Zhang *et al.* [175] developed a tube electrode for a high-speed ECD drilling machine setup having a precise five-axis motion table and high precision positioning, which gives an improvement in surface quality as machining accuracy. Saranya and Sankar [176] analyzed the effects of feed rate and tool shape on quartz material by the ECDM process. They found that a tool with a spherical tip offers a profile with a decreased overcut and entrance diameter.

Behroozfar and Razfar [177] studied the influence of high voltage on the tool of three various tool materials, i.e., steel, tungsten carbide, and brass. The tool wear is mostly dependent on the chemical composition and melting point of the tool materials. Goud *et al.* [30] revealed that copper, stainless steel, and brass materials are mostly used as tool electrodes in ECDM. In recent, abrasive coated mild steel and coated tools like nickel-coated mild steel were used to enhance the process performance of the ECDM. Guo *et al.* [178] fabricated micro/nano-electrode for micro-ECDM using a focused-ion-beam chemical vapor deposition. The various 3D nano-electrodes containing pillar, horn, array types, and corners were fabricated. The experimental observation showed that the unilateral discharge gap and the hole diameter increased with increase in machining voltage. The complex 3D micro/nano-electrode for 3D micro-ECDM can be fabricated by using the FIB-CVD phenomenon at the submicron scale. Hajian *et al.* [179] evaluated the bending forces applied to the tool electrode. They found that increasing the electrolyte concentrate and maximum machining voltage causes reducing the bending force. Liu *et al.* [180] used a rotary helical tool in micro-WECDM during the machining of glass material. They achieved pattern structure and kerf having a high aspect ratio of complex closed structure. Tang *et al.* [181] stated that the side-insulated tool electrode gives greater surface integrity and smaller hole diameter without a noticeable heat-affected zone. Han *et al.* [182] used a surface textured tool in the ECDM process to produce high-precision microgrooves

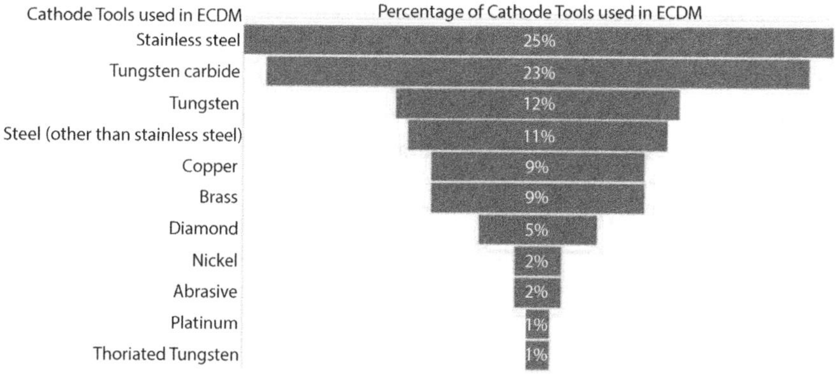

Figure 2.3 Percentage of cathode tools used in ECDM.

on soda-lime glass. This tool has given the improved surface quality and high edge linearity of microgrooves. Saranya and Sankar [183] coated thick insulating film around the tool for increasing the quality of microchannels by the SACE process. This insulated cathode tool restricts the spark discharges only at the tool's bottom surface. They fabricated a microchannel with the help of an insulated tool, which increase its machining depth by 19.53 % and overcut decreased by 57.8 % and had a smooth surface and regular edges.

Figure 2.3 shows the percentage of cathode tools material used in ECDM. So, according to the present observations stainless steel, Tungsten carbide is widely used as cathode tool followed by copper, brass, and so on. Same observations have been found by Goud *et al.* [30]. According to them, it has high wear and temperature resistance, chemical inertness, so the temperature of the electrode rises immediately, which resulted in greater discharge activity, so with this electrode material, removal is fast as compared to other materials.

2.6.6 Literature Review on Electrolyte Chemicals Used in ECDM

Yang *et al.* [20] observed better-overcut quality by the addition of SiC abrasive into the electrolyte. They observed that the addition of abrasive into the electrolyte decreases the surface roughness. Han *et al.* [184] studied the addition of the conductive particles in the electrolyte, which resulted in the enhancement of machined surface integrity and surface roughness. Laio *et al.* [185] investigated the influence of Sodium Dodecyl Sulfate surfactant mixed into electrolyte chemical for making microhole on quartz material by using ECDM. They observed that the current density was enhanced due to the addition of surfactant mixed into the electrolyte. Jawalkar *et al.* [186] observed the MRR was maximum when NaOH was utilized as an electrolyte as related to NaNO3 during the process of soda-lime glass material. The voltage was found to be a major parameter with a contribution of 70.14 % towards material removal. Dhanvijay *et al.* [187] investigated the machining of ceramics by electrolyte flow and stagnant method with stainless steel and copper tool. They found that the constant flow of electrolyte using stainless steel electrode provides the reduced diametric overcut and maximum MRR. Gupta *et al.* [10] compared the results of different electrolytes in ECDM, i.e., NaOH, KOH and NaCl during the drilling of glass material. They found that the maximum spark intensity was caused by the maximum peak voltage for NaOH electrolyte. Therefore, hole overcut, and material removal rate was significantly influenced by spark characteristics. They observed that the NaOH gives higher material removal rate than that KOH and NaCl electrolytes. Pawar *et al.* [153] stated that most researchers have used NaOH as an electrolyte followed by KOH, $NaNO_3$, NaCl, and HCl etc. Nguyen *et al.* [188] The electrolyte concentration, directly influences the creation of hydrogen films, was an important factor to enhance the machining resolution. Zhang *et al.* [189] the very high-pressure interior flushing rises the process efficiency, quality of surface, and material removal rate. Paul and Hiremath [190] successfully machined silicon wafers by using mixed electrolytes of KOH and NaOH of varying electrolyte concentrations with the help of a newly fabricated micro-ECDM setup. They observed that the improvement in material removal rate occurred because of a mixed electrolyte. Paul and Hiremath [191] revealed that many researchers have developed experimental setups, used various cathode tools and auxiliary electrodes, different workpieces and used NaOH and KOH as electrolytes. Saranya *et al.* [192] detected that the KOH needs a higher critical voltage than the NaOH solution due to its relatively smaller conductivity. Also, the precise channels with regular channel edges and smooth surfaces were obtained at an applied voltage of 4V and the tool travel rate was between 1 to 2 mm/min. Sabahi *et al.* [193] used anionic and cationic surfactant used in different concentrations was liquified in the 25 wt% NaOH and KOH alkaline chemicals for decreasing the thickness of the gas film by influencing the physical and chemical properties of the electrolyte. They observed that because of the existence of surfactant microchannels with more MRR, lower heat-affected zone and increased the surface quality. Sabahi and Razfar [194] investigated the effects of two alkaline electrolytes, i.e., NaOH and KOH individually and these two electrolytes mixed in equal proportion. The fabrication of a deeper microchannel having sharper sidewalls was created on the glass surface as compared to KOH and NaOH independently. Sabahi *et al.* [195]. The microchannels were produced on the glass surface by applying a magnetic field under NaOH and KOH electrolytes in the ECDM process. They observed that because of a magnetic field in NaOH 15 wt.%, the hardness of the microchannels edge was improved. Harugade and Waigaonkar [196] used three different electrolytic solutions, i.e., H_2SO_4, NaOH, and $NaNO_3$. They found that the MRR was greater for NaOH, and it was reduced for $NaNO_3$ and H_2SO_4. Also, they have used H_2SO_4 the workpiece material became cracked because of unstable spark generation, therefore, H_2SO_4 can't be used to machine soda-lime glass material. Gupta [197] investigated the effect of the electrolyte level during the process of glass by using ECDM. They concluded that at the lower electrolyte level conditions results obtained higher MRR, lower heat-affected zone, depth of cut and overcut. Varghese and Paul [198] carried out experiments on Polypropylene materials with graphite particle powder mixed in the electrolyte to enhance the machining performance. Therefore, MRR can be improved because of conductive powder particles in the electrolyte during the ECDM process.

Figure 2.4 shows a percentage pie chart for chemicals used as electrolytes in ECDM. It is used as working media in this process. During these observations total eight electrolytes and their combinations have been

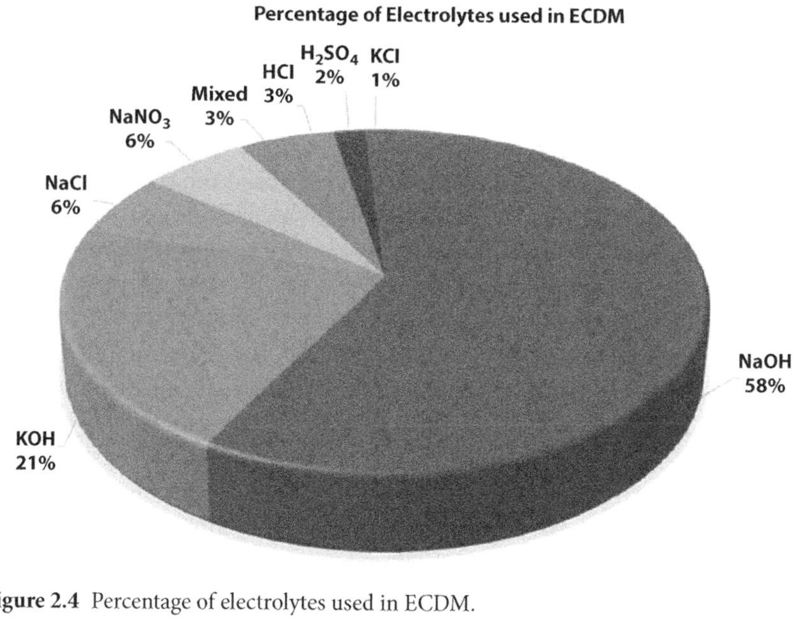

Figure 2.4 Percentage of electrolytes used in ECDM.

applied for the machining of different materials. The most often used electrolytes are NaOH and KOH. Same observations have been noticed by Goud *et al.* [30]. Due to the irritant nature and higher mobility of K+ ion in KOH and Na+ ion in NaOH more MRR was yielded than the other electrolytes. While other electrolytes did not get that much MRR as well as they produce toxic gases, which are harmful, they have less contribution to the ECDM process.

2.6.7 Literature Review on Optimization Techniques Used in ECDM

Doloi *et al.* [135] the Taguchi technique of parametric optimization was used for the design optimization of the process parameters. The maximum MRR obtained at the optimal level of parametric conditions was an electrolyte concentration of 25 % aqueous KOH, a voltage of 70 V, and an inter-electrode gap of 20 mm. Sarkar *et al.* [140] the response surface methodology (RSM) used for developing mathematical modeling and optimization method for processing on silicon nitride ceramics by using ECDM. They concluded that the voltage was a more significant parameter. Bhuyan and Yadava [199] used a weighted principal component methodology with RSM for multi-objective optimization of a fabricated traveling wire ECSM process. The Taguchi methodology combined with weighted principal component methodology approach has decreased surface roughness by 15 % and enhanced the material removal rate by 222 % from the initial value. Also, as compared to the Taguchi methodology, the hybrid approach improves the MRR by 46 % and decreases the surface roughness by 10 %. Razfar *et al.* [200] observed that applied voltage considerably increases the radial overcut, heat affected zone, roundness error and material removal rate. Paul and Hiremath [150] used the response surface modeling for optimization of input factors and developing a mathematical model for the MRR and TWR. They obtained optimum input process factors by the Taguchi method, and these were the voltage of 60 V, the duty factor of 70 % and electrolyte concentration of 30 % wt. for reduced radial overcut. Paul and Hiremath [114] The grey relational analysis results showed reducing diametrical overcut and heat-affected zone, the optimum parameters should be voltage 45 V, duty factor 60 % and electrolyte concentration 30 %. Paul and Hiremath [116] the experimental results were modeled using response surface modeling. They found that the MRR enhanced with the rise in voltage and temperature. Mallick *et al.* [201] used a multi-objective optimization technique to obtain the optimal parametric combination for high machining depth, MRR and minimum overcut by utilizing RSM. Ladeesh and Manu [125] investigated the effect of input process factors such as cycle time, duty cycle, electrolyte concentration and voltage on MRR using response surface methodology. The significant factors and their percentage contribution were identified by using the analysis of variance. They observed that the duty cycle was the most dominant factor followed by voltage, cycle time and concentration. Sarkar *et al.* [154] applied L9 array Taguchi method for carried out experiments. They used single and multi-objective optimal factors combination for higher MRR and lower radial overcut. Goud and Sharma [202] used Taguchi's L9 orthogonal array coupled with the Grey relational analysis method to investigate the optimal process parameter conditions for rising the MRR and decreasing the width overcut of microchannels. The optimal parameters for maximum MRR, lower width and overcut were concentration of electrolyte of 20 %, voltage of 40 V, and feed rate of 5 mm/min. Antil *et al.* [203] optimized parameters for SiC reinforced polymer matrix composite material by the grey relational analysis method. The results showed that the multi-response optimization technique using grey relational analysis contributes to increasing MRR with the least taper and overcut.

Figure 2.5 shows the percentage of input parameters used during ECDM. From the observations, it was noticed that the voltage and electrolyte concentration are the most frequently used parameters in ECDM followed by tool speed, feed rate and so on. As the applied voltage is important to discharge energy during machining and finely it resulted in vaporizing, melting, and thermally removal of the workpiece so that applied voltage is one of the most considered input parameters. The properties of electrolytes play a key role in the machining quality of this process. Cao *et al.* [22] fabricated glass having micro-structure features less than 100 μm by using ECDM process. As a result, they obtain high aspect ratio structures. They observed, KOH electrolyte gives less machining gap as compared to NaOH. According to Singh *et al.* [39] MRR is noticeably raised with the rise in voltage and concentration of electrolyte.

Liu *et al.* [57] studied the effect of pulse duration, duty cycle, current, and electrolyte concentration during the ECDM process. According to them increase in input parameters, the sparking rate increases. The electrolyte concentration produces a higher machining rate because of the rise in spark intensity and etching rate. So that the electrolyte concentration is also the most important input parameter and used in ECDM. The tool speed input parameter controls Surface characteristics of microchannels this observed by P. Maillard *et al.* [77]. Goud *et al.* [30] reviewed concept of material removal mechanism in ECDM. According them ECDM can successfully work on metal matrix materials and ceramics. They also stated that NaOH and KOH are frequently used electrolyte for this process because they are less

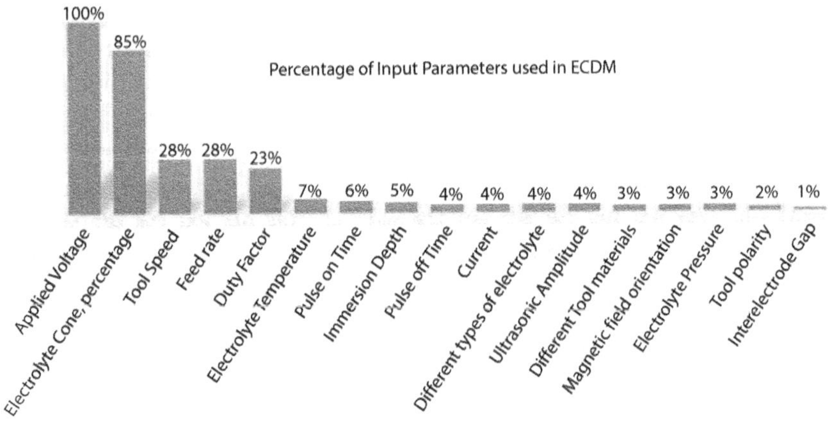

Figure 2.5 Percentage of input parameters used in ECDM.

hazardous. The tungsten carbide tools have high wear resistant properties so it is commonly used in tool electrode ECDM.

Figure 2.6 presents the output parameters of ECDM observed during the current review. The figure shows that the MRR, machining depth and hole overcut are the most valuable output parameters followed by surface roughness, surface quality etc. Most of the existing research studies

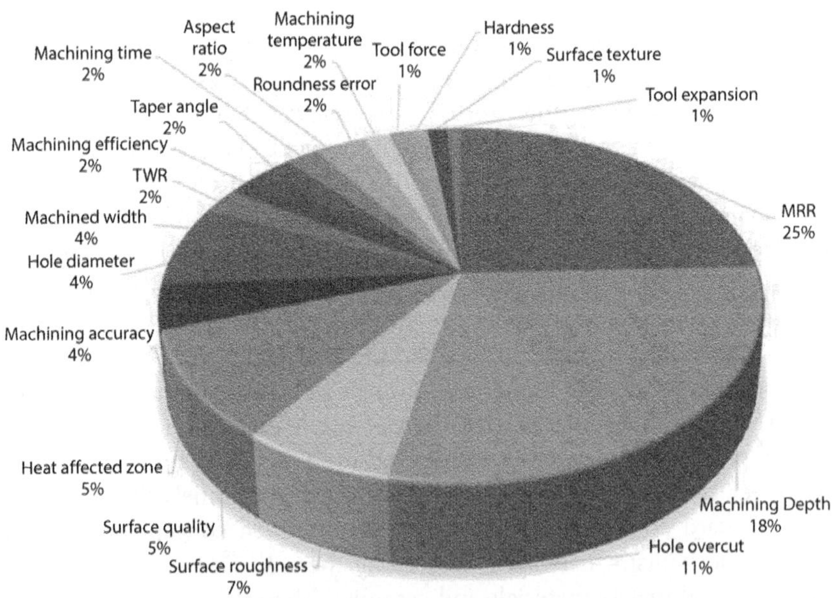

Figure 2.6 Percentage of output parameters used in ECDM.

have been focused on MRR and overcut. The machining performance is referred to as the dimension of component, MRR and machining quality. The MRR is raised with the rise in current supply of ECDM process. MRR is nothing but the final output result of the work done so that's directly connected to the input parameters used during experiments. So that whatever the research is done in ECDM in this the key achievement is MRR with some superior qualities. There are many evidences found during this investigation that shows the positive or negative correlation between input and output parameters. Singh *et al.* [1] discovered that MRR is noticeably raised with the rise in voltage and electrolyte concentration. Yang *et al.* [93] stated that alkaline electrolytes like NaOH and KOH produce maximum MRR as related to neutral electrolytes. Therefore, it is mostly used electrolytes. Didar *et al.* observed that MRR and overcut rate significantly improved with the use of spherical tool electrode. Yang *et al.* [165] studied those vibrations of tool electrode to improve MRR due to better flushing action. Cheng *et al.* [36] observed that MRR and overcut are directly related to the circulation of electrolyte. The electrolyte concentration was the most significant factor that will affect MRR and spark gap width. Zhang *et al.* [175] conclusions show a rise in MRR and improvement in taper angle with an increase in the inner diameter of the electrode. The decrease in the rate of electrolyte concentration of the ECDM process gives a low overcut rate. Hence the output parameters are the outcomes of input parameters in ECDM. These are characteristics of surface quality influenced by input parameters. In the present study, we observed many output parameters some of them are demanding but some are unwanted impacts that we can't avoid so that this can be a future The second most focused output parameter is machining depth it is nothing but one superior quality of microhole. Wuthrich *et al.* [19] stated that machining depth is a surface characteristic, and it is influenced by the change in voltage. Manna and Narang [148] stated that during microdrilling, the diameter of the hole at starting was equal to tool diameter and further it is reduced as the depth of the hole increases, which is also considered as machining depth. Coteata *et al.* [100] studied that machining at higher depths is possible with variation in the shape of a tool electrode.

The hole overcut is also one of the important output parameters, but it can be a negative impact on input parameters. According to Singh and Singh [39], hole overcut is output quality characteristic that is significantly affected by the applied voltage. According to him, overcut rate can be minimized using abrasive coated tool electrode and side insulated tool electrodes. The MRR and overcut rate is significantly improved with the use of

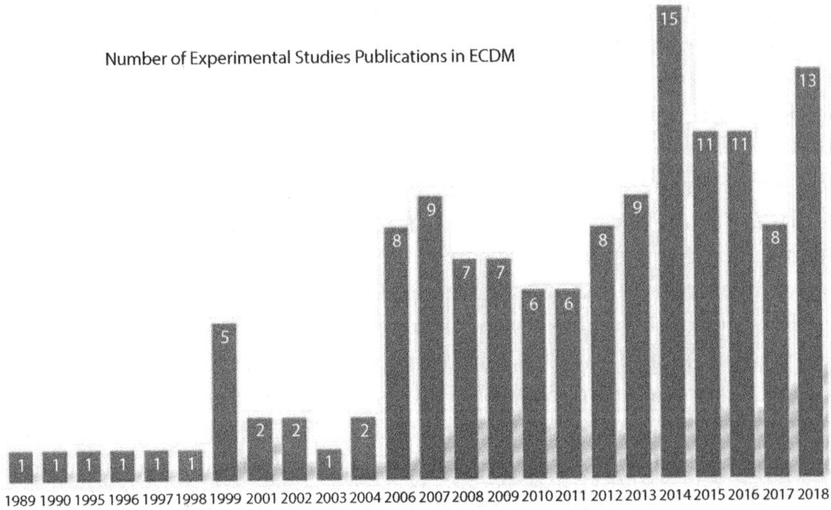

Figure 2.7 Number of experimental studies publications in ECDM from Table 2.2.

a spherical tool electrode. While according to Allagui and Wuthrich [56], side spark is an important aspect of controlling the overcut rate. Cheng *et al.* [36] stated that MRR and overcut are directly related to circulation work as a challenge in ECDM experiments.

Figure 2.7 presents the number of experimental studies publications in ECDM for the period 1989 to 2018. This analysis is based on Table 2.2. it shows very less experimental work done in past but after the year 2000, there is progress in this research area. It may be due to the lack of experimental literature, which includes detailed information on machining design, methodology and cost of material setup. ECDM is less time-consuming, efficient, and used for the microfabrication of various materials in different fields. As ECDM has a wide range of applications it gets high demand in various domains as well as its problems are a challenge for new research trends. Hence these things are a motivation to do the experimental study on ECDM and it will help to fulfill the lacunae in ECDM research.

Table 2.2 shows a brief overview of past research work with respect to workpiece material, tool material, electrolyte chemical, input process parameters, and its output responses of ECDM.

Table 2.2 Brief overview of past research work based on workpiece material, tool material, electrolyte chemical, input process parameters and its output responses of ECDM by various researchers.

S. no.	Name of researchers	Year	Workpiece material	Tool material	Electrolyte chemical	Input process parameters	Output process parameters
1	H. Tokura *et al.* [131]	1989	Alumina, ASN, MSN, YASN	Ni- wire	NaOH	Applied voltage, electrolyte conc., load, dipping, different material	Material removal rate (MRR)
2.	S. Tandon *et al.* [132]	1990	Glass-fiber epoxy, Kevlar-fiber-epoxy	Brass	NaCl	Specific conductance, fiber volume fraction, voltage	MRR, Average overcut, TWR, Relative tool wear
3	V. Raghuram *et al.* [68]	1995	Quartz glass	Thoriated tungsten	NaOH, KOH, HCl	Applied voltage, different types of electrolyte material, electrolyte concentration	The effect of different power supply configuration, Through holes obtained
4	Singh *et al.* [133]	1996	Piezoelectric ceramic, Carbon fiber epoxy	Brass	NaOH	Voltage, electrolyte concentration, feed rate	MRR, Average diametrical overcut

(Continued)

Table 2.2 Brief overview of past research work based on workpiece material, tool material, electrolyte chemical, input process parameters and its output responses of ECDM by various researchers. (*Continued*)

S. no.	Name of researchers	Year	Workpiece material	Tool material	Electrolyte chemical	Input process parameters	Output process parameters
5	I. Basak *et al.* [9]	1997	Glass	—	NaOH, KOH, NaCl, KCl	Applied voltage, types of electrolyte matl., electrolyte con., tool dia.	MRR
6	N. Gautam *et al.* [134]	1998	Borosilicate glass and Quartz	—	NaOH	Voltage, electrolyte conc., electrolyte temp., speed, tool tip depth in electrolyte	Machining depth, MRR, hole circularity
7	V. Jain *et al.* [37]	1999	Soda-lime glass and Borosilicate glass	—	NaOH	Voltage, electrolyte conc., tool diameter, heat transfer coefficient, depth of tool inside electrolyte	MRR, machining efficiency
8	B. Bhattacharyya *et al.* [40]	1999	aluminium oxide ceramic	Copper	NaOH	Voltage, electrolyte conc., shape of tool tip	MRR, overcut

(*Continued*)

Table 2.2 Brief overview of past research work based on workpiece material, tool material, electrolyte chemical, input process parameters and its output responses of ECDM by various researchers. (*Continued*)

S. no.	Name of researchers	Year	Workpiece material	Tool material	Electrolyte chemical	Input process parameters	Output process parameters
9.	R. Wuthrich *et al.* [92]	1999	Pyrex and Borosilicate glass material	Stainless steel	NaOH	Applied voltage, electrolyte conc., applied force, rotation of tool	MRR, machining depth, surface quality
10	B. Doloi *et al.* [135]	1999	Zirconium oxide ceramic	Copper	KOH	Voltage, interelectrode gap, electrolyte concentration,	MRR, radial overcut
11	V. Fascio *et al.* [136]	1999	Soda-lime glass, Glass M, Borosilicate	Stainless steel	NaOH	Applied voltage, electrolyte conc., working distance, the position of the tool tip, number of machining steps	Machining depth, surface quality

(*Continued*)

Table 2.2 Brief overview of past research work based on workpiece material, tool material, electrolyte chemical, input process parameters and its output responses of ECDM by various researchers. (*Continued*)

S. no.	Name of researchers	Year	Workpiece material	Tool material	Electrolyte chemical	Input process parameters	Output process parameters
12	C. Yang *et al.* [93]	2001	Cornings 7059 Glass	Tungsten	NaOH, KOH, NaCl, KCl	Applied voltage, different electrolyte chemicals, electrolyte conc., electrolyte temperature, speed of tool electrode, machine time	MRR, machined hole diameter, surface roughness, surface integrity, hole roundness,
13	H. Lim *et al.* [158]	2001	Ceramic material	Tungsten	KOH	Applied voltage, electrolyte concentration, tool electrode various positions, Diameter, and length of tools	Machining depths, various geometrical shapes
14	A. Kulkarni *et al.* [38]	2002	Silicon, Brass, Copper, and Tantalum	Copper	HCl	Applied voltage, electrolyte concentration, time-varying current	Signature of individual discharges, their size, discharge-affected region, material removal

(*Continued*)

Table 2.2 Brief overview of past research work based on workpiece material, tool material, electrolyte chemical, input process parameters and its output responses of ECDM by various researchers. (*Continued*)

S. no.	Name of researchers	Year	Workpiece material	Tool material	Electrolyte chemical	Input process parameters	Output process parameters
15	V. Jain *et al.* [137]	2002	Borosilicate glass, alumina	Mild steel & Abrasive tool	NaOH	Applied voltage, electrolyte concentration, electrolyte temperature, tool speed, tool eccentricity, machining time	Material removal and machined depth
16	V. Fascio *et al.* [94]	2003	Glass	Stainless Steel	NaOH	Applied voltage, electrolyte concentration, scan rate	Measured current/ voltage of tool electrode at different domain
17	E. Lee *et al.* [138]	2004	Borosilicate glass	Nickel, Platinum	NaOH	Applied voltage, electrolyte concentration, electrode diameter	Etch rate, microfluidic system developed

(*Continued*)

Table 2.2 Brief overview of past research work based on workpiece material, tool material, electrolyte chemical, input process parameters and its output responses of ECDM by various researchers. *(Continued)*

S. no.	Name of researchers	Year	Workpiece material	Tool material	Electrolyte chemical	Input process parameters	Output process parameters
18	W. Peng *et al.* [139]	2004	Optical glass and Quartz	Stainless steel wire	KOH	Applied voltage, electrolyte conc., duty factor, efficiency, wire tension, wire diameter	Slicing of workpiece material
19	C. Yang *et al.* [20]	2006	Pyrex glass	Brass wire	NaOH, KOH	Applied voltage, abrasive concentration and grit size, power frequency, duty factor, tension, wire feed rate	Effects on expansion, MRR, surface roughness
20	K. Bhondwe [55]	2006	Soda-lime glass, Alumina	—	NaOH	Voltage, duty factor electrolyte conc., energy partition	MRR

(Continued)

Table 2.2 Brief overview of past research work based on workpiece material, tool material, electrolyte chemical, input process parameters and its output responses of ECDM by various researchers. (*Continued*)

S. no.	Name of researchers	Year	Workpiece material	Tool material	Electrolyte chemical	Input process parameters	Output process parameters
21	R. Wuthrich *et al.* [72]	2006	Glass	316L stainless steel	NaOH	Applied voltage, electrolyte concentration, applied force, rotation of tool electrode	Drilling discharge regime identified, microhole obtained, drill depth
22	R. Wuthrich *et al.* [73]	2006	Pyrex Glass	Nickel	NaOH	Applied voltage, electrolyte conc., electrolyte conc. mixed with soap, rotation of tool	Reproducibility of machining, MRR, Machined surface quality
23	R. Wuthrich *et al.* [74]	2006	Glass	316L stainless steel	NaOH	Applied voltage, electrolyte concentration, electrolyte level and electrolyte temperature	MRR and study of current signal

(*Continued*)

Table 2.2 Brief overview of past research work based on workpiece material, tool material, electrolyte chemical, input process parameters and its output responses of ECDM by various researchers. (*Continued*)

S. no.	Name of researchers	Year	Workpiece material	Tool material	Electrolyte chemical	Input process parameters	Output process parameters
24	D. Kim *et al.* [75]	2006	Pyrex Glass	Copper	KOH	Applied voltage, pulse frequency, electrolyte concentration, duty ratio	TWR, MRR, surface roughness, hole roundness, microdrilled surface quality
25	R. Wuthrich *et al.* [95]	2006	Glass	Stainless steel	NaOH	voltage, Electrolyte Concentration, Amplitude, Frequency	Microhole depth, Mean MRR
26	B. Sarkar *et al.* [140]	2006	Silicon nitride	Stainless steel	NaOH	Voltage, inter-electrode gap, electrolyte conc.,	Heat affected zone, MRR, radial overcut
27	A. kulkarni *et al.* [44]	2007	Copper, Brass, Tantalum, and Silicon	Copper	HCl	Voltage, electrolyte conc., pulse current, pulse-on time	Current measurements, temperature measurement
28	J. west *et al.* [76]	2007	Borosilicate glass	Stainless steel	NaOH, KOH	Voltage, Electrolyte concentration, Tool polarity, Substrate thickness	Average entrance diameter, Average depth, Mean MRR

(*Continued*)

Table 2.2 Brief overview of past research work based on workpiece material, tool material, electrolyte chemical, input process parameters and its output responses of ECDM by various researchers. (*Continued*)

S. no.	Name of researchers	Year	Workpiece material	Tool material	Electrolyte chemical	Input process parameters	Output process parameters
29	P. Maillard *et al.* [77]	2007	Glass	316 L Stainless steel	NaOH	Voltage, electrolyte concentration, mean machining time	Various types of microhole, drilling depth, roundness error
30	S. Chak *et al.* [96]	2007	Aluminium oxide	Diamond embedded abrasive	NaOH, KOH	Applied voltage, duty factor, electrolyte conductivity, rotation of tool electrode	Material removed, machining depth, diametric overcut
31	Z. Zheng *et al.* [141]	2007	Pyrex glass	Tungsten carbide	KOH	Rectangular pulse voltage, electrolyte conc., tool travel rate, rotation of tool electrode,	Machining depth, groove width, 3D microstructure on Pyrex glass
32	Z. Zheng [142]	2007	Pyrex glass	Tungsten carbide rod	KOH	Voltage, electrolyte conc., tool rotational rate, tool travel velocity, pulse-off time, pulse-on time	Groove width, machining depth, 3D microstructure on Pyrex glass

(*Continued*)

Table 2.2 Brief overview of past research work based on workpiece material, tool material, electrolyte chemical, input process parameters and its output responses of ECDM by various researchers. (*Continued*)

S. no.	Name of researchers	Year	Workpiece material	Tool material	Electrolyte chemical	Input process parameters	Output process parameters
33	B. Yan *et al.* [160]	2007	Pyrex glass	Tungsten	KOH	Voltage, rotation speed of tool electrode, electrolyte conc.	Surface roughness, surface morphology, expansion of hole, hole roundness
34	Z. Zheng *et al.* [161]	2007	Pyrex glass	Tungsten carbide	KOH	Voltage, electrolyte conc., sidewall thickness, polarity, tool speed, machining depth	Effect on geometrical shape of tool electrode, microhole accuracy, machining time
35	M. Han *et al.* [184]	2007	Borosilicate glass	Tungsten carbide	NaOH	Applied voltage, peak current, electrolyte conc., graphite powder concentration	Surface roughness, surface integrity

(*Continued*)

Table 2.2 Brief overview of past research work based on workpiece material, tool material, electrolyte chemical, input process parameters and its output responses of ECDM by various researchers. (*Continued*)

S. no.	Name of researchers	Year	Workpiece material	Tool material	Electrolyte chemical	Input process parameters	Output process parameters
36	T. Didar *et al.* [21]	2008	Glass	316 L Stainless steel	NaOH	Voltage, electrolyte concentration, tool speed	Microchannel depth, obtained different types of microchannel, machined surface quality
37	S. Chak *et al.* [97]	2008	Aluminum oxide	Hollow Brass and Abrasive electrode.	NaOH, KOH	Applied voltage, duty factor, electrolyte conductivity, spring feed to tool electrode	Material removed, Improved quality of holes
38	M. Coteata *et al.* [100]	2008	High-speed steel	Chisel steel	Aqueous sodium Silicate	Applied voltage, working liquid density, electrode tool diameter, capacity	MRR, shape accuracy of the hole

(*Continued*)

Table 2.2 Brief overview of past research work based on workpiece material, tool material, electrolyte chemical, input process parameters and its output responses of ECDM by various researchers. (*Continued*)

S. no.	Name of researchers	Year	Workpiece material	Tool material	Electrolyte chemical	Input process parameters	Output process parameters
39	V. Jain *et al.* [143]	2008	Quartz	Copper	NaOH	Voltage, Electrolyte Conc., Electrolyte temperature, Feed rate to work-piece, polarity, Initial gap between workpiece and tool, Tool tip dipped into electrolyte	MRR, Overcut, Tool wear, Machined surface integrity
40	K. Furutani *et al.* [144]	2008	Soda lime glass	Tungsten	NaCl	Applied voltage, electrolyte conc., rotational speed of tool electrode, dipped depth of electrode, pressure	Surface roughness, groove width and depth

(*Continued*)

Table 2.2 Brief overview of past research work based on workpiece material, tool material, electrolyte chemical, input process parameters and its output responses of ECDM by various researchers. (*Continued*)

S. no.	Name of researchers	Year	Workpiece material	Tool material	Electrolyte chemical	Input process parameters	Output process parameters
41	M. Han *et al.* [162]	2008	Borosilicate glass	Tungsten carbide	NaOH	Applied, voltage, electrolyte conc., feed rate, conventional and Side-insulated tool electrode	Surface roughness, geometric accuracy, surface integrity
42	M. Han *et al.* [163]	2008	Borosilicate glass	Tungsten rod	NaOH	Voltage, electrolyte conc., feed rate, frequency, pulse on time, side-insulated tool	Machining depth, machining width, machined surface accuracy
43	X. Cao *et al.* [22]	2009	Soda-lime glass or Pyrex glass	Tungsten carbide	NaOH and KOH	Applied voltage, electrolyte concentration, machining feed rate, rotational speed of tool electrode	Fabricated 3D microstructure, microhole and groove machined, machined layer depth

(*Continued*)

Table 2.2 Brief overview of past research work based on workpiece material, tool material, electrolyte chemical, input process parameters and its output responses of ECDM by various researchers. (*Continued*)

S. no.	Name of researchers	Year	Workpiece material	Tool material	Electrolyte chemical	Input process parameters	Output process parameters
44	M. Jalali *et al.* [80]	2009	Glass	316 L stainless steel	NaOH	Applied voltage, electrolyte conc., rotational speed of tool electrode, force applied to tool electrode	Material removal mechanism, microhole depth, estimation of the machining temperature
45	M. coteata *et al.* [101]	2009	Steel	Rapid steel	Aqueous solution of sodium silicate	Voltage, electrodes diameters. Densities of the work liquid, tool speed	MRR, electrode tool wear
46	M. Han *et al.* [102]	2009	Borosilicate glass	Tungsten carbide	NaOH	Applied voltage, voltage type, electrolyte concentration, ultrasonic frequency, tool dia.	Machining-depth, geometrical accuracy, surface quality, overcut

(*Continued*)

Table 2.2 Brief overview of past research work based on workpiece material, tool material, electrolyte chemical, input process parameters and its output responses of ECDM by various researchers. (*Continued*)

S. no.	Name of researchers	Year	Workpiece material	Tool material	Electrolyte chemical	Input process parameters	Output process parameters
47	L. Kudla *et al.* [145]	2009	Borosilicate glass, Diamond	Tungsten, Platinum	NaOH, KNO or H_2SO_4	Applied voltage, electrolyte conc., tool shapes like sharp conical tip, sphere on the end	Microgroove, ring-shaped groove, machined depth
48	T. Didar *et al.* [146]	2009	Soda-lime glass	Stainless steel tool	NaOH	Applied Voltage, Electrolyte Concentration, Machining speed	Removed mass during machining, hardness and density of machined zones
49	M. Mousa *et al.* [164]	2009	Borosilicate glass	Steel 316L, high carbon steel, Tungsten, Copper	NaOH	Applied voltage, electrolyte concentration, rotational speed of tool electrode, force applied to tool electrode	Drilling depth, machining time, comparative study of different tool electrodes

(*Continued*)

Table 2.2 Brief overview of past research work based on workpiece material, tool material, electrolyte chemical, input process parameters and its output responses of ECDM by various researchers. *(Continued)*

S. no.	Name of researchers	Year	Workpiece material	Tool material	Electrolyte chemical	Input process parameters	Output process parameters
50	C. Cheng *et al.* [36]	2010	Pyrex glass	Tungsten carbide	KOH	Applied voltage, electrolyte Concentration, tool rotating speed, tool shape	Machining depth, Machining efficiency and accuracy
51	J. Liu *et al.* [57]	2010	Particulate reinforced aluminum alloy 359	Steel	$NaNO_3$	Applied voltage, current, electrolyte concentration, pulse duration, duty cycle	Material removal mechanism, comparative study of ECDM and EDM
52	C. Wei *et al.* [105]	2010	Soda-lime glass	Tungsten carbide	NaOH	Voltage, electrolyte conc., applied force, rotational and stationary tools	Hole, diameter, machining depth, radial overcut

(Continued)

Table 2.2 Brief overview of past research work based on workpiece material, tool material, electrolyte chemical, input process parameters and its output responses of ECDM by various researchers. (*Continued*)

S. no.	Name of researchers	Year	Workpiece material	Tool material	Electrolyte chemical	Input process parameters	Output process parameters
53	C. Cheng *et al.* [104]	2010	Pyrex glass	Tungsten carbide	KOH	Applied voltage, electrolyte concentration, tool rotating speed, magnetic orientations	Machining depth, machining time, geometry accuracy, efficiency
54	C. Yang *et al.* [165]	2010	Quartz	Stainless steel, Tungsten carbide, Tungsten	KOH	Applied voltage, electrolyte concentration, machining depth, rotational speed of tool electrode	Hole diameter, Surface roughness, Machining speed, Tool wear
55	J. Ziki *et al.* [166]	2010	Glass	Stainless steel	NaOH	Applied voltage, electrolyte concentration, constant velocity feed rate	Machining depth, investigated force exerted on tool electrode

(*Continued*)

Table 2.2 Brief overview of past research work based on workpiece material, tool material, electrolyte chemical, input process parameters and its output responses of ECDM by various researchers. (*Continued*)

S. no.	Name of researchers	Year	Workpiece material	Tool material	Electrolyte chemical	Input process parameters	Output process parameters
56	C. Yang *et al.* [23]	2011	Quartz	Tungsten carbide	KOH	Applied voltage, electrolyte conc., tool diameter, rotational speed, tool shape	Machining depth, machining time, hole diameter
57	C. Wei *et al.* [59]	2011	Soda-lime glass	Tungsten carbide	NaOH	Voltage, electrolyte conc., electrolyte level, machining time, tool electrode material	Machined hole depth
58	D. Nandi *et al.* [82]	2011	Glass	Copper	NaCl, NaOH and HCl	Voltage, electrolyte concentration, tool diameter, immersion depth	Voltage-current characteristics,
59	M. Coteata *et al.* [107]	2011	Steel	High speed steel	Aqueous solution of sodium silicate	Work liquid density, electrode tool diameter, electrode tool strokes per minute, voltage	Tool axial wear, machined hole depth

(*Continued*)

Table 2.2 Brief overview of past research work based on workpiece material, tool material, electrolyte chemical, input process parameters and its output responses of ECDM by various researchers. (*Continued*)

S. no.	Name of researchers	Year	Workpiece material	Tool material	Electrolyte chemical	Input process parameters	Output process parameters
60	J. Ozhikandathil *et al.* [147]	2011	Silicon substrate	Stainless steel	NaOH	Applied voltage, electrolyte conc., machining time	Hole diameter, machining hole depth
61	M. Han *et al.* [167]	2011	Soda-lime glass	Brass wire	NaOH	Voltage, tool speed, electrolyte conc., pulse frequency, duty cycle, feed rate	2D contour cutting, Surface roughness
62	J. Ziki *et al.* [43]	2012	Soda-lime glass	Stainless steel	NaOH	Tool speed, gap between tool and glass surface, duty cycle	Surface textures were obtained porous spongy-like textures to feathery like
63	M. Panda *et al.* [60]	2012	Silicon nitride	Stainless steel	NaOH	Supply voltage, energy partition, electrolyte conc., spark on-time, duty factor	MRR, Average surface roughness

(*Continued*)

Table 2.2 Brief overview of past research work based on workpiece material, tool material, electrolyte chemical, input process parameters and its output responses of ECDM by various researchers. *(Continued)*

S. no.	Name of researchers	Year	Workpiece material	Tool material	Electrolyte chemical	Input process parameters	Output process parameters
64	Y. Mochimaru *et al.* [84]	2012	Borosilicate glass	Tungsten rod	NaOH	Voltage, electrolyte conc., pulse frequency, duty factor	Microholes obtained
65	C. Lu *et al.* [110]	2012	Pyrex glass.	Tungsten rod	NaOH	Applied voltage, feed rate, electrolyte concentration	Machining efficiency, surface quality
66	M. Rusli *et al.* [111]	2012	Soda-lime glass	Tungsten	NaCl	Voltage, electrolyte conc., pressure, electrolyte level, time, ultrasonic amplitude	MRR, surface integrity
67	A. Manna *et al.* [148]	2012	E-glass–fiber–epoxy composite	IS-3748 steel	NaOH	Applied voltage, gap between tool and anode, electrolyte conc.	MRR and overcut

(Continued)

Table 2.2 Brief overview of past research work based on workpiece material, tool material, electrolyte chemical, input process parameters and its output responses of ECDM by various researchers. (*Continued*)

S. no.	Name of researchers	Year	Workpiece material	Tool material	Electrolyte chemical	Input process parameters	Output process parameters
68	C. Jawalkar *et al.* [149]	2012	Soda-lime glass	Stainless steel	NaOH	Applied voltage, work feed rate, electrolyte conc.	Material removal and tool wear
69	J. Ziki *et al.* [168]	2012	Al_2O_3	Stainless steel, Steel, Tungsten	NaOH	Applied voltage, electrolyte conc., electrode diameter, tool electrode vertical feed, duty cycle	Thermal expansion, TWR
70	J. Liu *et al.* [24]	2013	Particulate reinforced aluminum alloy 6061	Diamond coated tool	$NaNO_3$	Voltage, machining depth electrolyte conc., pulse duration, feed speed, duty cycle, spindle speed	MRR, surface roughness
71	H. Krotz *et al.* [61]	2013	$100Cr_6$ steel	Tungsten	$NaNO_3$	Voltage, pulse-on time, electrolyte conc., the pulse-off time	Heat-affected zone

(*Continued*)

Table 2.2 Brief overview of past research work based on workpiece material, tool material, electrolyte chemical, input process parameters and its output responses of ECDM by various researchers. (*Continued*)

S. no.	Name of researchers	Year	Workpiece material	Tool material	Electrolyte chemical	Input process parameters	Output process parameters
72	J. Ziki *et al.* [112]	2013	Soda-lime glass	Stainless steel	NaOH	Voltage, electrolyte conc., tool feed-rates, tool sizes	Investigated forces exerted on the tool-electrode
73	L. Paul *et al.* [150]	2013	Borosilicate glass	Tungsten carbide	NaOH	Voltage, electrolyte conc., duty factor	MRR, radial overcut, TWR
74	S. Jui *et al.* [169]	2013	Glass	Tungsten carbide	H_2SO_4	Voltage, forward duty cycle, pulse on-time, electrolyte conc., reverse pulse time, tool rotation	Aspect ratio, diametrical overcut
75	X. Cao *et al.* [170]	2013	Glass	Tungsten carbide	KOH	Voltage, electrolyte conc., pulse on/off-time, feed rate	Average MRR, depth of cut, surface quality, efficiency

(*Continued*)

Table 2.2 Brief overview of past research work based on workpiece material, tool material, electrolyte chemical, input process parameters and its output responses of ECDM by various researchers. (*Continued*)

S. no.	Name of researchers	Year	Workpiece material	Tool material	Electrolyte chemical	Input process parameters	Output process parameters
76	Y.S. Laio *et al.* [185]	2013	Quartz	Tungsten	Sodium Dodecyl Sulfate surfactant + KOH	Applied voltage, electrolyte concentration, frequency, duty factor, machining speed	Better quality microhole
77	B. Bhuyan *et al.* [199]	2013	Borosilicate glass	Brass	NaOH	Voltage, wire feed velocity, electrolyte concentration, workpiece thickness	MRR, Surface Roughness
78	M. Razfar *et al.* [200]	2013	Soda-lime glass	Brass	NaOH	Applied voltage, tool immersion depth, electrolyte conc.	Depth, average hole diameter, roundness error, radial overcut, heat-affected zone
79	V. Jain *et al.* [6]	2014	Quartz	Copper	NaOH	Applied voltage, electrolyte conc., table feed	MRR, width of channel

(*Continued*)

Table 2.2 Brief overview of past research work based on workpiece material, tool material, electrolyte chemical, input process parameters and its output responses of ECDM by various researchers. (*Continued*)

S. no.	Name of researchers	Year	Workpiece material	Tool material	Electrolyte chemical	Input process parameters	Output process parameters
80	B. Jiang *et al.* [25]	2014	Soda-lime glass	Tungsten	NaOH	Applied voltage, electrolyte conc., tapered tool electrode with different shape	MRR, machining depth
81	A. De Souza *et al.* [85]	2014	Stainless steel	Tungsten	NaCl, Na_2SO_4	Applied voltage, electrolyte conductivity, tool rotation, capacitor capacitance	Machining time, tool electrode wear
82	B. Jiang *et al.* [113]	2014	Soda-lime glass	Tungsten carbide	NaOH	Voltage, electrolyte conc., feed force, speed of tool, vibration waveform, duty ratio, amplitude, frequency	MRR, machining depth

(*Continued*)

Table 2.2 Brief overview of past research work based on workpiece material, tool material, electrolyte chemical, input process parameters and its output responses of ECDM by various researchers. (*Continued*)

S. no.	Name of researchers	Year	Workpiece material	Tool material	Electrolyte chemical	Input process parameters	Output process parameters
83	L. Paul *et al.* [114]	2014	Stainless steel wire	Silicon Wafer	NaOH	Voltage, electrolyte conc., duty factor	MRR, diametrical overcut, heat affected zone
84	L. Paul *et al.* [115]	2014	P-type silicon wafer	Musical string steel wire	NaOH	Voltage, electrolyte conc., tool feed rate, duty factor	MRR, heat-affected zone
85	L. Paul *et al.* [116]	2014	Soda-lime glass	Copper wire	NaOH	Voltage, electrolyte conc., duty factor, temp.	MRR, surface roughness, tool wear rate
86	C. Gao *et al.* [117]	2014	Pyrex glass	Tungsten carbide	NaOH and KOH	Voltage, electrolyte conc., contact force, rotary rate, rotary direction, specimen thickness, time	Hole diameter, tool wear, aspect ratio

(*Continued*)

Table 2.2 Brief overview of past research work based on workpiece material, tool material, electrolyte chemical, input process parameters and its output responses of ECDM by various researchers. (*Continued*)

S. no.	Name of researchers	Year	Workpiece material	Tool material	Electrolyte chemical	Input process parameters	Output process parameters
87	B. Bhuyan *et al.* [151]	2014	Borosilicate glass	Brass wire	NaOH	Voltage, electrolyte conc., workpiece feed rate, pulse-on time, pulse-off time	MRR, Kerf width, Surface finish
88	M. Coteata *et al.* [152]	2014	Spring steel 60CrMnSi	High speed steel	NaOH	Applied voltage, electrolyte density, tool diameter, capacitance	MRR, electrode wear, depth of the drilled hole
89	S. Huang *et al.* [171]	2014	304 stainless steel	Tungsten carbide	water	Applied voltage, tool speed and electrode diameters	Tool wear
90	M. Razfar *et al.* [174]	2014	Soda-lime glass	Tungsten carbide	NaOH	Applied voltage, electrolyte conc., vibration frequency, vibration amplitude	MRR, machining depth

(*Continued*)

Table 2.2 Brief overview of past research work based on workpiece material, tool material, electrolyte chemical, input process parameters and its output responses of ECDM by various researchers. *(Continued)*

S. no.	Name of researchers	Year	Workpiece material	Tool material	Electrolyte chemical	Input process parameters	Output process parameters
91	C. Jawalkar *et al.* [186]	2014	Soda-lime glass	Stainless steel (SS-304)	NaOH	Voltage, time of current flow, electrode spacing	Material removal, tool wear
92	M. Dhanvijay *et al.* [187]	2014	Alumina	Copper and stainless steel	NaOH	Voltage, electrolyte concentration, pulse time, duty factor	MRR, diametric overcut
93	B. Mallick *et al.* [201]	2014	Silica glass	Stainless steel	NaOH	Voltage, electrolyte conc., interelectrode, gap	Material removal rate, overcut, machining depth,
94	P. Gupta *et al.* [10]	2015	Glass	Stainless steel	NaCl, KOH, NaOH	Voltage, pulse on time, electrolyte concentration, pulse off time	MRR, hole overcut

(Continued)

Table 2.2 Brief overview of past research work based on workpiece material, tool material, electrolyte chemical, input process parameters and its output responses of ECDM by various researchers. (*Continued*)

S. no.	Name of researchers	Year	Workpiece material	Tool material	Electrolyte chemical	Input process parameters	Output process parameters
95	J. Abou Ziki *et al.* [26]	2015	Glass	Stainless steel	KOH	Voltage, electrolyte conc., force applied, pulse-off-time, pulse-on-time, rotation speed of tool	Machining temperature
96	A. Kamaraj *et al.* [62]	2015	Borosilicate glass	Tungsten	NaOH	Voltage, immersion depth, electrolyte conc., tool diameter, machining time, tool feed rate	Overcut, aspect ratio
97	J. Abou Ziki *et al.* [86]	2015	Glass	Steel	NaOH	Applied voltage, electrolyte concentration, feed rate	Obtained tool-glass bond during drilling

(*Continued*)

Table 2.2 Brief overview of past research work based on workpiece material, tool material, electrolyte chemical, input process parameters and its output responses of ECDM by various researchers. (*Continued*)

S. no.	Name of researchers	Year	Workpiece material	Tool material	Electrolyte chemical	Input process parameters	Output process parameters
98	J. Abou Ziki *et al.* [87]	2015	Glass	Stainless-steel	NaOH	Voltage, duty cycle, electrolyte conc., feed rate	Tool-electrode temp., tool expansion, drilling depth
99	P. Gupta *et al.* [90]	2015	Glass	Stainless steel	NaCl, NaOH	Voltage, pulse on time, electrolyte concentration, pulse off time	Surface damage, machining depth
100	H. Krotz *et al.* [118]	2015	Steel 100Cr6	Tungsten rod	Sodium nitrate	Voltage, pulse-on-time, conductivity of electrolyte, pulse-off-time	Machining efficiency, surface and heat-affected zone
101	Y. Zhang *et al.* [175]	2015	Cobalt-based superalloy, Nickel-based superalloy	Brass	Salt solution	Applied voltage, solution conductivity, pulse-off-time, working fluid pressure, pulse-on-time, tool rotation	MRR, machining speed, Machining accuracy

(*Continued*)

Table 2.2 Brief overview of past research work based on workpiece material, tool material, electrolyte chemical, input process parameters and its output responses of ECDM by various researchers. (*Continued*)

S. no.	Name of researchers	Year	Workpiece material	Tool material	Electrolyte chemical	Input process parameters	Output process parameters
102	S. Saranya *et al.* [176]	2015	Quartz	Tungsten carbide	NaOH	Voltage, Electrolyte conc., Feed rate, Tool shape, Tool geometry	Machined profile, Hole overcut
103	K. Nguyen *et al.* [188]	2015	Quartz	Tungsten carbide alloy	KOH	Voltage, Electrolyte conc., Feed rate, Pulse on/off-time, Tool rotation	Surface roughness, Groove width and depth
104	Y. Zhang *et al.* [189]	2015	Nickel-based super-alloys	Brass	Sodium nitrate water solution	Voltage, Pulse duration, Pulse interval, Solution conductivity, Tube electrode diameter, Tube electrode rotation, Work fluid pressure	Average hole diameter, Material removal rate, Taper angle

(*Continued*)

Table 2.2 Brief overview of past research work based on workpiece material, tool material, electrolyte chemical, input process parameters and its output responses of ECDM by various researchers. (*Continued*)

S. no.	Name of researchers	Year	Workpiece material	Tool material	Electrolyte chemical	Input process parameters	Output process parameters
105	A. Behroozfar *et al.* [89]	2016	Soda-lime glass	Tungsten carbide	NaOH	Voltage, Electrolyte conc., pulse off time, duty cycles, pulse on time,	removed material, diameter and depth of the crater
106	P. Gupta *et al.* [91]	2016	Alumina ceramic	Stainless steel	NaOH	Voltage, Pulse-off-time, Electrolyte conc., Pulse-on-time, Pulse frequency, Duty ratio	Depth of penetration, MRR, Heat affected zone Surface damage
107	K. Furutani *et al.* [120]	2016	Silica glass, Piezoelectric ceramics	Tungsten	NaCl, NaOH	Voltage, Electrolyte conc., Feed rate, Machining width, Revolution of workpiece, Machining time	Material removal rate, Machined Surface Quality

(*Continued*)

Table 2.2 Brief overview of past research work based on workpiece material, tool material, electrolyte chemical, input process parameters and its output responses of ECDM by various researchers. (*Continued*)

S. no.	Name of researchers	Year	Workpiece material	Tool material	Electrolyte chemical	Input process parameters	Output process parameters
108	Z. Zhang *et al.* [121]	2016	Soda-lime glass	tungsten carbide	$NaNO_3$, KOH, and NaOH	Voltage, electrolyte conc., tool rotation, dia. of electrode, pulse frequency, duty cycle	Investigated critical voltage, Machining accuracy
109	M. Hajian *et al.* [122]	2016	Soda-lime glass	HSS	NaOH	Voltage, electrolyte conc., tool speed, magnetic field orientations	Machining depth, surface roughness
110	J. Madhavi *et al.* [123]	2016	Borosilicate glass	Tungsten carbide	NaOH	Applied voltage, electrolyte conc., duty factor, feed rate	MRR, tool wear rate, Taper angle, Radial overcut
111	V. Ladeesh *et al.* [124]	2016	Borosilicate glass	Diamond	KOH	Voltage, electrolyte conc., pulse on time, tool speed	Material removal rate, tool wear

(*Continued*)

Table 2.2 Brief overview of past research work based on workpiece material, tool material, electrolyte chemical, input process parameters and its output responses of ECDM by various researchers. *(Continued)*

S. no.	Name of researchers	Year	Workpiece material	Tool material	Electrolyte chemical	Input process parameters	Output process parameters
112	V. Ladeesh *et al.* [125]	2016	Soda-lime glass	Diamond coated tool	KOH	Applied voltage, electrolyte concec., duty cycle, cycle	Material removal, tool wear
113	A. Behroozfar *et al.* [177]	2016	Ceramic	Brass, Steel, Tung. carbide	NaOH	Applied voltage, electrolyte concentration, different types of tool electrodes	Tool wear
114	D. Guo *et al.* [178]	2016	304 austenite stainless steel	Amorphous Pt deposited microtool	Deionized water	Open voltage, pulse duration, pulse separation, working depth, working step	Hole diameter, The unilateral discharge gap
115	L. Paul *et al.* [190]	2016	Silicon Wafers	Tungsten carbide	NaOH + KOH	Voltage, electrolyte conc., mixed electrolyte, duty factor	MRR

(Continued)

Table 2.2 Brief overview of past research work based on workpiece material, tool material, electrolyte chemical, input process parameters and its output responses of ECDM by various researchers. (*Continued*)

S. no.	Name of researchers	Year	Workpiece material	Tool material	Electrolyte chemical	Input process parameters	Output process parameters
116	V. Ladeesh *et al.* [126]	2017	Borosilicate glass	Diamond coated solid tool	KOH	Voltage, electrolyte conc., duty factor, feed rate, rotation speed of tool electrode, frequency	MRR, surface roughness
117	B. Sarkar *et al.* [154]	2017	Silicon carbide	Stainless steel	NaOH	Voltage, interelectrode gap, electrolyte conc.	Material removal rate, radial overcut
118	M. Hajian *et al.* [179]	2017	Soda lime glass	HSS	NaOH	Voltage, electrolyte conc., tool, feed rate, magnetic field, immersion depth	Tool bending force

(*Continued*)

Table 2.2 Brief overview of past research work based on workpiece material, tool material, electrolyte chemical, input process parameters and its output responses of ECDM by various researchers. (*Continued*)

S. no.	Name of researchers	Year	Workpiece material	Tool material	Electrolyte chemical	Input process parameters	Output process parameters
119	W. Tang *et al.* [181]	2017	Quartz	Tungsten carbide, diamond coated	NaOH	Voltage, electrolyte conc., different types of tool electrodes, tool electrode diameter, tool rotation, pulse on time, pulse off time	Machining depth, heat affected zone, taper angle, surface integrity
120	M. Han *et al.* [182]	2017	Soda-lime glass	Brass	NaOH	Applied voltage, electrolyte concentration, tool diameter, shape of tool, feed rate, tool speed	Different texture of microgroove, surface quality, aspect ratio
121	S. Saranya *et al.* [192]	2017	Glass	Tungsten carbide	NaOH, KOH	Voltage, electrolyte conc., immersion depths, tool diameters, tool travel rate	Geometric accuracy, Surface smoothness, channel outlines

(*Continued*)

Table 2.2 Brief overview of past research work based on workpiece material, tool material, electrolyte chemical, input process parameters and its output responses of ECDM by various researchers. (*Continued*)

S. no.	Name of researchers	Year	Workpiece material	Tool material	Electrolyte chemical	Input process parameters	Output process parameters
122	N. Sabahi *et al.* [193]	2017	Soda-lime glass	Tungsten carbide	Surfactant-mixed KOH/ NaOH	Voltage, electrolyte conc., electrolyte type, surfactant type, surfactant conc., electrolyte temp., tool speed, tool immersion	Depth of microchannel, surface quality, surface roughness, heat-affected zone, hardness of microchannel
123	M. Goud *et al.* [202]	2017	Quartz glass	Stainless steel	NaOH	Voltage, feed rate, electrolyte conc.	Material removal rate, width overcut
124	V. Ladeesh [64]	2018	Soda-lime glass	Diamond core drill bit	NaOH	Voltage, rotation speed of tool electrode, electrolyte conc., feed rate	MRR, edge-chipping thickness
125	M. Hajian *et al.* [65]	2018	Soda lime glass	Stainless steel	NaOH	Voltage, feed rate, electrolyte conc., rotation tool speed	Machining depth

(*Continued*)

Table 2.2 Brief overview of past research work based on workpiece material, tool material, electrolyte chemical, input process parameters and its output responses of ECDM by various researchers. (*Continued*)

S. no.	Name of researchers	Year	Workpiece material	Tool material	Electrolyte chemical	Input process parameters	Output process parameters
126	T. Singh *et al.* [128]	2018	Borosilicate glass	Stainless steel coated with SiC	NaOH	Voltage, electrolyte conc., exerted pressure, pulse-on-time, inter electrode gap, pulse-off-time,	MRR, hole overcut, taper, machining depth
127	S. Elhami *et al.* [129]	2018	Soda lime glass	Tungsten carbide	NaOH	Voltage, electrolyte conc., electrolyte temp., amplitude	Material removal, tool wear
128	Y. Xu *et al.* [130]	2018	Soda-lime glass	Tungsten carbide	NaOH	Voltage, electrolyte conc., feeding method, tool feed force, rotation speed of tool	Machining depth, roundness, machining speed

(*Continued*)

Table 2.2 Brief overview of past research work based on workpiece material, tool material, electrolyte chemical, input process parameters and its output responses of ECDM by various researchers. (*Continued*)

S. no.	Name of researchers	Year	Workpiece material	Tool material	Electrolyte chemical	Input process parameters	Output process parameters
129	S. He *et al.* [155]	2018	ZrO_2 Ceramic	Tungsten	NaOH	Applied voltage, electrolyte conc., tool electrode diameter, machining gap, spindle speed, pulse duration, pulse interval	Investigated debris as crystalline and amorphous forms, different color presence of the processed debris
130	S. Saranya *et al.* [183]	2018	Fused quartz	Carbon alloy steel	NaOH	Voltage, electrolyte conc., tool diameter, initial working gap, tool travel rate	Depth of microchannel, width of microchannel, surface quality
131	N. Sabahi *et al.* [194]	2018	Soda-lime glass	Tungsten carbide	NaOH, KOH, Mixed	Voltage, electrolyte conc., mixed electrolyte, electrolyte temp., speed	Tool wear, Thermal damages, surface quality, and overcut

(*Continued*)

Table 2.2 Brief overview of past research work based on workpiece material, tool material, electrolyte chemical, input process parameters and its output responses of ECDM by various researchers. (*Continued*)

S. no.	Name of researchers	Year	Workpiece material	Tool material	Electrolyte chemical	Input process parameters	Output process parameters
132	N. Sabahi *et al.* [195]	2018	Soda-lime glass	Tungsten carbide	KOH/ NaOH	Voltage, electrolyte conc., speed, Lorentz force direction, immersion depth	Hardness of microchannel edges, heat-affected zone
133	M. Harugade *et al.* [196]	2018	Soda-lime glass	Copper	NaOH, $NaNO_3$, H_2SO_4	Voltage, electrolyte conc., inter-electrode gap,	Material removal rate, hole diameter, machining depth
134	P. Gupta *et al.* [197]	2018	Borosilicate glass	Stainless steel	NaOH	Voltage, machining time, pulse-on-time, electrolyte conc., pulse-off-time	MRR, overcut, depth of cut, heat-affected zone

(*Continued*)

Table 2.2 Brief overview of past research work based on workpiece material, tool material, electrolyte chemical, input process parameters and its output responses of ECDM by various researchers. (*Continued*)

S. no.	Name of researchers	Year	Workpiece material	Tool material	Electrolyte chemical	Input process parameters	Output process parameters
135	A. Varghese *et al.* [198]	2018	Polypropylene		Graphite powder + NaOH	Voltage, graphite powder conc., duty factor	Material removal rate
136	P. Antil *et al.* [203]	2018	SiC reinforced polymer matrix	Stainless steel	NaOH. NaOH+ HCl	Inter-electrode gap, applied voltage, duty factor, electrolyte concentration	MRR, overcut, taper

2.7 Conclusion

The present review includes the analytical study of various fields of ECDM, from this some major conclusions have been reported. It is very easy to obtain a precise result of ECDM on glass material due to its chemical resistance and transparency as well as it is less time-consuming and has high demand in various fields as compared to other workpiece materials. The lower depth and smooth surface on glass obtained for lower voltage and at higher voltage, the higher depth and heat-affected zone with cracks obtained on a glass surface.

The stainless steel and Tungsten carbide is widely used as cathode tool as they have significantly low specific heat capacity, high wear resistance, chemical inertness and temperature resistance, which resulted in higher discharge activity than the other tool material. The KOH and NaOH have irritant nature and higher mobility of K+ ion and Na+ ion, which help to yield more MRR, so these are the most used electrolytes in ECDM. The high electrolyte concentration gives high MRR, overcut and machined depth. The high applied voltage leads to high MRR, high tool wear rate, increases heat-affected zone, reduce bending force, increases hole diameter and radial overcut

Acknowledgments

The authors are thankful to Department of Production and Industrial Engineering, NIT Jamshedpur for all the support for making this research happen.

References

1. Singh, T. and Dvivedi, A., Developments in electrochemical discharge machining: A review on electrochemical discharge machining, process variants and their hybrid methods. *Int. J. Mach. Tools Manuf.*, 105, 1, 2016.
2. Dixit, P.M. and Dixit, U.S., Metal forming and machining processes, in: *Modeling of Metal Forming and Machining Processes: By Finite Element and Soft Computing Methods*, pp. 1–32, Springer, London, 2008.
3. Jain, V.K., Advanced (non-traditional) machining processes, in: *Machining: Fundamentals and Recent Advances*, pp. 299–327, Springer, London, 2008.
4. Ghosh, A., Electrochemical discharge machining: Principle and possibilities. *Sadhana*, 22, 435, 1997.

5. Yadav, R.N., Electro-chemical spark machining–based hybrid machining processes: Research trends and opportunities. *Proc. Inst. Mech. Eng. B J. Eng. Manuf.*, 233, 1037, 2019.
6. Jain, V.K. and Priyadarshini, D., Fabrication of microchannels in ceramics (quartz) using electrochemical spark micromachining (ECSMM). *J. Adv. Manuf. Syst.*, 13, 5, 2014.
7. McGeough, J.A., Khayry, A.B.M., Munro, W. *et al.*, Theoretical and experimental investigation of the relative effects of spark erosion and electrochemical dissolution in electrochemical ARC machining. *CIRP Ann. Manuf. Technol.*, 32, 113, 1983.
8. Sarkar, B.R., Doloi, B., Bhattacharyya, B., Electrochemical Discharge Micromachining of Engineering Materials. In: Kibria, G., Bhattacharyya, B., Davim, J. (eds), Non-traditional Micromachining Processes, Materials Forming, Machining and Tribology, pp. 367–392, Springer International Publishing AG, Cham, 2017.
9. Basak, I. and Ghosh, A., Mechanism of material removal in electrochemical discharge machining: A theoretical model and experimental verification. *J. Mater. Process. Technol.*, 71, 350, 1997.
10. Gupta, P.K., Dvivedi, A., Kumar, P., Effect of electrolytes on quality characteristics of glass during ECDM. *Key Eng. Mater.*, 658, 141, 2015.
11. Rudorff, D.W., Principles and applications of spark machining. *Proc. Inst. Mech. Eng.*, 171, 495, 1957.
12. Wuthrich, W., *Spark Assisted Chemical Engraving-A Stochastic Modeling Approach*, Ph. D. Thesis, Ecole Polytechniqsue Federale de Lausanne, Switzerland, 2003.
13. Wuthrich, R., and Ziki, J., Machining with Electrochemical Discharges- An Overview, in: *Micromachining using Electrochemical Discharge Phenomenon: Fundamentals and Application of Spark Assisted Chemical Engraving*, William Adrew, pp. 1–9, Norwich, NY, USA, 2014.
14. Kurafuji, H. and Suda, K., Electrical discharge drilling of glass. *Ann. CIRP*, 16, 415, 1968.
15. Cook, N.H., Foote, G.B., Jordan, P. *et al.*, Experimental studies in electromachining. *J. Eng. Ind.*, 95, 945, 1973.
16. Tsuchiya, H., Inoue, T., Miyazaiki, M., Wire electrochemical discharge machining of glasses and ceramics. *Bull. Japan Soc Precis. Eng.*, 19, 73, 1985.
17. Esashi, M., Matsumoto, Y., Shoji, S., Absolute pressure sensors by air-tight electrical feedthrough structure. *Sens. Actuators A Phys.*, 23, 1048, 1990.
18. Skrabalak, G., Zybura-Skrabalak, M., Ruszaj, A., Building of rules base for fuzzy-logic control of the ECDM process. *J. Mater. Process. Technol.*, 149, 530, 2004.
19. Wuthrich, R. and Fascio, V., Machining of non-conducting materials using electrochemical discharge phenomenon—An overview. *Int. J. Mach. Tools Manuf.*, 45, 1095, 2005.

20. Yang, C.T., Song, S.L., Yan, B.H. *et al.*, Improving machining performance of wire electrochemical discharge machining by adding SiC abrasive to electrolyte. *Int. J. Mach. Tools Manuf.*, 46, 2044, 2006.
21. Didar, T.F., Dolatabadi, A., Wuthrich, R., Characterization and modeling of 2D-glass micro-machining by spark-assisted chemical engraving (SACE) with constant velocity. *J. Micromech. Microeng.*, 18, 065016, 2008.
22. Cao, X.D., Kim, B.H., Chu, C.N., Micro-structuring of glass with features less than 100 μm by electrochemical discharge machining. *Precis. Eng.*, 33, 459, 2009.
23. Yang, C.K., Wu, K.L., Hung, J.C. *et al.*, Enhancement of ECDM efficiency and accuracy by spherical tool electrode. *Int. J. Mach. Tools Manuf.*, 51, 528, 2011.
24. Liu, J.W., Yue, T.M., Guo, Z.N., Grinding-aided electrochemical discharge machining of particulate reinforced metal matrix composites. *Int. J. Mach. Tools Manuf.*, 68, 2349, 2013.
25. Jiang, B., Lan, S., Ni, J. *et al.*, Experimental investigation of spark generation in electrochemical discharge machining of non-conducting materials. *J. Mater. Process. Technol.*, 214, 892, 2014.
26. Abou Ziki, J.D., Hof, L.A., Wuthrich, R., The machining temperature during spark assisted chemical engraving of glass. *Manuf. Lett.*, 3, 9, 2015.
27. Goud, M. and Sharma, A.K., A three-dimensional finite element simulation approach to analyze material removal in electrochemical discharge machining. *Proc. Inst. Mech. Eng. C J. Mech. Eng. Sci.*, 231, 2417, 2016.
28. Hof, L.A., Guo, X., Seo, M. *et al.*, Glass imprint templates by spark assisted chemical engraving for microfabrication by hot embossing. *Micromachines*, 8, 29, 2017.
29. Hof, L.A. and Wuthrich, R., Industry 4.0–towards fabrication of mass-personalized parts on glass by spark assisted chemical engraving (SACE). *Manuf. Lett.*, 15, 76, 2018.
30. Goud, M., Sharma, A.K., Jawalkar, C., A review on material removal mechanism in electrochemical discharge machining (ECDM) and possibilities to enhance the material removal rate. *Precis. Eng.*, 45, 1, 2016.
31. Crichton, I.M. and McGeough, J.A., Studies of the discharge mechanisms in electrochemical arc machining. *J. Appl. Electrochem.*, 15, 113, 1985.
32. McGeough, J.A. and Rasmussen, H., A theoretical analysis of electrochemical are machining. *Proc. R. Soc A Math. Phys. Eng. Sci.*, 429, 429, 1990.
33. Pawar, P., Ballav, R., Kumar, A., FEM analysis of different materials based on explicit dynamics ANSYS in electrochemical discharge Machine, in: *Simulations for Design and Manufacturing, Lecture Notes on Multidisciplinary Industrial Engineering*. Dixit, U., Kant, R. (eds), pp. 231–258, Springer, Singapore, 2018.
34. Kulkarni, A.V., Electrochemical spark micromachining process, in: *Micromachining Techniques for Fabrication of Micro and Nano Structures*, pp. 235–252, InTech Open, London, 2012.
35. Kellogg, H.H., Anode effect in Aqueous Electrolysis. *J. Electrochem. Soc*, 97, 133, 1950.

36. Cheng, C.P., Wu, K.L., Mai, C.C., Study of gas film quality in electrochemical discharge machining. *Int. J. Mach. Tools Manuf.*, 50, 689, 2010.
37. Jain, V.K., Dixit, P.M., Pandey, P.M., On the analysis of the electrochemical spark machining process. *Int. J. Mach. Tools Manuf.*, 39, 165, 1999.
38. Kulkarni, A., Sharan, R., Lal, G.K., An experimental study of discharge mechanism in electrochemical discharge machining. *Int. J. Mach. Tools Manuf.*, 42, 1121, 2002.
39. Singh, M. and Singh, S., Electrochemical discharge machining: A review on preceding and perspective research. *Proc. Inst. Mech. Eng. B J. Eng. Manuf.*, 233, 1425, 2019.
40. Bhattacharyya, B., Doloi, B.N., Sorkhel, S.K., Experimental investigations into electrochemical discharge machining (ECDM) of non-conductive ceramic materials. *J. Mater. Process. Technol.*, 95, 145, 1999.
41. Jiang, B., Lan, S., Ni, J., On modeling and simulation of the discharging activity in electrochemical discharge machining. *Proceedings of the ASME 2015 International Manufacturing Science and Engineering Conference*, vol. 1, p. V001T02A050, 2015.
42. Wuthrich, R., Fujisaki, K., Couthy, P. *et al.*, Spark assisted chemical engraving (SACE) in microfactory. *J. Micromech. Microeng.*, 15, S276, 2005.
43. Abou Ziki, J.D., Didar, T.F., Wuthrich, R., Micro-texturing channel surfaces on glass with spark assisted chemical engraving. *Int. J. Mach. Tools Manuf.*, 57, 66, 2012.
44. Kulkarni, A.V., Electrochemical discharge machining process. *Def. Sci. J.*, 57, 765, 2007.
45. Tsutsumi, C., Okano, K., Suto, T., High quality machining of ceramics. *J. Mater. Process. Technol.*, 37, 639, 1993.
46. Pawar, P., Kumar, A., Ballav, R., Development and manufacturing of arduino based electrochemical discharge machine. *J. Mach. Eng.*, 18, 45, 2018.
47. Pawar, P., Ballav, R., Kumar, A., Review on material removal technology of soda-lime glass material. *Indian. J. Sci. Technol.*, 10, 1, 2017.
48. Pawar, P., Ballav, R., Kumar, A., Machining processes of silicon carbide: A review. *Rev. Adv. Mater. Sci.*, 51, 62, 2017.
49. Khairy, A.B.E. and McGeough, J.A., Die-sinking by electroerosion-dissolution machining. *CIRP Ann. Manuf. Technol.*, 39, 191, 1990.
50. Basak, I., *Electrochemical Discharge Machining: Mechanism and A Scheme For Enhancing Material Removal Capacity*, Ph.D. Thesis, I.I.T. Kanpur, India, 1991.
51. Basak, I. and Ghosh, A., Mechanism of spark generation during electrochemical discharge machining: A theoretical model and experimental verification. *J. Mater. Process. Technol.*, 62, 46, 1996.
52. Mediliyegedara, T.K.K.R., De Silva, A.K.M., Harrison, D.K. *et al.*, System identification and controller design for the tool position control system of an electro chemical discharge machining (ECDM) machine. *Proceedings of the 34th International MATADOR Conference*, Springer, pp. 141–146, 2004.

53. Mediliyegedara, T.K.K.R., De Silva, A.K.M., Harrison, D.K. *et al.*, A Fuzzy logic approach for the pulse classification of electro chemical discharge machining (ECDM). *Proceedings of the 34th International MATADOR Conference*, Springer, pp. 161–166, 2004.
54. Fascio, V., Wuthrich, R., Bleuler, H., Spark assisted chemical engraving in the light of electrochemistry, Electrochim. *Acta*, 49, 3997, 2004.
55. Bhondwe, K.L., Yadava, V., Kathiresan, G., Finite element prediction of material removal rate due to electro-chemical sparks machining. *Int. J. Mach. Tools Manuf.*, 46, 1699, 2006.
56. Allagui, A. and Wuthrich, R., Gas film formation time and gas film lifetime during electrochemical discharge phenomenon. *Electrochim. Acta*, 54, 5336, 2009.
57. Liu, J.W., Yue, T.M., Guo, Z.N., An analysis of the discharge mechanism in electrochemical discharge machining of particulate reinforced metal matrix composites. *Int. J. Mach. Tools Manuf.*, 50, 86, 2010.
58. El-Haddad, R. and Wuthrich, R., A mechanistic model of the gas film dynamics during the electrochemical discharge phenomenon. *J. Appl. Electrochem.*, 40, 1853, 2010.
59. Wei, C., Xu, K., Ni, J. *et al.*, A finite element-based model for electrochemical discharge machining in discharge regime. *Int. J. Adv. Manuf. Technol.*, 54, 987, 2011.
60. Panda, M.C. and Yadava, V., Intelligent modeling and multiobjective optimization of die sinking electrochemical spark machining process. *Mater. Manuf. Process.*, 27, 10, 2012.
61. Krotz, H., Roth, R., Wegener, K., Experimental investigation and simulation of heat flux into metallic surfaces due to single discharges in micro-electrochemical arc machining (micro-ECAM). *Int. J. Adv. Manuf. Technol.*, 68, 1267, 2013.
62. Kamaraj, A.B., Jui, S.K., Cai, Z., Sundaram, M.M., A mathematical model to predict overcut during electrochemical discharge machining. *Int. J. Adv. Manuf. Technol.*, 81, 685, 2015.
63. Paul, L. and Korah, L.V., Effect of power source in ECDM process with FEM modeling. *Procedia Manuf.*, 25, 1175, 2016.
64. Ladeesh, V.G. and Manu, R., Effect of machining parameters on edge-chipping during drilling of glass using grinding-aided electrochemical discharge machining (G-ECDM). *Adv. Manuf.*, 6, 215, 2018.
65. Hajian, M., Razfar, M.R., Movahed, S., Etefagh, A.H., Experimental and numerical investigations of machining depth for glass material in electrochemical discharge milling. *Precis. Eng.*, 51, 521, 2018.
66. Crichton, I.M., McGeough, J.A., Munro, W., White, C., Comparative studies of ecm, edm and ecam. *Precis. Eng.*, 3, 155, 1981.
67. De Silva, A.K., *Process Developments In Electrochemical Arc Machining*, Ph.D. Thesis, University of Edinburgh, 1988.

68. Raghuram, V., Pramila, T., Srinivasa, Y.G., Narayanasamy, K., Effect of the circuit parameters on the electrolytes in the electrochemical discharge phenomenon. *J. Mater. Process. Technol.*, 52, 301, 1995.
69. Kulkarni, A., Sharan, R., Lal, G.K., Measurement of temperature transients in the electrochemical discharge machining process. *AIP Conf. Proc.*, 684, 1069, 2003.
70. Kulkarni, A.V. and Karnik, M.G., Experimental measurements and theoretical estimation of temperature in ECDM process. *Proceedings of the International Conference on MEMS, NANO and Smart Systems (ICMENS 2004)*, IEEE, pp. 243–247, 2004.
71. Mediliyegedara, T.K.K.R., De Silva, A.K.M., Harrison, D.K., McGeough, J.A., New developments in the process control of the hybrid electro chemical discharge machining (ECDM) process. *J. Mater. Process. Technol.*, 167, 338, 2005.
72. Wuthrich, R., Spaelter, U., Wu, Y., Bleuler, H., A systematic characterization method for gravity-feed micro-hole drilling in glass with spark assisted chemical engraving (SACE). *J. Micromech. Microeng.*, 16, 1891, 2006.
73. Wuthrich, R. and Hof, L.A., The gas film in spark assisted chemical engraving (SACE)–a key element for micro-machining applications. *Int. J. Mach. Tools Manuf.*, 46, 828, 2006.
74. Wuthrich, R., Spaelter, U., Bleuler, H., The current signal in spark-assisted chemical engraving (SACE): What does it tell us? *J. Micromech. Microeng.*, 16, 779, 2006.
75. Kim, D.J., Ahn, Y., Lee, S.H., Kim, Y.K., Voltage pulse frequency and duty ratio effects in an electrochemical discharge micro drilling process of pyrex glass. *Int. J. Mach. Tools Manuf.*, 46, 1064, 2006.
76. West, J. and Jadhav, A., ECDM methods for fluidic interfacing through thin glass substrates and the formation of spherical microcavities. *J. Micromech. Microeng.*, 17, 403, 2007.
77. Maillard, P., Despont, B., Bleuler, H., Wuthrich, R., Geometrical characterization of micro-holes drilled in glass by gravity-feed with spark assisted chemical engraving (SACE). *J. Micromech. Microeng.*, 17, 1343, 2007.
78. CHAK, S.K., *Experimental Investigations on ECDM Process Improvement While Machining Al2O3 and SiC*, Ph.D Thesis, I.I.T. Delhi, India, 2007.
79. Sandison, M.E., Zagnoni, M., Abu-Hantash, M., Morgan, H., Micromachined glass apertures for artificial lipid bilayer formation in a microfluidic system. *J. Micromech. Microeng.*, 17, S189, 2007.
80. Jalali, M., Maillard, P., Wuthrich, R., Toward a better understanding of glass gravity-feed micro-hole drilling with electrochemical discharges. *J. Micromech. Microeng.*, 19, 045001, 2009.
81. Tastekin, D., Krotz, H., Gerlach, C., Roth-Stielow, J., A novel electrical power supply for electrothermal and electrochemical removal machining methods. *Energy Conversion Congress and Exposition*, IEEE, p. 2682, 2009.
82. Nandi, D., Puri, A.B., Basak, I., Behaviour of bubbles generated in electro-chemical discharge machining. *Int. J. Eng. Sci. Technol.*, 3, 8274, 2011.

83. Sankar, A.R., Bindu, V.S.S., Das, S., Coupled effects of gold electroplating and electrochemical discharge machining processes on the performance improvement of a capacitive accelerometer. *Microsyst. Technol.*, 17, 1661, 2011.
84. Mochimaru, Y., Ota, M., Yamaguchi, K., Micro hole processing using electro-chemical discharge machining. *J. Adv. Mech. Des. Syst. Manuf.*, 6, 949, 2012.
85. De Souza, A. and Ingpeed, A., Analyzing machining time, geometric form and tool-electrode wear as work result of the ECDM-process producing microholes in stainless steel. *Dev. J. Integr. Eng.*, 1, 1, 2014.
86. Abou Ziki, J.D. and Wuthrich, R., Nature of drilling forces during spark assisted chemical engraving. *Manuf. Lett.*, 4, 10, 2015.
87. Abou Ziki, J.D. and Wuthrich, R., The machining gap during constant velocity-feed glass micro-drilling by spark assisted chemical engraving. *J. Manuf. Process.*, 19, 87, 2015.
88. Jiang, B., Lan, S., Wilt, K., Ni, J., Modeling and experimental investigation of gas film in micro electrochemical discharge machining process. *Int. J. Mach. Tools Manuf.*, 90, 8, 2015.
89. Behroozfar, A. and Razfar, M.R., Experimental and numerical study of material removal in electrochemical discharge machining ECDM. *Mater. Manuf. Process.*, 31, 495, 2016.
90. Gupta, P.K., Dvivedi, A., Kumar, P., Effect of pulse duration on quality characteristics of blind hole drilled in glass by ECDM. *Mater. Manuf. Process.*, 31, 1740, 2016.
91. Gupta, P.K., Bhamu, J.P., Rajoria, C.S., Lautre, N.K., Agarwal, V., Effect of duty ratio at different pulse frequency during hole drilling in ceramics using electrochemical discharge machining. *MATEC Web of Conferences*, vol. 77, p. 1, 2016.
92. Wuthrich, R., Fascio, V., Viquerat, D., Langen, H., In situ measurement and micromachining of glass. *International Symposium on Micromechatronics and Human Science*, IEEE, pp. 185–191, 1999.
93. Yang, C.T., Ho, S.S., Yan, B.H., Micro hole machining of borosilicate glass through electrochemical discharge machining (ECDM). *Key Eng. Mater.*, 196, 149, 2001.
94. Fascio, V., Langen, H.H., Bleuler, H., Comninellis, C., Investigations of the spark assisted chemical engraving. *Electrochem. Commun.*, 5, 203, 2003.
95. Wuthrich, R., Despont, B., Maillard, P., Bleuler, H., Improving the material removal rate in spark-assisted chemical engraving (SACE) gravity-feed micro-hole drilling by tool vibration. *J. Micromech. Microeng.*, 16, N28, 2006.
96. Chak, S.K. and Rao, P.V., Trepanning of Al_2O_3 by electro-chemical discharge machining (ECDM) process using abrasive electrode with pulsed DC supply. *Int. J. Mach. Tools Manuf.*, 47, 2061, 2007.

97. Chak, S.K. and Rao, P.V., The drilling of Al_2O_3 using a pulsed DC supply with a rotary abrasive electrode by the electrochemical discharge process. *Int. J. Adv. Manuf. Technol.*, 39, 633, 2008.
98. Morrison, A., Rodrigues, L., Wuthrich, R., Reducing variability in spark assisted chemical engraving gravity feed drilling of glass. *Microsystems and Nanoelectronics Research Conference*, IEEE, p. 161, 2008.
99. Lal, A., Bleuler, H., Wuthrich, R., Fabrication of metallic nanoparticles by electrochemical discharges. *Electrochem. Commun.*, 10, 488, 2008.
100. Coteata, M., Slatineanu, L., Dodun, O., Ciofu, C., Electrochemical discharge machining of small diameter holes. *Int. J. Mater. Form.*, 1, 1327, 2008.
101. Coteata, M., Ciofu, C., Slatineanu, L., Munteanu, A., Dodun, O., Establishing the electrical discharges weight in electrochemical discharge drilling. *Int. J. Mater. Form.*, 2, 673, 2009.
102. Han, M.S., Min, B.K., Lee, S.J., Geometric improvement of electrochemical discharge micro-drilling using an ultrasonic-vibrated electrolyte. *J. Micromech. Microeng.*, 19, 065004, 2009.
103. Liu, J.W., *Grinding-Aided Electrochemical Discharge Machining of Metal Matrix Composites*, Ph.D. Thesis, The Hong Kong Polytechnic University, Hong Kong, 2009.
104. Cheng, C.P., Wu, K.L., Mai, C.C., Hsu, Y.S., Yan, B.H., Magnetic field-assisted electrochemical discharge machining. *J. Micromech. Microeng.*, 20, 075019, 2010.
105. Wei, C., Ni, J., Hu, D., Electrochemical discharge machining using micro-drilling tools. *Trans. NAMRI/SME*, 38, 105, 2010.
106. Wei, C., Hu, D., Xu, K., Ni, J., Electrochemical discharge dressing of metal bond micro-grinding tools. *Int. J. Mach. Tools Manuf.*, 51, 165, 2011.
107. Coteata, M., Schulze, H.P., Slatineanu, L., Drilling of difficult-to-cut steel by electrochemical discharge machining. *Mater. Manuf. Process.*, 26, 1466, 2011.
108. Kulkarni, A.V., Jain, V.K., Misra, K.A., Electrochemical spark micromachining: Present scenario. *Int. J. Autom. Technol.*, 5, 52, 2011.
109. Doan, C.X., *Spark-Assisted Micro Machining of Glass, Ph.D*, Thesis, Seoul National University, 2012.
110. Lu, C., Gu, A., Li, M., Yang, S., The micro-milling machining of pyrex glass using the electrochemical discharge machining process. *Adv. Mat. Res.*, 403-408, 738, 2012.
111. Rusli, M. and Furutani, K., Performance of micro-hole drilling by ultrasonic-assisted electro-chemical discharge machining. *Adv. Mat. Res.*, 445, 865, 2012.
112. Abou Ziki, J.D. and Wuthrich, R., Forces exerted on the tool-electrode during constant-feed glass micro-drilling by spark assisted chemical engraving. *Int. J. Mach. Tools Manuf.*, 73, 47, 2013.
113. Jiang, B., Lan, S., Ni, J., Experimental investigation of drilling incorporated electrochemical discharge machining. *Proceedings of the ASME International Manufacturing Science and Engineering Conference*, vol. 2, p. 1, 2014.

114. Paul, L. and Hiremath, S., Evaluation of process parameters of ECDM using grey relational analysis. *Procedia Mater. Sci.*, 5, 2273, 2014.
115. Paul, L., Hiremath, S.S., Ranganayakulu, J., Experimental investigation and parametric analysis of electro chemical discharge machining. *Int. J. Manuf. Technol. Manage.*, 28, 57, 2014.
116. Paul, L. and Hiremath, S.S., Characterisation of micro channels in electrochemical discharge machining process. *Appl. Mech. Mater.*, 490, 238, 2014.
117. Gao, C., Liu, Z., Li, A., Study of micro drilling on pyrex glass using spark assisted chemical engraving. *Micro Nanosyst.*, 6, 26, 2014.
118. Krotz, H. and Wegener, K., Sparc assisted electrochemical machining: a novel possibility for microdrilling into electrical conductive materials using the electrochemical discharge phenomenon. *Int. J. Adv. Manuf. Technol.*, 79, 1633, 2015.
119. Kulkarni, A.V. and Jain, V.K., Design and development of an electrochemical spark micro manufacturing equipment. *Int. J. Mech. Eng. Robot. Res.*, 4, 368, 2015.
120. Furutani, K. and Kojima, S., Prototyping of acceleration sensor by using lathe-type electro-chemical discharge machine. *Procedia CIRP*, 42, 772, 2016.
121. Zhang, Z., Huang, L., Jiang, Y., Liu, G., Nie, X., Lu, H., Zhuang, H., A Study to explore the properties of electrochemical discharge effect based on pulse power supply. *J. Adv. Manuf. Technol.*, 85, 2107, 2016.
122. Hajian, M., Razfar, M.R., Movahed, S., An experimental study on the effect of magnetic field orientations and electrolyte concentrations on ECDM milling performance of glass. *Precis. Eng.*, 45, 322, 2016.
123. Madhavi, B.J. and Hiremath, S.S., Investigation on machining of holes and channels on borosilicate and sodalime glass using μ-ECDM setup. *Procedia Technol.*, 25, 1257, 2016.
124. Ladeesh, V.G. and Manu, R., Grinding aided electrochemical discharge drilling (G-ECDD) of borosilicate glass and its performance evaluation. *Procedia Technol.*, 25, 1122–1128, 2016.
125. Ladeesh, V.G. and Manu, R., Experimental investigation on the performance of grinding assisted electrochemical discharge drilling of glass. *MATEC Web Conf.*, 51, 03001, 2016.
126. Ladeesh, V.G. and Manu, R., A mathematical model for surface roughness of fluidic channels produced by grinding aided electrochemical discharge machining (G-ECDM). *MATEC Web Conf.*, 104, 02008, 2017.
127. Jiang, B., *Micro Machining Using the Electrochemical Discharge Assisted Cutting*, Ph.D. Thesis, University of Michigan, 2017.
128. Singh, T. and Dvivedi, A., A pressurized feeding approach for effective control on working gap in ECDM of borosilicate glass. *Mater. Manuf. Process.*, 33, 462, 2018.

129. Elhami, S. and Razfar, M.R., Effect of ultrasonic vibration on the single discharge of electrochemical discharge machining. *Mater. Manuf. Process.*, 33, 444, 2018.
130. Xu, Y., Chen, J., Jiang, B., Ni, J., Investigation of micro-drilling using electrochemical discharge machining with counter resistant feeding. *J. Mater. Process. Technol.*, 257, 141, 2018.
131. Tokura, H., Kondoh, I., Yoshikswa, M., Ceramic material processing by electrical discharge in electrolyte. *J. Mater. Sci.*, 24, 991, 1989.
132. Tandon, S., Jain, V.K., Kumar, P., Rajurkar, K.P., Investigations into machining of composites. *Precis. Eng.*, 12, 227, 1990.
133. Singh, Y.P., Jain, V.K., Kumar, P., Agrawal, D.C., Machining piezoelectric (PZT) ceramics using an electrochemical spark machining (ECSM) process. *J. Mater. Process. Technol.*, 58, 24, 1996.
134. Gautam, N. and Jain, V.K., Experimental investigations into ECSD process using various tool kinematics. *Int. J. Mach. Tools Manuf.*, 38, 15, 1998.
135. Doloi, B., Bhattacharyya, B., Sorkhel, S.K., Electrochemical discharge machining of non-conducting ceramic. *Def. Sci. J.*, 49, 4, 331–338, 1999.
136. Fascio, V., Wuthrich, R., Viquerat, D., Langen, H., 3D microstructuring of glass using electrochemical discharge machining (ECDM). *International Symposium on Micromechatronics and Human Science*, IEEE, pp. 179–183, 1999.
137. Jain, V.K., Choudhury, S.K., Ramesh, K.M., On the machining of alumina and glass. *Int. J. Mach. Tools Manuf.*, 42, 1269, 2002.
138. Lee, E.S., Howard, D., Liang, E., Collins, S.D., Smith, R.L., Removable tubing interconnects for glass-based micro-fluidic systems made using ECDM. *J. Micromech. Microeng.*, 14, 535, 2004.
139. Peng, W.Y. and Liao, Y.S., Study of electrochemical discharge machining technology for slicing non-conductive brittle materials. *J. Mater. Process. Technol.*, 149, 363, 2004.
140. Sarkar, B.R., Doloi, B., Bhattacharyya, B., Parametric analysis on electrochemical discharge machining of silicon nitride ceramics. *Int. J. Adv. Manuf. Technol.*, 28, 873, 2006.
141. Zheng, Z.P., Cheng, W.H., Huang, F.Y., Yan, B.H., 3D microstructuring of Pyrex glass using the electrochemical discharge machining process. *J. Micromech. Microeng.*, 17, 5, 960–966, 2007.
142. Zheng, Z.P., Wu, K.L., Hsu, Y.S., Huang, F.Y., Yan, B.H., Feasibility of 3D surface machining on pyrex glass by electrochemical discharge machining (ECDM). *Proc. AEMS07*, Japan, pp. 98–103, 2007.
143. Jain, V.K. and Adhikary, S., On the mechanism of material removal in electrochemical spark machining of quartz under different polarity conditions. *J. Mater. Process. Technol.*, 200, 460–470, 2008.
144. Furutani, K. and Maeda, H., Machining a glass rod with a lathe-type electro-chemical discharge machine. *J. Micromech. Microeng.*, 18, 065006, 2008.
145. Kudla, L., Investigation into electrochemical discharge machining of microholes. *J. Autom. Mob. Robot. Intell. Syst.*, 3, 21, 2009.

146. Didar, T.F., Dolatabadi, A., Wuthrich, R., Local hardness and density variation in glass substrates machined with spark assisted chemical engraving (SACE). *Mater. Lett.*, 63, 51, 2009.
147. Ozhikandathil, J., Morrison, A., Packirisamy, M., Wuthrich, R., Low resistive silicon substrate as an etch-stop layer for drilling thick SiO_2 by spark assisted chemical engraving (SACE). *Microsyst. Technol.*, 17, 373, 2011.
148. Manna, A. and Narang, V., A study on micro machining of e-glass–fibre–epoxy composite by ECSM process. *Int. J. Adv. Manuf. Technol.*, 61, 1191, 2012.
149. Jawalkar, C.S., Sharma, A.K., Kumar, P., Micromachining with ECDM: Research potentials and experimental investigations. *World Acad. Sci. Eng. Technol.*, 61, 90, 2012.
150. Paul, L. and Hiremath, S.S., Response surface modelling of micro holes in electrochemical discharge machining process. *Procedia Eng.*, 64, 1395, 2013.
151. Bhuyan, B.K. and Yadava, V., Experimental study of traveling wire electrochemical spark machining of borosilicate glass. *Mater. Manuf. Process.*, 29, 298, 2014.
152. Coteata, M. and Cretu, G., ECDM drilling process analysis via taguchi method. *Appl. Mech. Mater.*, 657, 337, 2014.
153. Pawar, P., Ballav, R., Kumar, A., Revolutionary developments in ECDM process: An overview. *Mater. Today Proc.*, 2, 3188, 2015.
154. Sarkar, B.R., Doloi, B., Bhattacharyya, B., Investigation on electrochemical discharge micro-machining of silicon carbide. *IJMFMP*, 4, 29, 2017.
155. He, S., Tong, H., Liu, G., Spark assisted chemical engraving (SACE) mechanism on ZrO_2 ceramics by analyzing processed products. *Ceram. Int.*, 44, 7967, 2018.
156. Wang, J., Guo, Y.B., Fu, C., Jia, Z., Surface integrity of alumina machined by electrochemical discharge assisted diamond wire sawing. *J. Manuf. Process.*, 31, 96, 2018.
157. Schopf, M., Beltrami, I., Boccadoro, M., Kramer, D., Schumacher, B., ECDM (electro chemical discharge machining), a new method for trueing and dressing of metal bonded diamond grinding tools. *CIRP Ann. Manuf. Technol.*, 50, 125, 2001.
158. Lim, H.J., Lim, Y.M., Kim, S.H., Kwak, Y.K., Self-aligned micro tool and electrochemical discharge machining (ECDM) for ceramic materials. *Proceedings of SPIE Optical Engineering for Sensing and Nanotechnology*, vol. 4416, pp. 348–353, 2001.
159. Wuthrich, R., Hof, L.A., Lal, A., Fujisaki, K., Bleuler, H., Mandin, P., Picard, G., Physical principles and miniaturization of spark assisted chemical engraving (SACE). *J. Micromech. Microeng.*, 15, S268, 2005.
160. Yan, B.H., Yang, C.T., Huang, F.Y., Lu, Z.H., Electrophoretic deposition grinding (EPDG) for improving the precision of microholes drilled via ECDM. *J. Micromech. Microeng.*, 17, 376, 2007.

161. Zheng, Z.P., Su, H.C., Huang, F.Y., Yan, B.H., The tool geometrical shape and pulse-off time of pulse voltage effects in a pyrex glass electrochemical discharge microdrilling process. *J. Micromech. Microeng.*, 17, 265, 2007.
162. Han, M.S., Min, B.K., Lee, S.J., Modeling gas film formation in electrochemical discharge machining processes using a side-insulated electrode. *J. Micromech. Microeng.*, 18, 045019, 2008.
163. Han, M.S., Min, B.K., Lee, S.J., Reduction of overcut in the electrochemical discharge machining process using a side-insulated electrode. *International Conference on Smart Manufacturing Application*, IEEE, pp. 251–254, 2008.
164. Mousa, M., Allagui, A., Ng, H.D., Wuthrich, R., The effect of thermal conductivity of the tool electrode in spark-assisted chemical engraving gravity-feed micro-drilling. *J. Micromech. Microeng.*, 19, 015010, 2009.
165. Yang, C.K., Cheng, C.P., Mai, C.C., Wang, A.C., Hung, J.C., and Yan, B.H., "Effect of surface roughness of tool electrode materials in ECDM performance. *Int. J. Mach. Tools Manuf.*, 50, 1088, 2010.
166. Abou Ziki, J.D., Dhawale, N.K.M., Wuthrich, R., Modeling the forces exerted on the tool-electrode during spark assissted chemical engraving constant velocity feed-drilling. *Proceedings of the ASME 3rd Joint US-European Fluids Engineering Summer Meeting and 8TH Joint US-European Fluids Engineering Summer Meeting and 8th International Conference on Nanochannels, Microchannels, and Minichannels*, pp. 1303–1311, 2010.
167. Han, M.S., Min, B.K., Lee, S.J., Micro-electrochemical discharge cutting of glass using a surface-textured tool. *CIRP J. Manuf. Sci. Technol.*, 4, 362, 2011.
168. Abou Ziki, J.D. and Wuthrich, R., Tool wear and tool thermal expansion during micro-machining by spark assisted chemical engraving. *Int. J. Adv. Manuf. Technol.*, 61, 481, 2012.
169. Jui, S.K., Kamaraj, A.B., Sundaram, M.M., High aspect ratio micromachining of glass by electrochemical discharge machining (ECDM). *J. Manuf. Process.*, 15, 460, 2013.
170. Cao, X.D., Kim, B.H., Chu, C.N., Hybrid micromachining of glass using ECDM and micro grinding. *Int. J. Precis. Eng. Manuf.*, 14, 5, 2013.
171. Huang, S.F., Liu, Y., Li, J., Hu, H.X., Sun, L.Y., Electrochemical discharge machining micro-hole in stainless steel with tool electrode high-speed rotating. *Mater. Manuf. Process.*, 29, 634, 2014.
172. Abou Ziki, J.D., *Spark Assisted Chemical Engraving: A Novel Approach for Quantifying the Machining Zone Parameters Using Drilling Forces*, Ph.D. Thesis, Concordia University, Montreal, Canada, 2014.
173. Furutani, K., Shintani, H., Murase, Y., Arakawa, S., Performance of electrochemical discharge machining by forced discharge dispersion. *Int. J. Electr. Mach.*, 19, 9, 2014.
174. Razfar, M.R., Behroozfar, A., Ni, J., Study of the effects of tool longitudinal oscillation on the machining speed of electrochemical discharge drilling of glass. *Precis. Eng.*, 38, 885, 2014.

175. Zhang, Y., Xu, Z., Zhu, D., Xing, J., Tube electrode high-speed electrochemical discharge drilling using low-conductivity salt solution. *Int. J. Mach. Tools Manuf.*, 92, 10, 2015.
176. Saranya, S. and Sankar, A.R., Effect of tool shape and tool feed rate on the machined profile of a quartz substrate using an electrochemical discharge machining process. *Proceedings of the 2nd International Symposium on Physics and Technology of Sensors*, IEEE, pp. 313–316, 2015.
177. Behroozfar, A. and Razfar, M.R., Experimental study of the tool wear during the electrochemical discharge machining (ECDM). *Mater. Manuf. Process.*, 31, 574, 2016.
178. Guo, D., Wu, X., Lei, J., Xu, B., Kometani, R., Luo, F., Fabrication of micro/nanoelectrode using focused-ion-beam chemical vapor deposition, and its application to micro-ECDM. *Procedia CIRP*, 42, 733, 2016.
179. Hajian, M., Razfar, M.R., Etefagh, A.H., Experimental study of tool bending force and feed rate in ECDM milling. *Int. J. Adv. Manuf. Technol.*, 91, 1677, 2017.
180. Liu, Y., Wei, Z., Wang, M., Zhang, J., Experimental investigation of micro wire electrochemical discharge machining by using a rotating helical tool. *J. Manuf. Process.*, 29, 265, 2017.
181. Tang, W., Kang, X., Zhao, W., Enhancement of electrochemical discharge machining accuracy and surface integrity using side-insulated tool electrode with diamond coating. *J. Micromech. Microeng.*, 27, 065013, 2017.
182. Han, M.S., Chae, K.W., Min, B.K., Fabrication of high-aspect-ratio microgrooves using an electrochemical discharge micromilling process. *J. Micromech. Microeng.*, 27, 055004, 2017.
183. Saranya, S. and Sankar, A.R., Fabrication of precise microchannels using a side-insulated tool in a spark assisted chemical engraving process. *Mater. Manuf. Process.*, 33, 1422, 2018.
184. Han, M.S., Min, B.K., Lee, S.J., Improvement of surface integrity of electro-chemical discharge machining process using powder-mixed electrolyte. *J. Mater. Process. Technol.*, 191, 224, 2007.
185. Laio, Y.S., Wu, L.C., Peng, W.Y., A study to improve drilling quality of electrochemical discharge machining (ECDM) process. *Procedia CIRP*, 6, 609, 2013.
186. Jawalkar, C.S., Sharma, A.K., Kumar, P., Investigations on performance of ECDM process using NaOH and $NaNO_3$ electrolytes while micro machining soda lime glass. *Int. J. Manuf. Technol. Manage.*, 28, 80, 2014.
187. Dhanvijay, M.R. and Ahuja, B.B., Micromachining of ceramics by electrochemical discharge process considering stagnant and electrolyte flow method. *Procedia Technol.*, 14, 165, 2014.
188. Nguyen, K.H., Lee, P.A., Kim, B.H., Experimental investigation of ECDM for fabricating microstructures of quartz. *Int. J. Precis. Eng. Manuf.*, 16, 5, 2015.
189. Zhang, Y., Xu, Z., Xing, J., Zhu, D., Enhanced machining performance of micro holes using electrochemical discharge machining with super-high-pressure interior flushing. *Int. J. Electrochem. Sci.*, 10, 8465, 2015.

190. Paul, L. and Hiremath, S.S., Improvement in machining rate with mixed electrolyte in ECDM process. *Procedia Technol.*, 25, 1250, 2016.
191. Paul, L. and Hiremath, S.S., Experimental and theoretical investigations in ECDM process–an overview. *Procedia Technol.*, 25, 1242, 2016.
192. Saranya, S., Nair, A., Sankar, A.R., Experimental investigations on the electrical and 2D-machining characteristics of an electrochemical discharge machining (ECDM) process. *Microsyst. Technol.*, 23, 1453, 2017.
193. Sabahi, N., Razfar, M.R., Hajian, M., Experimental investigation of surfactant-mixed electrolyte into electrochemical discharge machining (ECDM) process. *J. Mater. Process. Technol.*, 250, 190, 2017.
194. Sabahi, N. and Razfar, M.R., Investigating the effect of mixed alkaline electrolyte (NaOH + KOH) on the improvement of machining efficiency in 2D electrochemical discharge machining (ECDM). *Int. J. Adv. Manuf. Technol.*, 95, 643, 2018.
195. Sabahi, N., Hajian, M., Razfar, M.R., Experimental study on the heat-affected zone of glass substrate machined by electrochemical discharge machining (ECDM) process. *Int. J. Adv. Manuf. Technol.*, 97, 1557, 2018.
196. Harugade, M.L. and Waigaonkar, S.D., Effect of different electrolytes on material removal rate, diameter of hole, and spark in electrochemical discharge machining, in: *Advances in Manufacturing, Lecture Notes in Mechanical Engineering*, pp. 427–437, Springer, 2018.
197. Gupta, P.K., Effect of electrolyte level during electro chemical discharge machining of glass. *J. Electrochem. Soc*, 165, E279, 2018.
198. Varghese, A. and Paul, L., Effect of powder mixed electrolyte in ECDM process. *Mater. Today Proc.*, 5, 11864, 2018.
199. Bhuyan, B.K. and Yadava, V., Experimental modeling and multi-objective optimization of traveling wire electrochemical spark machining (TW-ECSM) process. *J. Mech. Sci. Technol.*, 27, 2467, 2013.
200. Razfar, M.R., Ni, J., Behroozfar, A., Lan, S., An investigation on electrochemical discharge micro-drilling of glass, in: *ASME 2013 International Manufacturing Science and Engineering Conference collocated with the 41st North American Manufacturing Research Conference*, pp. V002T03A013–V002T03A013, 2013.
201. Mallick, B., Sarkar, B.R., Doloi, B., Bhattacharyya, B., Multi criteria optimization of electrochemical discharge micromachining process during micro-channel generation on glass. *Appl. Mech. Mater.*, 592-594, 525, 2014.
202. Goud, M. and Sharma, A.K., On performance studies during micromachining of quartz glass using electrochemical discharge machining. *J. Mech. Sci. Technol.*, 31, 1365, 2017.
203. Antil, P., Singh, S., Manna, A., Electrochemical discharge drilling of SiC reinforced polymer matrix composite using Taguchi's Grey relational analysis. *Arab. J. Sci. Eng.*, 43, 1257, 2018.

3

Laser-Assisted Micromilling

Asma Perveen[1]*, Sandip Kunar[2], Golam Kibria[3] and Prasenjit Chatterjee[4]

[1]Department of Mechanical & Aerospace Engineering, School of Engineering & Digital Sciences, Nazarbayev University, Republic of Kazakhstan
[2]Department of Mechanical Engineering, Aditya Engineering College, Surampalem, India; Jawaharlal Nehru Technological University Kakinada, Kakinada, East Godavari, India
[3]Department of Mechanical Engineering, Aliah University, Kolkata, India
[4]Department of Mechanical Engineering, MCKV Institute of Engineering, Howrah, India

Abstract

With the development of materials with superior properties and miniaturization requirement, conventional machining processes face additional challenges due to tool wear, tool stiffness, which gives the motivation for exploring hybrid/sequential machining techniques. In this respect, laser machining has turned out to be one of the wonders in the manufacturing area; however, it also comes with issues, such as heat-affected zone. On the other hand, laser-assisted machining techniques seem to elevate most of the heat-affected zone issue along with severe tool wear issue faced by conventional machining. This chapter discusses motivation, and overview of laser-assisted micromilling. Hard to machine materials, such as, steel alloy, Ti alloy, Ni alloy, cementite carbides, and other ceramics, are investigated using laser-assisted micromilling.

Keywords: Laser, fiber laser, difficult to cut, oxidation, ball end mill, hardness, dimensional accuracy

**Corresponding author*: asma.perveen@nu.edu.kz

Sandip Kunar, Golam Kibria, Prasenjit Chatterjee and Asma Perveen (eds.) Hybrid Micromachining and Microfabrication Technologies: Principles, Varieties and Applications, (101–114) © 2023 Scrivener Publishing LLC

3.1 Introduction

Difficult to cut materials have found their applications in several important industries, such as aerospace, biomedical, molding due to their exotic properties. However, when it comes to applications, superior properties are not the only thing to consider, but also cost involved with materials processing has important roles to play. Most of these desired but hard-to-cut materials come with the expenses of higher materials processing cost due to the unavailability of the cost-effective processing technology [1].

Nowadays, due to large-scale miniaturization, development in micromachining becomes a top priority, which offers the fabrication of microscale/mesoscale structures made of large variety of engineering materials. One of the micromachining techniques that has been well researched compared to other techniques, such as photolithography, is micromilling. Micromilling has proven its competency in machining structures like micromolds [2]. However, micromolds/dies undergo thermal and mechanical cycles, which is repetitive in nature during the molding process, therefore material choice must satisfy this requirement of withstanding thermal loading cycle [3]. Eventually, hard to cut materials, such as nickel alloy, hardened tool steel is one of the major choice for such kind of application [4]. Nevertheless, dimensional accuracy and material removal rate of such materials suffers negatively due to limited machine tools and microtool stiffness, as well as rapid tool wear [5].

Therefore, external energy-assisted machining has become a popular research trends for such kind of materials where work materials undergo softening before machining using some kind of energy. The whole idea is to reduce the yield strength, as well as hardness of the concerned materials using some kind of heat source. Literature has reported some of the energy-assisted machining, such as induction preheating [6, 7], gas torch preheating [8], and plasma-assisted machining [9], and laser-assisted machining [10, 11]. Among all of these assisted machining, Laser-assisted machining has got special attention due to controlled laser power, as well as controlled spot size and speed, which provides rapid local rise of temperature when it comes to microscale [12].

This chapter will address laser-assisted micromilling of application oriented but difficult to machine materials, associated cutting force, tool wear, and heat-affected zone. Contemporary research trends and challenges will also be discussed.

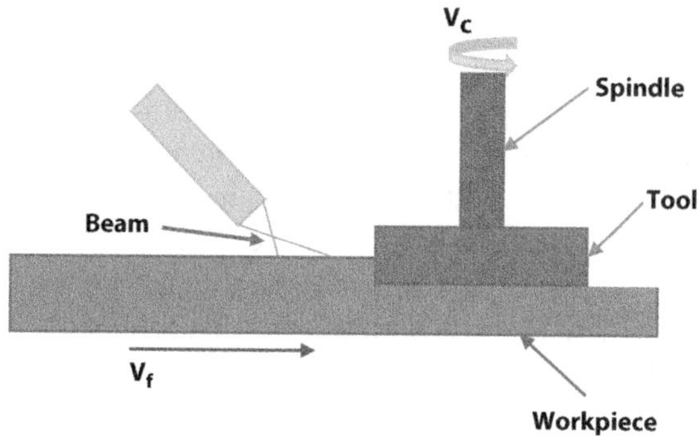

Figure 3.1 Illustration of laser-assisted micromilling.

3.2 Laser-Assisted Micromilling

Integration of continuous fiber laser with low power to micromilling process aids in heating up localized area of workpiece intensely well ahead of the cutting zone and therefore softens the cutting zone. Machinability of difficult to machine materials gets enhanced in this way due to strength reduction. Although it results in reduced cutting force along with improved surface finish and accuracy, appearance of crack in heat-affected zone may occur. Increasing temperature not only increases the density of crack, but also deteriorates the tool life leading to premature tool failure instead of tool life due to the accelerated diffusion and adhesion tool wear. Figure 3.1 provides the illustrations of laser-assisted micromilling setup.

3.2.1 Laser-Assisted Micromilling of Steel Alloys

It is found that A-286 Steel is heavily used in aerospace industries since it has favorable tensile strength and high-temperature oxidation resistance property. However due to high nickel content its machinability is not very worthy. Micromilling of A-286 Steel comes with challenges, such as size effect, stiffness of miniature microtools, ploughing, and rubbing. Therefore, micromilling of such alloys experiences stresses and tool wear, which is extraordinary, compared to traditional steel [13]. In order to elevate these shortcomings, laser heating can be used to soften such high strength alloy to improve its machinability. Here the main idea is to induce local softening to facilitate the mechanical micromilling while reducing not only the cutting force but also the tool wear.

Melkote *et al.* [14] in their study, proposed a unique laser-assisted micromilling method where comparison of laser-assisted micromilling with traditional micromilling of hardened A2 steel was investigated. Their setup comes with low power (7.5 W) continuous wave fiber laser without assisting gas ahead of the cutting tool to gain localized thermal softening. Surface roughness and dimensional accurateness of the micromilled grooves were measured as performance parameters for variation of cutting speed. Experimental findings suggested that rate of change in groove depth is reduced for laser-assisted micromilling along the line of cutting with respect to traditional micromilling which might be because of the lower rate of tool wear. This fact is also verified using the optical image of tool where microtool without laser demonstrated severe wear including loose of round shape due to initial tool chipping followed by the mechanical wearing. On the other hand, laser-assisted microtool experienced less wear and wear mode is mostly gradual wearing. In addition, there is also initial depth of groove difference between both cases; due to predominant tool deflection which is less in case of laser-assisted micromilling. It appears that assistive laser induced thermal extension of tool as well as workpiece contravene the tool deflection effect on groove depth and therefor aids in achieving better dimensional accuracy [15]. Moreover, with increase of cutting speed, laser-assisted micromilling offers better groove depth accuracy. Cutting speed of 32 m/min demonstrated highest dimensional accuracy for laser-assisted micromilled groove, which might be due to the reduced tool wear because of decreased tool cutting time. Whereas cutting speed of 14 m/min offers better groove accuracy for traditional micromilling, which may be attributed to the decreased tool wear, associated with the lower cutting speed. Groove width also presented insignificant change along the cutting line for laser-assisted case compared to traditional micromilling. Laser-assisted micromilling also demonstrated more consistent and reduced surface roughness along the line of cutting. Increasing cutting speed comes with increasing surface roughness for laser-assisted micromilling due to burnishing effect of microtool whereas traditional micromilling shows no clear trends [14]. Kumar *et al.* [16] conducted similar experiment on hardened A2 tool steel using continuous wave fiber laser and reported on reduced cutting force, surface roughness and tool wear with assisted micromilling. Ozel *et al.* [17] reported on reduced cutting forces and increased cutting tool temperature by conducting pulsed laser-assisted micromilling on AISI 4340 steel. Ding *et al.* [18] attempted to model laser-assisted micromilling for difficult to machine alloy and reported increased workpiece temperature and flow stress along with reduced tool wear.

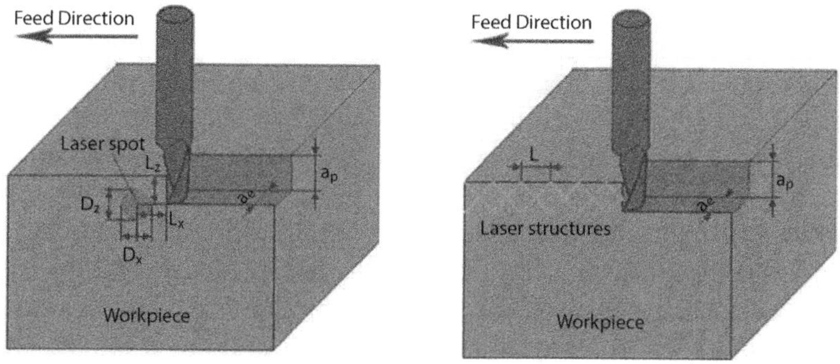

Figure 3.2 Illustration of laser-assisted micromilling along with (a) softened area (b) structured area.

Mohammadali Kadivar *et al.* [19] also conducted two stage laser-assisted micromilling of austenitic stainless steel X5CrNi18-10 where ultra-short pulsed laser is employed for microstructure creation in the first stage and then micromilling is used to get rid of the laser induced structured area of the workpiece subsequently (Figure 3.2). Issue related to heat dissipation can be elevated due to the ultrashort pulse since heat accumulation will be insignificant. Their study suggested reduced cutting force and temperature with respect to conventional milling. The reason behind this is twofold. First, due to laser structuring, smaller volume of materials is removed by second stage micromilling, which refers to lower contact length and process that is more intermittent and therefore reduced force as well as temperature. Second, comparatively shorter thinner and segmented chips produced in laser-assisted milling may also contribute to lower cutting forces. When the cutting speed rises from 50 to 250 m/min, it causes 50% reduction in normal and tangential cutting force, which is due to the higher stacking fault energy. This eventually causes reduced strain hardening because of impedance to the martensitic transformation. However, increasing feed per tooth worsens not only the surface roughness but also the cutting forces.

3.2.2 Laser-Assisted Micromilling of Titanium Alloys

Shelton *et al.* [20] has demonstrated the capabilities of laser-assisted micromilling for difficult to cut materials, such as Ti6Al4V, AISI 422, and AISI 316 using microslotting operation. In their study, CO_2 laser having 30 kW peak power was used along with the inert assist gases, which aid in blowing away the chips and prevent the oxidation of the workpiece. Acoustic emission (AE) sensors were also used for process monitoring purpose. As per their thermal modeling result, increasing initial uniform temperature may aid in reducing cutting force and flow stress for all conditions due to increased shear temperature. It is also suggested that increasing cutting speed also increases the shear plane temperature. Experimental results imply that rate of increase in acoustic emission value is reduced along with the increasing value of machining time for laser-assisted micromilling (LAMM) case which suggests reduced tool wear and it is also agreed with simulation result due to reduced contact pressure and low tool temperature. LAMM provides reduced AE signal (15.9%), which corresponds to reduced thrust force data (16.8%) given by finite element model. In addition, significant surface finish improvement was reported for LAMM for both Ti and steel alloy.

Micromilling with ball end mill tool can experience severe rubbing phenomena due to the radial shape geometry of tool and it can get worsen for ductile materials, such as Ti alloy. Therefore, laser-assisted micromilling of titanium alloy(Ti6Al4V) by means of ball end shaped milling tool was investigated by Mohid *et al.* [21]. Laser heating parameters were chosen not to exceed the phase transformation temperature of Ti6Al4V. A gap of 0.1 mm is maintained between heated area and tool outer diameter to avoid tool heating. Their study also simulated the tool-workpiece interface temperature, which suggested the topmost surface area temperature to be between 128°C and 178°C for various feed rate, cutting speed and depth of cut. Experimental results reported reduced average surface roughness for depth of cut of 0.02 mm along with feed per tooth of 0.003 and 0.0042 mm, which gives about 12% to 20% reduction of roughness for assisted milling. In addition, higher cutting speed results in less reduction in surface roughness due to lower temperature. Higher value of depth of cut also results in large fluctuation in average roughness value due to larger volume of produced chip, which is associated with large cutting force and eventual rapid flank wear. Most consistent average roughness value is achieved for depth of cut of 0.007 mm when feed rate increases from 0.0021 mm/flute to 0.0042 mm/flute due to dominating rubbing and ploughing mechanism at lower feed rate.

Xia *et al.* [22] proposed another novel method called laser-induced oxidation-assisted micromilling for successful micromachining of Ti6Al4V alloy where laser heating induced porous oxide layer along with the subsequent dense sub-surface layer are removed with less cutting force and reduced tool wear (Figure 3.3). Nanosecond pulse Glass fiber laser with average power output of 20 W is used for laser irradiation purpose. Microstructure

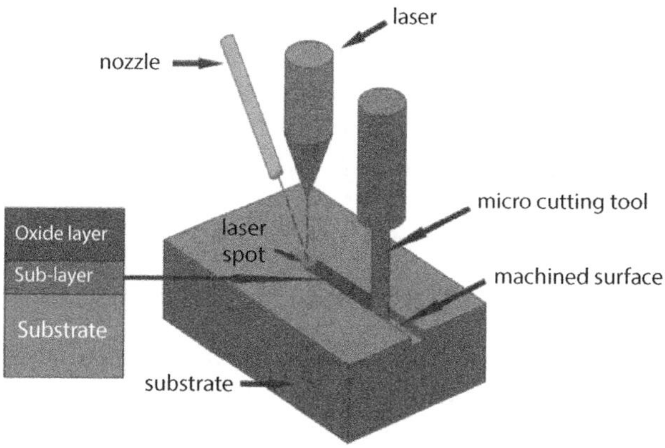

Figure 3.3 Schematic of laser-induced oxidation-assisted micromilling.

suggested surface layer of 18.2 μm and subsurface layer of 2.3 μm contains anatase and rutile TiO_2. Their study also reported 50% to 65% reduction in cutting force with laser-induced oxidation-assisted micromilling when it is compared against traditional micromilling for same cutting parameters. In addition, reduced tool wear as well as reduced top burr width is reported for this novel method. Tool wear mechanism related to the proposed method is coating spalling, whereas micromilling tool for conventional process experiences nose breakage, adhesive wear in addition to spalling. This is due to reduced cutting load associated with the removal of porous oxide layer compared to conventional micromilling where greater cutting load causes both adhesive wear and tool nose breakage. For the very same reason of less cutting load and tool wear, proposed assisted micromilling also offers reduced surface roughness compared to conventional case.

Hojati *et al.* [23] also investigated on laser microstructuring of Ti6Al4V using picosecond laser before micromilling operation and reported on the significant reduction of cutting force along x direction. Their study also revealed higher reduction in cutting force for higher density laser structures, which was done by reducing the gap between the lines. Line gap and cutting force seemed to have logarithmic relationship. It is also suggested that for higher feed rate, tool deflection can be minimized by laser structuring. The reasons behind the reduction in cutting forces are twofold. First, amount of materials removed by micromilling becomes smaller due to laser structuring and therefore reduced chip thickness can be expected. Secondly, laser structuring induced materials alteration, which contributes to around 30% reduction in materials strength. However, contribution percentage from both factor is still subjected to future investigation.

Shen *et al.* [24] proposed a 2D finite element model to predict the flow stress and temperature distribution of LAAM of Ti6Al4V alloy. Their results suggested less burr free surface with no heat-affected zone for LAAM due to thermal softening. For tool rotation angle of 0° to 90°, flow stress varies from 2464 MPa to 1922 MPa and temperature varies from 401°C to 580°C.

3.2.3 Laser-Assisted Micromilling of Ni Alloys

Ni based alloys are considered super alloys although their applications are limited due to the large tool wear and shortened tool life experienced during conventional micromilling. Rahim *et al.* [25] investigated Inconel 718 alloy using LAMM where AlTiN coated two flute carbide microball end mill was used. Their study reported on the reduced cutting force for LAMM when it is benchmarked against traditional micromilling which is due to the reduced materials strength and shear resistance. It is also revealed that workpiece reached to temperature of 800K where workpiece experiences thermal softening. Although LAMM process reduces tool adhesion wear; however, it also induces coating delamination due to temperature exceeding coating melting temperature.

Kim *et al.* [26] derived prediction equations for estimating the cutting force and preheating temperature associated with the LAAM of Inconel 718 alloy using factorial and Box-Behnken design using response surface method. Here absorptivity for Ni alloy was confirmed using Diode laser. Their study showed good agreement with experimental result and predicted result. Maximum error reported for cutting force was 6% to 4%, and for preheating temperature was about 0.5–1% respectively. Authors reported on the applicability of prediction equations for similar process using other materials. Alahmari *et al.* [27] also reported mathematical modeling based on response surface method for predicting the microchannel size micromilled on Inconel 718 using Nd:YAG pulsed laser. Their study suggested optimized parameters for accurate microchannel of 100 μm width and they are 82% current intensity, 37.07 kHz pulse frequency and scan speed 300 mm/s. Optimized setup yields in microchannel size of 102.34 μm in Inconel 718. Low current density is associated with narrow microchannel and vice versa. On the other hand, increasing pulse frequency results in more accurate size of microchannel. Experimental results are in good agreement with predicted result, which yields in less than 2% of error.

Ahmed *et al.* [28] reported on dimensional variation of microchannel fabricated by LAAM on Inconel 718 while deploying Nd:YAG pulsed laser. Their approach involved full factorial face-centered central composite design and response surface methodology. Smaller microchannel seems to have large dimensional variation compared to large microchannel. While fabricating small microchannel, conical shape microchannel was reported instead of flat bottom surface due to wrong combination of laser parameters. In addition, average taper angle of 10° to 15° was found for microchannel fabricated on Inconel 718.

3.2.4 Laser-Assisted Micromilling of Cementite Carbide

Although cementite carbide has excellent applications in mold making and wear resistance part industries, machining of cementite poses a greater challenge due to the generation of excessive tool wear and subsequent reduced tool life. To address that issue, laser-assisted micromilling was proposed where laser energy is exploited to oxidize the part of the surface area of the workpiece ahead of micromilling cutting path which in turn generates fully porous oxide layer. The oxidized layer is micromilled mechanically easily in a subsequent operation. Xian *et al.* [29] investigated LAAM on cementite carbide as shown in Figure 3.4 and reported on improved tool wear, reduced cutting force and better surface finish. As per their study, laser beam irradiating vertically on the cementite carbide surface well ahead of the micromilling tool made of diamond in a smaller spot diameter facilitates the oxidation with the help of supplied oxygen from a tube. This additional oxygen supply enhances the oxidation more than

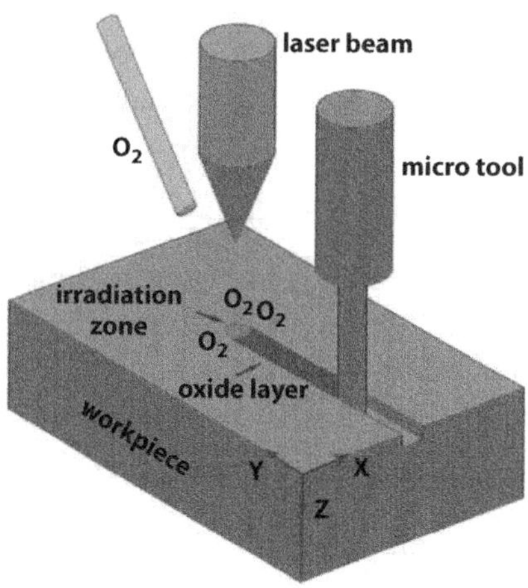

Figure 3.4 Schematic representing material removal mechanism of micromilling hybrid process.

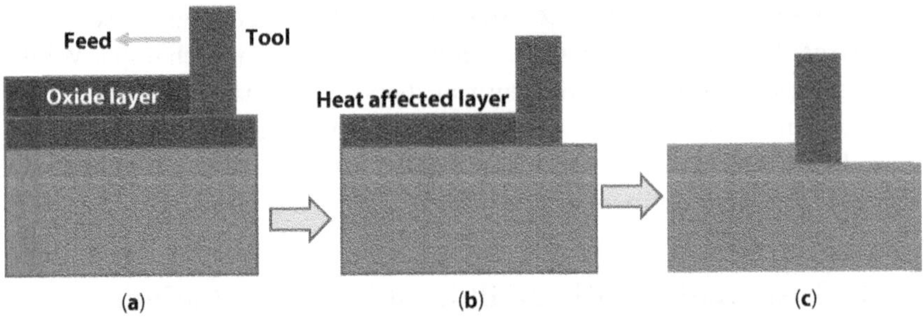

Figure 3.5 The Illustration of material removal procedure. (a) Removal of oxide layer, (b) removal of heat-affected layer removal, (c) Removal of base material layer removal.

that occurs in normal atmosphere. This oxide layer offers less resistance in terms of hardness due to less density compared to bulk materials and therefore enhances the machinability of carbide material.

Figure 3.5 demonstrates the associated three stages material removal mechanism. During first stage, micromilling tool removes oxidized layer and then heat-affected layer is removed without further heating in second stage. During last stage, very small layer is removed in a ductile manner to generate better surface finish. Since only small layer of materials is removed directly by the tool, this method offers reduced tool wear, and cutting force in addition to decreased surface roughness. Wu *et al.* [30] conducted further investigation and reported on 80% reduction of hardness of cementite carbide workpiece along with more than 52% reduction in cutting force. It was also suggested to keep some distance between laser spot and outer diameter of the cutting tool to avoid laser heating and thermal damage to the diamond micromilling tool, which would otherwise experiences graphitizing at temperature exceeding 800 °C.

3.2.5 Laser-Assisted Micromilling of Ceramics

Brecher *et al.* [31] investigated laser-assisted milling for advanced materials, such as Si_3N_4, using ytterbium YAG laser where local softening was induced by laser prior to milling process and reported on the reduction of 73%, 90%, and 90% cutting force along x, y, and z direction respectively. They also reported on significant improvement in material removal rate. Yang *et al.* [32] also investigated on sintered TiB_2–SiC ceramic using laser induced oxidation micromilling with diamond coated micromill and reported on 104% machining efficiency improvement with respect to conventional micromilling. This is due to laser-induced oxidation, which reduces the hardness of the sublayer to the half of the original hardness.

This in turn lessens the cutting forces and improves the surface finish; however, increases of both depth of cut and feed per tooth increases the thrust as well as cutting force. Xia *et al.* [33] also investigated on TiB_2–TiC ceramic using similar technique and their finding supported the reduction of hardness of the porous layer to half of original material. They also reported 106% improvement in machining efficiency. Surface quality along with perpendicularity of sidewall of microgroove resulted from laser-assisted micromilling provides relatively superior outcomes.

3.3 Conclusion

This chapter deals with laser-assisted micromilling of difficult to cut materials. Some of the interesting research highlights are given as follows:

- It appears that laser-assisted milling generally aids in reduced cutting force, tool life and enhanced surface integrity. On average 15–50% reduction in cutting force are demonstrated by several researchers.
- In case of two-stage laser-assisted milling, the reasons behind the reduction in cutting forces are twofold. First, amount of materials removed by micromilling becomes smaller due to laser structuring and therefore reduced chip thickness can be expected. Second, laser structuring induced materials alteration, which may contribute to around 30% reduction in materials strength.
- Laser energy assists in increasing the machinability of the difficult to cut materials by reducing both the yields strength and hardness of materials.
- Overall, lower tool wear compared to conventional milling was reported; however, laser to tool tip distance must be maintained to avoid the melting of tool tip.
- Dimensional deviation using laser-assisted milling was reported much more for small size microchannel compared to large size microchannel.
- Predictive statistical model can provide preheating temperature and cutting force value with error percentage of 0.5–1% and 6–4%, respectively.
- While machining of cementite carbide, lasing action results in oxidized layer which can be subsequently removed by diamond micromilling tool due to reduced hardness without

difficulty and therefore this process offers reduced tool wear and cutting force in addition to improved surface finish.

- Laser-assisted micromilling of silicon nitride ceramics results in 50% reduction of hardness for porous oxidized layer, improved machinability by 104%. In case of TiB^2–TiC ceramics, improvement in terms of perpendicularity of sidewall and surface quality was reported.

References

1. Jeon, Y., Park, H.W., Lee, C.M., Current research trends in external energy assisted machining. *Int. J. Precis. Eng. Manuf.*, 14, 337–342, 2013.
2. Weule, H., Hüntrup, V., Tritschler, H., Micro-cutting of steel to meet new requirements in miniaturization. *CIRP Ann.*, 50, 61–64, 2001.
3. Dornfeld, D., Min, S., Takeuchi, Y., Recent advances in mechanical micromachining. *CIRP Ann.*, 55, 745–768, 2006.
4. Cheng, X., Nakamoto, K., Sugai, M., Matsumoto, S., Wang, Z., Yamazaki, K., Development of ultra-precision machining system with unique wire EDM tool fabrication system for micro/nano-machining. *CIRP Ann.*, 57, 415–420, 2008.
5. Bissacco, G., Hansen, H.N., De Chiffre, L., Micromilling of hardened tool steel for mould making applications. *J. Mater. Process. Technol.*, 167, 201–207, 2005.
6. Lajis, M., Amin, A.N., Karim, A.M., Radzi, H., Ginta, T., Hot machining of hardened steels with coated carbide inserts. *Am. J. Appl. Sci.*, 2, 421–427, 2009.
7. Rudnev, V., Loveless, D., Cook, R.L., *Handbook of Induction Heating*, CRC Press, Boca Raton, Florida, United States, 2017.
8. Özler, L., Inan, A., Özel, C., Theoretical and experimental determination of tool life in hot machining of austenitic manganese steel. *Int. J. Mach. Tools Manuf.*, 41, 163–172, 2001.
9. Leshock, C.E., Kim, J.-N., Shin, Y.C., Plasma enhanced machining of Inconel 718: modeling of workpiece temperature with plasma heating and experimental results. *Int. J. Mach. Tools Manuf.*, 41, 877–897, 2001.
10. Kim, K.-S., Kim, J.-H., Choi, J.-Y., Lee, C.-M., A review on research and development of laser assisted turning. *Int. J. Precis. Eng. Manuf.*, 12, 753–759, 2011.
11. Attia, H., Tavakoli, S., Vargas, R., Thomson, V., Laser-assisted high-speed finish turning of superalloy inconel 718 under dry conditions. *CIRP Ann.*, 59, 83–88, 2010.
12. Jeon, Y. and Pfefferkorn, F., Effect of laser preheating the workpiece on micro end milling of metals. *J. Manuf. Sci. Eng. Trans.*, 130, 011004(1-9), 2008.

13. Câmara, M., Rubio, J.C., Abrão, A., Davim, J., State of the art on micromilling of materials, a review. *J. Mater. Sci. Technol.*, 28, 673–685, 2012.
14. Melkote, S., Kumar, M., Hashimoto, F., Lahoti, G., Laser assisted micro-milling of hard-to-machine materials. *CIRP Ann.*, 58, 45–48, 2009.
15. Singh, R. and Melkote, S.N., Characterization of a hybrid laser-assisted mechanical micromachining (LAMM) process for a difficult-to-machine material. *Int. J. Mach. Tools Manuf.*, 47, 1139–1150, 2007.
16. Kumar, M. and Melkote, S.N., Process capability study of laser assisted micro milling of a hard-to-machine material. *J. Manuf. Process.*, 14, 41–51, 2012.
17. Ozel, T. and Pfefferkorn, F., Pulsed laser assisted micromilling for die/mold manufacturing, in: *International Manufacturing Science and Engineering Conference*, pp. 337–342, 2007.
18. Ding, H., Shen, N., Shin, Y.C., Thermal and mechanical modeling analysis of laser-assisted micro-milling of difficult-to-machine alloys. *J. Mater. Process. Technol.*, 212, 601–613, 2012.
19. Kadivar, M., Azrhoushang, B., Zahedi, A., Müller, C., Laser-assisted micro-milling of austenitic stainless steel X5CrNi18-10. *J. Manuf. Process.*, 48, 174–184, 2019.
20. Shelton, J.A. and Shin, Y.C., Experimental evaluation of laser-assisted micro-milling in a slotting configuration. *J. Manuf. Sci. Eng. Trans.*, 132, 021008 (1-9) 2010.
21. Mohid, Z. and Rahim, E., Chip pattern, burr and surface roughness in laser assisted micro milling of Ti6Al4V using micro ball end mill. *J. Mech. Eng. Sci.*, 12, 3410, 2018.
22. Xia, H., Zhao, G., Yan, J., Li, L., He, N., Hao, X., Study on laser-induced oxidation assisted micro milling of Ti6Al4V alloy. *Int. J. Adv. Manuf. Technol.*, 103, 1579–1591, 2019.
23. Hojati, F., Azarhoushang, B., Daneshi, A., Biermann, D., Laser pre-structure-assisted micro-milling of Ti6Al4V titanium alloy. *Int. J. Adv. Manuf. Technol.*, 120, 1765–1776, 2022.
24. Shen, N. and Ding, H., Thermo-mechanical coupled analysis of laser-assisted mechanical micromilling of difficult-to-machine metal alloys used for bio-implant. *Int. J. Precis. Eng. Manuf.*, 14, 1677–1685, 2013.
25. Rahim, E., Warap, N., Mohid, Z., Ibrahim, R., Investigation on laser assisted micro ball milling of Inconel 718. *Appl. Mech. Mater. Trans. Tech Publications*, 79–83, 660, 2014.
26. Kim, D.-H. and Lee, C.-M., A study of cutting force and preheating-temperature prediction for laser-assisted milling of inconel 718 and AISI 1045 steel. *Int. J. Heat Mass Transf.*, 71, 264–274, 2014.
27. Alahmari, A.M., Darwish, S., Ahmed, N., Laser beam micro-milling (LBMM) of selected aerospace alloys. *Int. J. Adv. Manuf. Technol.*, 86, 2411–2431, 2016.

28. Ahmed, N., Alahmari, A.M., Darwish, S., Naveed, M., Laser beam micro-milling of nickel alloy: Dimensional variations and RSM optimization of laser parameters. *Appl. Phys. A*, 122, 12, 1–6, 2016.
29. Wu, X., Li, L., He, N., Zhao, G., Shen, J., Laser induced oxidation of cemented carbide during micro milling. *Ceram. Int.*, 45, 15156–15163, 2019.
30. Wu, X., Zeng, K., Zhong, L., Shen, J., Li, L., Hybrid micro-milling assisted with laser oxidation based on the hardness reduction that caused by cemented carbide oxidation. *Ceram. Int.*, 47, 35144–35151, 2021.
31. Brecher, C., Rosen, C.-J., Emonts, M., Laser-assisted milling of advanced materials. *Phys. Procedia*, 5, 259–272, 2010.
32. Yang, Y., Zhao, G., Hu, M., Li, L., He, N., Laser-induced oxidation assisted micro milling of spark plasma sintered TiB2-SiC ceramic. *Ceram. Int.*, 45, 12780–12788, 2019.
33. Xia, H., Zhao, G., Mao, P., Hao, X., Li, L., He, N., Improved machinability of TiB2–TiC ceramic composites via laser-induced oxidation assisted micro-milling. *Ceram. Int.*, 47, 11514–11525, 2021.

4

Ultrasonic-Assisted Electrochemical Micromachining

Sandip Kunar[1,6*], Itha Veeranjaneyulu[1], S. Rama Sree[2], Asma Perveen[3], Norfazillah Talib[4], Sreenivasa Reddy Medapati[1] and K.V.S.R. Murthy[5,6]

[1]Department of Mechanical Engineering, Aditya Engineering College, Surampalem, India
[2]Department of Computer Science & Engineering, Aditya Engineering College, Surampalem, India
[3]Mechanical & Aerospace Engineering Department, School of Engineering & Digital Sciences, Nazarbayev University, Republic of Kazakhstan
[4]Department of Manufacturing Engineering, Faculty of Mechanical and Manufacturing Engineering, Universiti Tun Hussein Onn Malaysia, Batu Pahat, Johor, Malaysia
[5]Department of Electrical and Electronics Engineering, Aditya Engineering College, Surampalem, India
[6]Jawaharlal Nehru Technological University Kakinada, Kakinada, East Godavari, India

Abstract
In this study, 301 stainless steel microholes are drilled using an electrode with an ultrasonic vibration-integrated array. It is described how various processing properties are affected by machining factors including electrode feed rate, operating voltage, and ultrasonic vibration amplitude. The outcomes of the experiment demonstrate that the periodic pressure difference for the electrolyte is produced by the vibrating electrode array supported by ultrasonics. Both a cavitation effect and a pumping effect are created by the periodic pressure difference. The two effects can efficiently remove the reaction heat, reactive products, and gas from the machining gap, as well as replenish the electrolyte in the machining gap. Machining speed is enhanced by boosting ultrasonic amplitude.

Keywords: Ultrasonic aided, EMM, micropattern, pumping effect, cavitation

**Corresponding author*: sandip.sandip.kunar@gmail.com

Sandip Kunar, Golam Kibria, Prasenjit Chatterjee and Asma Perveen (eds.) Hybrid Micromachining and Microfabrication Technologies: Principles, Varieties and Applications, (115–124) © 2023 Scrivener Publishing LLC

4.1 Introduction

Electrochemical micromachining (EMM), etching micromachining, abrasive jet micromachining, electro discharge micromachining, water jet micromachining, etc. are examples of nontraditional micromachining techniques. Electrochemical micromachining characteristics involve machining performance that is unrestricted by the material's mechanical performance, nearly no tool electrode wear, high machining speed, good surface quality after machining, no heat-affected zone, no burrs, and minimal residual stress. Consequently, it is frequently utilized in industrial operations in the automotive, biomedical, and semiconductor industries. Research into electrochemical micromachining has grown because of the rise in processing requirements for sophisticated materials. Because the tool electrode for electrochemical micromachining is so small, it is challenging to build an electrolyte supply flow channel system in the electrode. Electrolytes find it challenging to remove the gas bubbles, metal oxide, and reaction heat in the small machining gap caused by the electrochemical reaction. As a result, surface quality and machining precision both decline. Additionally, the workpiece and tool electrode may be harmed by the two poles' short circuit. According to the review of past researchers, researchers suggest utilizing ultra-voltage mode [1–4], special feed mode [5], electrode vibration [3, 6, 7], or a revolving electrode [8, 9] to resolve the utilization of electrolyte in the machining gap. These techniques can only slightly improve accuracy and surface quality and require more expensive equipment. Additionally, these techniques result in less practicability and economy.

However, it is not possible to rotate several electrodes at once. Additionally, the electrode cannot be constructed with a flow channel system. Renewal of the electrolyte is challenging [10–13]. In the interelectrode gap, reactive by-products, such as metal oxide, and reactive heat cannot be promptly discharged. The interelectrode gap's electric and flow fields vary, which reduce surface quality and machining accuracy. As a result, two poles may short circuit, making machining impossible. Studies about the benefits of ultrasonic-assisted machining have been investigated to increase productivity in the traditional machining of advanced materials [14–17]. Therefore, a unique electrochemical micromachining technique is used to create a microhole array using a high-frequency ultrasonic vibration assisted tool electrode. The electrolyte experiences a pumping and cavitation action because of ultrasonic vibration. With the former, the overall machining gap experiences an extrusion effect, and with the latter, a microrange of microjets is produced. Two effects refresh the electrolyte. As the electrolyte is replenished, additional reactive ions are added. In the interelectrode gap, reaction by-products like metal oxide, gas bubbles, and reaction heat can be promptly discharged. It is possible to increase both machining accuracy and speed. Several studies are carried out in this research.

4.2 Ultrasonic Effect

In this study, an ultrasonic tool holder vibrates the integrated electrode array. A cavitation and pumping effects in the electrolyte are encouraged by the vibrated electrode array. The following provides an explanation for the first two effects.

4.2.1 Pumping Effect

The tool electrode extrudes the electrolyte while vibrating up and down in the machining gap because the volume of the electrolyte is incompressible. The electrolyte is drawn into the cutting gap and released. The pumping effect is the name of this phenomena.

4.2.2 Cavitation Effect

An ultrasound's propagation medium is fluid. Rapid pressure fluctuations in the medium are caused by ultrasonography. Microbubbles are created when a medium transitions from a boiling liquid to a gas at a pressure lower than the saturated vapour pressure. The bubbles burst, and the surrounding liquid rushes to the center of the bubbles as the medium's pressure rises. Consequently, tiny but potent microjets are produced.

4.3 Experimental Procedure

For study, the investigational system is demonstrated in Figure 4.1. A stepper motor is used to position the ultrasonic electrode holder on the Z-axis, creating a high frequency vibration in that direction. The workpiece and array of tool electrodes are both attached to the pulsed power supply. An oscilloscope is used to track the waveforms of the working voltage and current during the micromachining. The electrolyte must cover the workpiece surface during the experimentation process. The electrolyte is supplied by

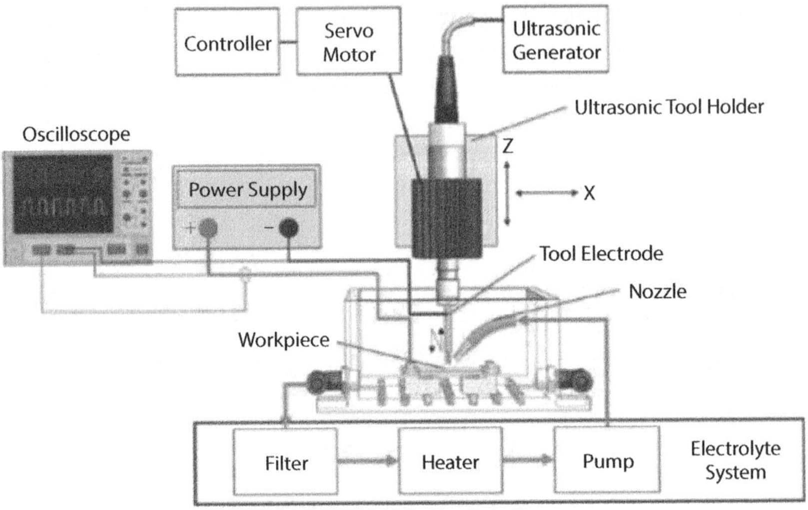

Figure 4.1 Experimental setup [18].

a snake tube with a flat nozzle, and as a result, the entire array of electrodes is covered by the range of electrolyte jet flow. The electrolyte circulation system filters the electrolyte. Before being fed into the electrolytic tank, the electrolyte's temperature is kept under control at room temperature. SUS 301 stainless steel is utilized in this investigation. Varying input parameters are machining gap, amplitude, feed rate, voltage, and frequency. Using a wire electrical discharge machine, the integrated tool electrode array used in this experiment is produced having tool electrode tip size of 0.8 mm × 0.8 mm and distance between two tips of 0.8 mm. Sodium nitrate is utilized as electrolyte for investigation.

4.4 Results and Discussion

4.4.1 Effect of Traditional Electrochemical Micromachining

The investigation is conducted by electrochemical micromachining during generation of microhole array without ultrasonic aid to investigate the impact of ultrasonic help on the electrochemical micromachining of various quality attributes. A picture of the electrochemically machined workpiece without the use of ultrasonics is shown in Figure 4.2. Without the help of ultrasonic technology, it was revealed that all the parameters failed to completely build exact microhole array. On the surface of the machined workpiece, a build-up

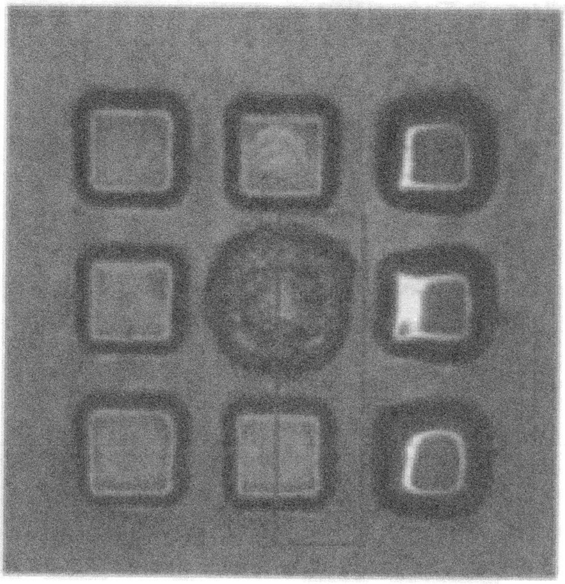

Figure 4.2 Machined micropattern without ultrasonic assistance [18].

of dark brown metal oxide is visible. On the surface of the workpiece after it had been cleaned by an ultrasonic cleaner, a unique geometric feature is seen. Metal oxide builds up in the machining gap at the tips of some electrodes during the electrochemical micromachining of microhole array without the help of an ultrasonic tool. It is challenging for the flow of electrolytes to carry away the metal oxide in the core zone of these electrodes. Due to this occurrence, the central zone of the single electrode has a relatively larger electrolyte resistance than the periphery area. The current moves along the least-resistance path. The single electrode's center zone lacks the necessary current density to dissolve the substance. The workpiece surface exhibits a central protrusion. In the machining gap, metal oxide finally thins out and forms a thin coating. The thin coating causes an increase in the machining gaps between the electrodes and the workpiece. These electrodes' feed-direction electric field strength is less intense. Reduced material dissolution occurs in the feed direction. Finally, after the electrode array's stroke is complete, it is impossible to perform a whole microhole array.

4.4.2 Effect of Electrolyte Jet During Micropatterning

In this study, a microhole array is electrochemically machined using an integrated tool electrode array vibrated by an ultrasonic tool holder. It is challenging to get rid of a flow channel for providing the electrolyte in the tool electrode because of how small they are. Additionally, no two electrodes can be simultaneously rotated. The flow of electrolytes and tool electrode vibration carry away the machined products in the machining zone. An electrolyte jet provides the new electrolyte. Ions are added to the cutting zone by the vibration of the tool electrode. However, because the electrolyte jet is supplied from the right, each row of the microhole array will have a different electrochemical machining condition. As a result, the array's side with the microholes has varied diagonal lengths. An experiment on the electrochemical micromachining of a microhole array without an electrolyte jet is conducted first to examine the impact of the electrolyte jet. There is no set distribution for the length of microholes without using an electrolyte jet. Figure 4.3 depicts the outcome of the machining process when the electrolyte jet is used to provide the electrolyte. The array's microholes' diagonal lengths roughly decrease from right to left. To discharge the reaction products more efficiently from the machining gap, an electrolyte jet flushes the right side of the microhole array. At these holes, the rate of electrochemical disintegration is accelerated.

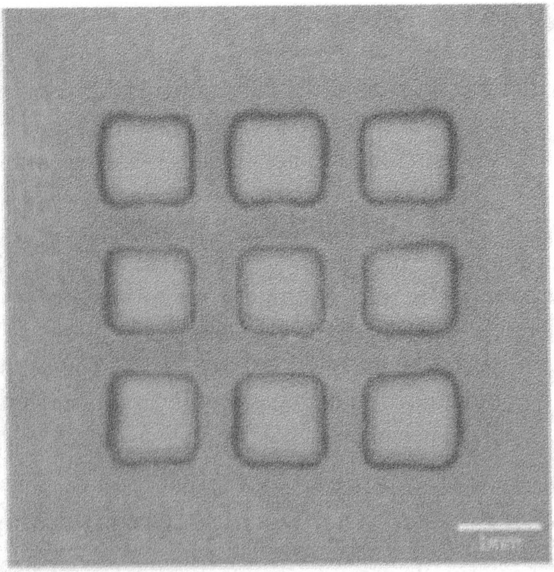

Figure 4.3 Array of microholes using electrolyte jet [18].

4.4.3 Effect of Ultrasonic Assistance During Micropatterning

In this study, the electrode feed stroke is fixed, therefore the machining time is shorter, and the electrode feed rate is greater. As machining time is reduced, the diagonal length of microholes typically reduces as well. The renewal of electrolyte in the machining gap when ultrasonic amplitude level 1 is employed for the experiment is poorer than the renewal of electrolyte under the other ultrasonic level. This occurrence results in a short circuit or perforation failure when the feed rate is greater than 1 μm/s. The ultrasonic vibrated tool electrode has a stronger ability to discharge the reaction products from the machining gap as the ultrasonic amplitude increases from level 1 to level 3, therefore, the electrode feed rate is permitted to be 3 μm/s. The electrode feed rate is permitted to be 5 μm/s because the ultrasonic amplitude is level 9. Better machining capabilities can be found at ultrasonic amplitude level 9.

4.4.4 Effect of Ultrasonic Amplitude During Micropatterning

The cavitation impact increases as the ultrasonic amplitude increases. The electrolyte's electrical conductivity will decrease as the volume of bubbles in the electrolyte increases. Additionally, the rate of electrochemical disintegration is slowed down, which shortens the array's average diagonal length.

4.4.5 Influence of Working Voltage During Micropatterning

The average diagonal length rises as the voltage increases. The sidewall of microhole is disintegrated by a continuous electrochemical reaction since the side face of the tool electrode utilized in this experiment is not insulated. As the voltage rises, the average diagonal length grows. Because the electrolyte has a lower electrical conductivity than the voltage, the minimum value of diagonal length slightly increases with voltage. When the operating voltage rises, the maximum value of the diagonal length also rises along with the amount of variation in the diagonal length.

4.4.6 Influence of Pulse-Off Time During Micropatterning

Since the experiment's pulse-on time is set at 50 seconds, the duty ratio decreases as the pulse-off duration increases. This indicates that there is less energy available from the power source for electrochemical reactions. Consequently, when the pulse-off time increases, the average diagonal length decreases. The amount of fluctuation in diagonal length decreases when the pulse-off duration rises from 20 μs to 50 μs. The pulse duty ratio increases as the pulse-off time decreases. With more machining time, there is more diagonal length variance.

4.4.7 Influence of Electrode Feed Rate During Micropatterning

When the feed rate is increased, the influence of stray currents is lessened. The average diagonal length decreases as the electrode input rate rises. The machining time in this experiment is inversely proportional to the electrode feed rate because the cutting depth is fixed. The average diagonal length and the extent of the stray current influence area grow when the electrode feed rate falls for the reason. The amount of variation in diagonal length decreases as the electrode feed rate rises. As a result, the electrochemical process requires less total energy as the electrode feed rate is greater. It is possible to lessen the variation in the dissolving rate by the electrolyte. Additionally, there is less variance in the diagonal length.

4.5 Conclusions

This study investigates the electrochemical micromachining of a microhole array with ultrasonic assistance. The outcomes demonstrate that ultrasonic-assisted electrochemical micromachining can significantly increase machining capability and speed. It is also possible to enhance the microaverage hole's diagonal length and tapered angles. The conclusions are as follows:

- The electrolyte can be successfully refilled in the machining gap and the reactive products can be discharged from the gap during electrochemical micromachining of the microhole array with ultrasonic assistance.
- One of the important factors affecting the machining gap in the feed direction and the lateral machining gap is the working voltage. The average diagonal length rises from when the working voltage rises, while the range of the diagonal length increases.
- Machining accuracy can be raised by quickening the electrode feed rate and shortening the time needed for an electrochemical reaction. The average diagonal length reduces and the amount of variance in diagonal length drops as the electrode input rate rises.

References

1. Rajurkar, K.P., Levy, G., Malshe, A., Sundaram, M., McGeough, J., Hu, X., Resnick, R., DeSilva, A., Micro and nano machining by electro-physical and chemical processes. *CIRP Ann.*, 55, 643–666, 2006.
2. Schuster, R., Kirchner, V., Allongue, P., Ertl, G., Electrochemical micromachining. *Science*, 289, 98–101, 2000.
3. Kozak, J., Rajurkar, K.P., Makkar, Y., Selected problems of micro-electrochemical machining. *J. Mater. Process. Technol.*, 149, 426–431, 2004.
4. Ahn, S.H., Ryu, S.H., Choi, D.K., Chu, C.N., Electro-chemical micro drilling using ultra short pulses. *Precis. Eng.*, 28, 129–134, 2004.
5. Hewidy, M., Ebeid, S., Rajurkar, K.P., El-Safti, M., Electrochemical machining under orbital motion conditions. *J. Mater. Process. Technol.*, 109, 339–346, 2001.
6. Natsu, W., Nakayama, H., Yu, Z., Improvement of ECM characteristics by applying ultrasonic vibration. *Int. J. Precis. Eng. Manuf.*, 13, 1131–1136, 2012.
7. Bradley, C. and Samuel, J., Controlled phase interactions between pulsed electric fields, ultrasonic motion, and magnetic fields in an anodic dissolution cell. *J. Manuf. Sci. Eng.*, 140, 041010, 2018.
8. Tsui, H.P., Hung, J.C., You, J.C., Yan, B.H., Improvement of electrochemical microdrilling accuracy using helical tool. *Mater. Manuf. Process.*, 23, 499–505, 2008.
9. Yang, Y.K., *A Study on Magnetic Field Assisted Micro Electro-Chemical Milling*, National Central University, Taoyuan, Taiwan, 2009.
10. Park, M.S. and Chu, C.N., Micro-electrochemical machining using multiple tool electrodes. *J. Micromech. Microeng.*, 17, 1451, 2007.
11. Wang, M. and Zhu, D., Fabrication of multiple electrodes and their application for micro-holes array in ECM. *Int. J. Adv. Manuf. Technol.*, 41, 42–47, 2009.
12. Skrabalak, G., Stwora, A., Electrochemical, electrodischarge and electrochemical-discharge hole drilling and surface structuring using batch electrodes. *Procedia CIRP*, 42, 766–771, 2016.
13. Arab, J., Adhale, P., Mishra, D.K., Dixit, P., Micro-hole array formation in glass using electrochemical discharge machining. *Procedia Manuf.*, 34, 349–354, 2019.
14. Wu, B., Zhao, B., Ding, W., Su, H., Investigation of the wear characteristics of microcrystal alumina abrasive wheels during the ultrasonic vibration-assisted grinding of PTMCs. *Wear*, 477, 203844, 2021.
15. Zhang, X., Yang, L., Wang, Y., Lin, B., Dong, Y., Shi, C., Mechanism study on ultrasonic vibration assisted face grinding of hard and brittle materials. *J. Manuf. Process.*, 50, 520–527, 2020.
16. Suárez, A., Veiga, F., de Lacalle, L.N.L., Polvorosa, R., Lutze, S., Wretland, A., Effects of ultrasonics-assisted face milling on surface integrity and fatigue life of Ni-alloy 718. *J. Mater. Eng. Perform.*, 25, 5076–5086, 2016.

17. Ainhoa, C., Luis, N.L.L., Francisco, J.C., Aitzol, L., Ultrasonic assisted turning of mild steels. *Int. J. Mater. Prod. Technol.*, 37, 60–70, 2010.
18. Shen, Z.Y. and Tsui, H.P., An investigation of ultrasonic-assisted electrochemical machining of micro-hole array. *Processes*, 9, 1615, 2021.

5

Micro-Electrochemical Piercing on SS 204

Manas Barman, Premangshu Mukhopadhyay* and Goutam Kumar Bose

Department of Mechanical Engineering, Haldia Institute of Technology, West Bengal, India

Abstract

Micro-electrochemical piercing (micro-ECP) is a nonconventional micromachining process used to obtain desired hole impression on the material with accurate dimensional features. Materials get eroded by the chemical reaction process in the machining zone. It is essential to get low tool wear and greater metal removal for assembly practices of miniature products. The need of proper chosen values of process parameters is essential to get the desired performance characteristics for micromachining of alloys. In most of the micromachining processes, one of the most important performance characteristic is the tool wear rate. With excessive tool wear, it will affect the accuracy and shape of microholes. So a need has evolved out to decrease the tool wear and radial overcut related to micro-ECM process. In this research work, an effort has been made to reduce the tool wear and overcut without addition of any external force. Concentration of NaCl electrolyte played a vital role for improving the performance features of microelectrochemical machining of SS 204. Highest MRR of 26.5 μg/min has been recorded for 25 V and 60 Hz. Lowest TWR of 0.035 μg/min and radial overcut of 10 μm of have been depicted for 0.5 M concentration of NaCl solution.

***Keywords*:** Micromachining, μECM, MRR, TWR, overcut, RSM

5.1 Introduction

Micro-electro chemical machining (μECM) is a chemical erosion method in which metals erode through chemical reaction with a very low current

**Corresponding author*: drpremangshu@gmail.com

Sandip Kunar, Golam Kibria, Prasenjit Chatterjee and Asma Perveen (eds.) Hybrid Micromachining and Microfabrication Technologies: Principles, Varieties and Applications, (125–136) © 2023 Scrivener Publishing LLC

density in the machining zone. An electrolytic salt solution is decomposed into cations and anions, thus separating the material from the workpiece. It is a crucial machining process concerning the micromilling of metals and its alloys with the necessary dimensional accuracy. On the other hand, hindrances are there like tool wear, radial overcut, etc., which openly affect the accuracy and precision of machined surface during micro-ECM punching on SS 204. There must have provision to remove such obstacles partially or completely. Madhusudan *et al.* [1] performed developed and performed experiments on micro-ECM setup with applied voltage, electrolytic concentration, and tool feed rate to examine their effects on performances measures and validated furthermore. Aravind *et al.* [2] conducted micro-ECM experiments on copper plate of 300 µm with the stainless-steel tool and $NaNO_3$ electrolyte with the developed provision to maintain constant inter electrode gap. The input parameters were voltage, electrolytic concentration, and duty factor. Madhankumar *et al.* [3] have put their prime focus on overcut to compare between Inconel 625 and Inconel 718 alloys in the micro-ECM process. He recorded the best possible value of overcut value of 0.021 µm by Inconel 718 alloy. Jung-Chou *et al.* [4] generated slot on nickel-based superalloy Rene 77 which is a difficult-to-cut material using micro-ECM process. Tayade *et al.* [5] provided the impact on improvement of performance features during micro-drilling on SS 304 including the micro-EDM and micro-ECM processes. Chuang *et al.* [6] investigated the machining effects of micro-ECM process in terms of material removal rate and accuracy of microholes on SS 304 stainless steel with tungsten carbide tool electrode. Natsu *et al.* [7] incorporated the researches on micro-ECM consisting of ultrashort pulse, wire ECM and electrolytic jet machining. Kibria *et al.* [8] provided a brief history and fundamentals of micromachining processes viz. micro-EDM, micro-ECM, laser micromachining, etc.

It is understood that very few works have been carried out on drilling-based micro-ECM process on SS grade alloys in the past. A short exploration of the effects of process parameters, like voltage, frequency, and electrolytic Concentration on performance characteristics, i.e., metal removal rate (MRR), tool wear rate (TWR), and radial overcut (DOC) for micro-ECM piercing of SS 204 using pure copper as tool electrode has been represented in the study related to research work.

5.2 Experimentation on SS 204 Plates With Cu Tool Electrodes

The experiments were designed for SS 204 (300 microns) plates, Cu (φ 600 microns), and NaCl solution as workpiece, tool and electrolyte respectively using unlocked central composite design-based response surface methodology technique. At the very beginning, the limits of process parameters have been set by undergoing some initial experiments using voltage (V), frequency (Hz), and concentration (M) as variable parameters. Peak current of 5 A and duty factor of 50% were considered as fixed parameters.

A sum of 60 (20 × 3) the number of experiments was designed and performed using the abovementioned technique for the desired tool-workpiece combination. Weights of SS 204 plates and copper electrode tool were measured using a weighing unit and dimensions of tool and SS 204 plates were measured by the optical profile projector.

The considered values of process parameters for the design of experiments with CCD-based RSM method has been shown in Table 5.1.

5.3 Results and Discussions

With respect to the design of experiments, experiments were conducted to find the values of MRR, TWR, and radial overcut. The experimental results for micro-ECM milling of SS 204 with Cu as tool electrode are shown in Table 5.2.

The formulae for calculations of MRR, TWR, and ROC are given as below in Eq. 5.1, Eq. 5.2, and Eq. 5.3.

$$MRR = \frac{W_{bw} - W_{aw}}{T_{m/c}} \tag{5.1}$$

$$TWR = \frac{W_{bt} - W_{at}}{T_{m/c}} \tag{5.2}$$

$$ROC = R_t - R_w \tag{5.3}$$

where W_{bw}, initial weight of workpiece before erosion; W_{aw}, final weight of workpiece after erosion; W_{bt}, initial weight of tool before wear; W_{at}, final weight of tool after wear; $T_{m/c}$, machining time; R_t, tool tip radius; R_w, workpiece radius.

Figure 5.1 shows a photograph of the micro-ECM machining setup used in the experimentation on SS 204. Outcomes of the above experimental values have been analyzed using surface plots obtained for MRR, TWR, and ROC as depicted in Figure 5.2, Figure 5.3, and Figure 5.4, respectively.

Table 5.1 Values of process parameters for designing the experiments on ss 204 plates.

Sl no.	Voltage	Frequency	Concentration	Voltage (V)	Frequency (Hz)	Concentration (M)
	Coded Values			Uncoded Values		
1	−1.68			5	20	0.3
2	−1			10	30	0.35
3	0			15	40	0.4
4	1			20	50	0.45
5	1.68			25	60	0.5

Table 5.2 Experimental Results for Micro-ECM Milling on SS 204.

Sl no.	Voltage (V)	Frequency (Hz)	Concentration (M)	MRR (g/min)	TWR (g/min)	ROC (mm)
1	15	40	0.5	0.01373	0.000097	0.28
2	15	40	0.4	0.01523	0.000085	0.18
3	15	40	0.3	0.01092	0.000084	0.11
4	15	40	0.4	0.01502	0.000085	0.11
5	25	40	0.4	0.02656	0.000458	0.01
6	20	50	0.45	0.02423	0.000198	0.11
7	20	50	0.35	0.01668	0.000344	0.15
8	20	30	0.45	0.01674	0.000049	0.12
9	15	40	0.4	0.01534	0.000085	0.17
10	15	40	0.4	0.01510	0.000127	0.18

(Continued)

Table 5.2 Experimental Results for Micro-ECM Milling on SS 204. (*Continued*)

Sl no.	Voltage (V)	Frequency (Hz)	Concentration (M)	MRR (g/min)	TWR (g/min)	ROC (mm)
11	15	20	0.4	0.01369	0.000085	0.39
12	15	60	0.4	0.01348	0.000212	0.32
13	5	40	0.4	0.00164	0.000035	0.2
14	10	50	0.45	0.00675	0.000046	0.11
15	15	40	0.4	0.01521	0.000085	0.16
16	10	50	0.35	0.00671	0.000190	0.04
17	20	30	0.35	0.01272	0.000622	0.01
18	10	30	0.35	0.00845	0.000125	0.08
19	10	30	0.45	0.00740	0.000120	0.07
20	15	40	0.4	0.01523	0.000127	0.18

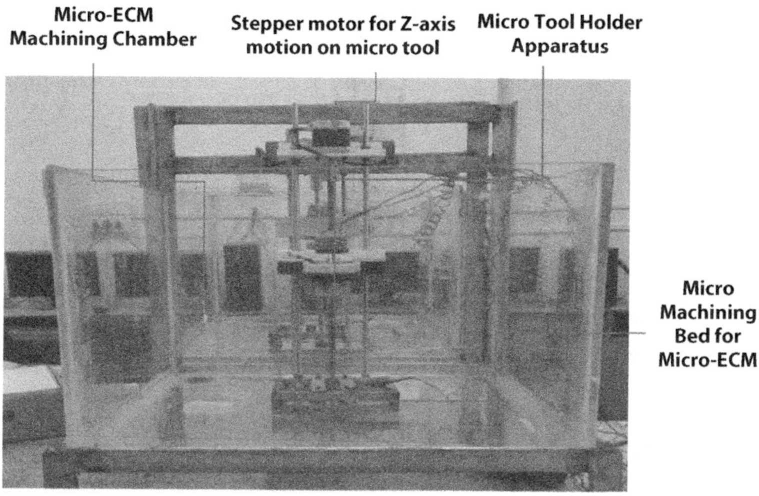

Figure 5.1 Micro-ECM machining setup for SS 204.

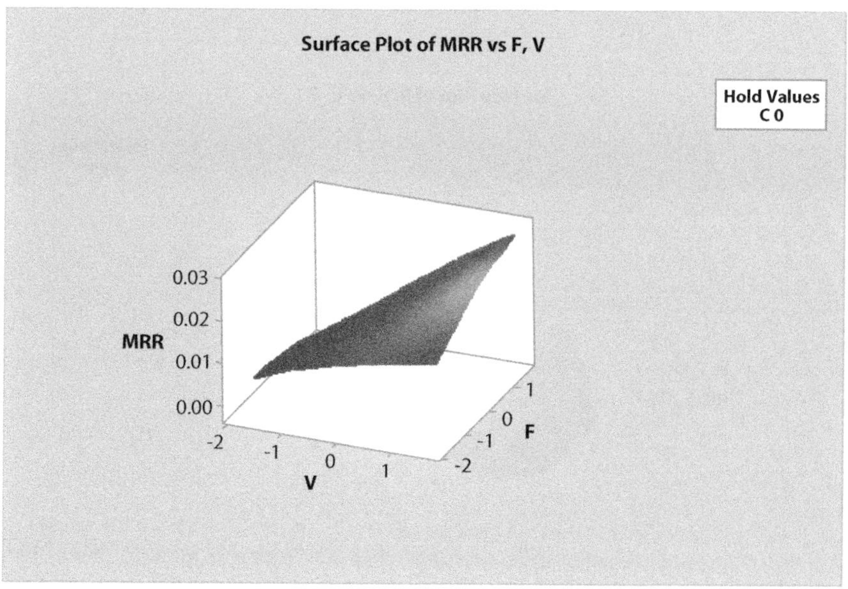

Figure 5.2 Surface plot for MRR.

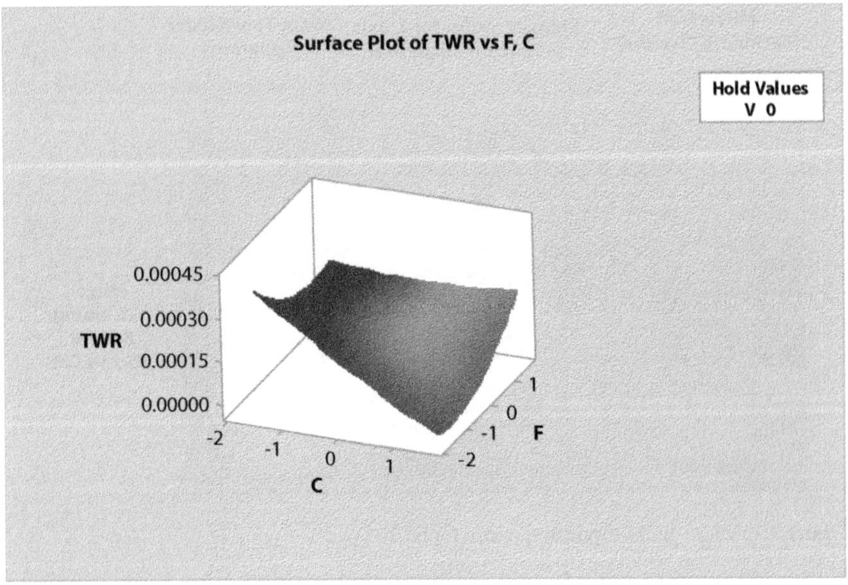

Figure 5.3 Surface plot for TWR.

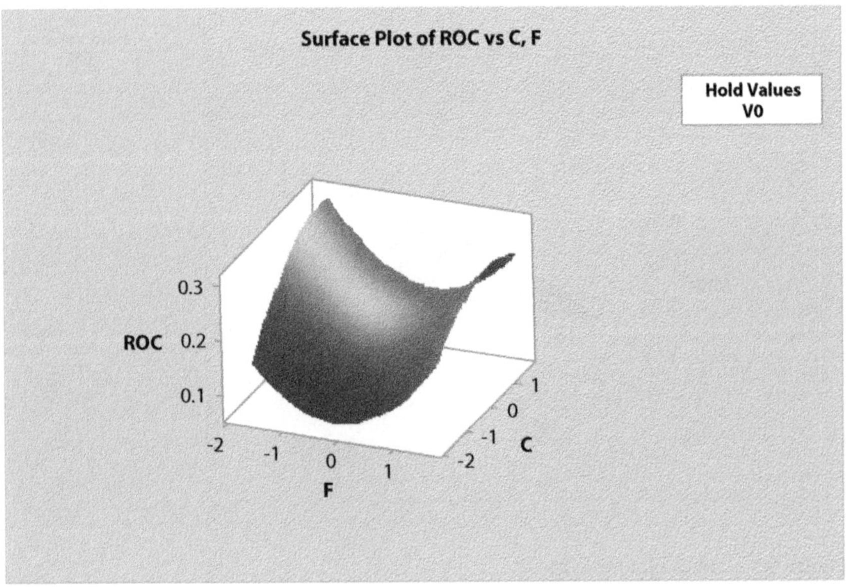

Figure 5.4 Surface plot for ROC.

The regression analysis was carried out using the exploration of RSM analysis by a suitable software, thereby forming equations for the performance features, such as MRR, TWR, and radial overcut as shown below.

$$\textbf{MRR},\ Y_u = 0.015213 + 0.006075\ V + 0.000638\ F + 0.001119\ C - 0.000546\ V{*}V\ 0.000728\ F{*}F - 0.001174\ C{*}C + 0.001730\ V{*}F + 0.001573\ V{*}C + 0.000577\ F{*}C$$

$$\textbf{TWR},\ Y_u = 0.000098 + 0.000106\ V + 0.000006\ F - 0.000062\ C + 0.000062\ V{*}V + 0.000027\ F{*}F + 0.000007\ C{*}C - 0.000015\ V{*}F - 0.000071\ V{*}C + 0.000036\ F{*}C$$

$$\textbf{ROC},\ Y_u = 0.1677 - 0.0168\ V + 0.0009\ F + 0.0305\ C - 0.0492\ V{*}V + 0.0392\ F{*}F - 0.0174\ C{*}C + 0.0162\ V{*}F + 0.0013\ V{*}C - 0.0087\ F{*}C$$

where V, voltage; F, frequency; C, concentration.

It is understood from Figure 5.2 that metal removal rate (MRR) increases uniformly with increase in voltage, as well as frequency by keeping the value of concentration fixed at 0.4 M. The reason behind this escalation is with increase in voltage and frequency of pulses in the machining area, intensity of pressure and thermal energy increases, resulting in wearing out of metals due to pressure force and temperature produced in that region. Also, the transitive value of 0.4 M concentration acts as a key role in the above extrapolation of slope.

From Figure 5.3, it is depicted that tool wear rate (TWR) started declining with the increase in concentration, whereas it remains constant for low range of frequency and started rising with the increase in frequency by keeping voltage fixed at 15 V. This is due to the reason that with the rise in concentration, a greater number of cations and anions of salt solution are being evolved which forms a oxidized coating at the tip surface of tool instantaneously during the duration of chemical reaction thus erosion of material gets reduced from tool. With the rise in the values of frequency the behaviour of coating gets disturbed, which affect the bonding between tool-tip surface and oxidized coating thus causing increase in wear rate of tool.

Radial overcut (ROC) as determined from Figure 5.4 decreases for lower ranges of frequency and higher values of concentration, i.e., at 0.5 M.

However, it increases for higher values of frequency and transitive values of concentration. The main cause surrounding this behavior is with rise in concentration of NaCl, a type of protective coating has been formed at the tip of tool after electrolysis of NaCl salt solution, which prevents the further erosion of metals from the lowest periphery of tooltip.

5.4 Conclusions

Based on the analysis from the experimental results of experimentation for micro-ECM milling of SS 204 with Cu tool electrode in the NaCl electrolytic solution, the conclusions have been noted down. It is seen that MRR is maximum at its peak of voltage, i.e., 25 V and frequency of 60 Hz due to the increased current density in the machining zone. On the other hand, TWR is at its minimum value of 0.5 M concentration and frequency of 20 Hz. Also, radial overcut is minimum at 0.5 M concentration, whereas it is at its peak of 60 Hz frequency.

So it can be concluded specifically that concentration of NaCl electrolyte played an important role in minimizing effects of tool wear rate and radial overcut. Whereas as usual the similar from earlier research works, metal removal rate changes drastically with increase in voltage, as well as pulse waveform frequency.

Therefore, the outcomes found hereby will help the researchers in the field of micro-ECM and other hybridized process for SS 204 as workpiece and Cu as tool electrode.

References

1. Painuly, M., Singh, R.P., Trehan, R., In-house development of μ-ECM setup and its experimental validation. *Adv. Mater. Process. Technol.*, 8, 2022.
2. Aravind, S. and Hiremath, S.S., Design and development of IEG control and characterization of micro-holes generated using in-house developed μ-ECM setup. *Arab. J. Sci. Eng.*, 47, 8877–8898, 2022.
3. Kumar, S.M., Kumaar, K.A., Arunachalam, S., Suhirtharaj, J.B., Anil, K.A., Aadhavan, P., Narayanan, K.R.H., Akshey, P.B., Assessments and comparison of Inconel 625 and Inconel 718 alloys from overcut in micro ECM. *Mater. Today Proc.*, 45, 6459–6464, 2021.
4. Hung, J.C. and Chen, P.C., Characteristics research of different electrode tip shapes by electrodeposition for micro slot ECM. *Proc. CIRP*, 95, 798–802, 2020.

5. Tayade, R.M., Doloi, B., Sarkar, B.R., Bhattacharyya, B., Investigations into micro-hole drilling on SS-304 by sequential electro-micro machining, in: *Advances in Micro and Nano Manufacturing and Surface Engineering, Lecture Notes on Multidisciplinary Industrial Engineering*, pp. 129–140, 2019.
6. Chen, C., Li, J., Zhan, S., Yu, Z., Xu, W., Study of micro groove machining by micro-ECM. *Proc. CIRP*, 42, 418–422, 2016.
7. Natsu, W., Micro electrochemical machining, in: *Micro and Nano Fabrication Technology*, pp. 807–855, Springer, Singapore, 2018.
8. Kibria, G., Bhattacharyya, B., Davim, J.P., *Non-Traditional Micromachining Processes: Fundamentals and Applications, Materials Forming Machining and Tribology*, Springer Cham, UK, 2017.

6

Laser-Assisted Electrochemical Discharge Micromachining

Sandip Kunar[1,6*], Kagithapu Rajendra[1], V. V. D. Praveen Kalepu[1], Prasenjit Chatterjee[2], Asma Perveen[3], Norfazillah Talib[4] and K.V.S.R. Murthy[5,6]

[1]*Department of Mechanical Engineering, Aditya Engineering College, Surampalem, India*
[2]*Department of Mechanical Engineering, MCKV Institute of Engineering, Howrah, India*
[3]*Mechanical & Aerospace Engineering Department, School of Engineering & Digital Sciences, Nazarbayev University, Republic of Kazakhstan*
[4]*Department of Manufacturing Engineering, Faculty of Mechanical and Manufacturing Engineering, Universiti Tun Hussein Onn Malaysia, Batu Pahat, Johor, Malaysia*
[5]*Department of Electrical and Electronics Engineering, Aditya Engineering College, Surampalem, India*
[6]*Jawaharlal Nehru Technological University Kakinada, Kakinada, East Godavari, India*

Abstract

Emerging nontraditional micromachining techniques like electrochemical discharge micromachining (ECDM) and laser micromachining are ideal for microprocessing of nonconducting brittle materials like glass. Glass' low machinability is a significant obstacle, though, and it needs to be overcome. The bottom surface of the microgrooves produced by ECDM is typically irregular and correlated with protrusion structures, while the frames, which frequently have overcutting and apparent wave formed HAZs, are not straight. Additionally, the cross segment created by the laser has a considerable taper and a V-shape. This study suggested the laser aided ECDM for glass microgrooving, which unites ECDM and laser micromachining, to address these issues. The single and hybrid processing techniques were examined for their morphological characteristics in this study. The outcomes

**Corresponding author*: sandip.sandip.kunar@gmail.com

Sandip Kunar, Golam Kibria, Prasenjit Chatterjee and Asma Perveen (eds.) Hybrid Micromachining and Microfabrication Technologies: Principles, Varieties and Applications, (137–150) © 2023 Scrivener Publishing LLC

demonstrate that ECDM produced tubular protrusions at the microgrooves' base. The cylindrical protrusions were repositioned after the laser processing of these microgrooves. The microgrooves' edge quality was still subpar, though. So, to obtain microgrooves with superior edge quality, the laser is employed to create them first. Then, ECDM is employed to change the cross-sectional shape from a V-shape to a U-shape, improving the taper of the microgrooves.

***Keywords*:** ECDM, LBM, microgroove

6.1 Introduction

Glass is a common example of an insulating and hard-brittle material that is utilized extensively in a variety of industries, including communications, nuclear engineering electronics, and instrumentation. The components composed of glass materials exhibit remarkable rust resistance and high temperature steadiness, particularly in some extreme, or complicated working situations. Even though glass has many outstanding qualities, it is often hard and brittle and hence challenging to mill using traditional methods, which impacts the product quality and restricts its additional application [1]. Water jet machining, mechanical grinding, and microabrasive air jet are examples of conventional techniques that are stated for glass machining. In these methods, the removal of material is happened owing to the creation of brittle flaws brought on by the squeeze or impingement of the cutting tool. Surface quality and processing accuracy suffer due to defects, like chipping and flaws, that are certainly created through machining [2, 3].

In recent years, unconventional technologies have been presented for glass [4], and the usual techniques are LBM [5–7] and ECDM [8–12], which are unconventional processes and remove materials without stress. As a result, noncontact approaches can reduce chipping and cracking [13], avoid stress damage, and considerably improve surface quality and processing precision. The two unconventional strategies each have unique traits. For LBM, the beam concentrates on a minor space, the pulsation energy releases inside a brief or ultrashort pulse duration, and as a result, the energy intensity is extremely high, reaching up to even PW/cm^2, resulting in distinctive nonlinear light [14, 15]. When using an ultrashort laser for processing, like a picosecond laser, the material's ability to absorb electrons determines how much laser energy is absorbed and by how quickly. The workpiece's ability to absorb energy is further enhanced by the electrons that grip photons, move, and then become excited to the conduction band. A Coulomb explosion may happen when the conduction band's electron density rises above a certain level, which aids in the removal of more material.

The auxiliary and tool electrodes were submerged in the solution of ECDM. A current loop developed once the DC power was switched on. The bubbles would combine to form a gas layer as the reaction continued, isolating the tool from the solution. The sparking is started, discharging an incredible quantity of energy, when the voltage grew up to the discharging that starts breakdown in the gas film. The material may melt or even vaporise because of the high pressure and temperature created by the discharge energy that was released, and it is subsequently hurled away by high-pressure influence. The alkaline solution on the workpiece was also chemically corroded because of the temperature increase [16–18].

Research on the micromachining of nonconductive materials has been done by academics from different nations, and a variety of hybrid processing techniques have been put forth in the literature. It was experimentally demonstrated [19] that a hybrid system combines microgrinding and ECDM for diminution of surface quality of structures. The research is carried out on the basic material removal mechanisms of embrittle materials and the material removal mechanisms between the hole wall and tool [20]. The viability of ECDM on nonconductive quartz glass materials with an assisting flow jet is also investigated [21]. An assist nozzle is increased penetrating depth by confirming that there is an adequate electrolyte flow for sparking and removal of debris. A magnetic field was applied to ECDM [22]. Research was done on the magnetohydrodynamic effect in ECDM. The performance of ECDM microhole drilling was greatly enhanced by adopting this hybrid technique. But there is still a lot of work to be done in the early stages of exploration and experimental study. For brittle materials, the hybrid processing approach is even based on spark discharge or laser, with vibration, inflation, or rotation added as an assistance. To increase the processing quality, there is still minimal study on etching glass using a combination of two primary energies of spark discharge and laser. As a result, there are numerous theoretical and technical challenges that must be researched to process glass in an optoelectronic composite in steps.

This study concentrated on the sequential processing of glass utilizing various laser and ECDM sequences. In this study, the morphology characteristics of various machining techniques were compared and studied based on thermal effect of laser and ECDM impact. Additionally, it examined how laser preprocessing affected the formation of an ECDM gas layer and material ablation.

6.2 Experimental Procedure

An ECDM system, including pulse power supply, auxiliary electrode, tool electrode, oscilloscope, four-axis linkage processing platform, current probe, microscope camera, response tank, etc., has been built for this investigation, as shown in Figure 6.1. A computer-operated motion controller device allows the solution tank, which is composed of acrylic sheet and mounted to the machining platform, to operate accurately along the X and Y axes. An electric motor spun the vertical chuck holding the tool electrode while the main spin moved the electrode up and down. The tool is attached to the negative end, and the auxiliary electrode is attached to the positive terminal of the DC power source. The oscilloscope is linked to greater frequency current probe utilized to observe the current through ECDM was used to show and save the signal acquired from the current. The material selection for electrodes in ECDM [23–25] is explored carefully. The tool must be able to survive strong alkali corrosion and high temperatures associated with electric spark discharge in addition to having adequate electrical conductivity. To create the tool electrode for this analysis, tungsten carbide (WC) was selected. Additionally, a cylindrical tool is employed in this experiment because it is superior to a smooth electrode at ejecting machining products and restoring the electrolyte close to the cutting spot. Block graphite with high conductivity and great alkali resistance was used to create the auxiliary electrode.

The computer control system, picosecond laser, four-axes accuracy motion system, optical measurement system, optical path system, and

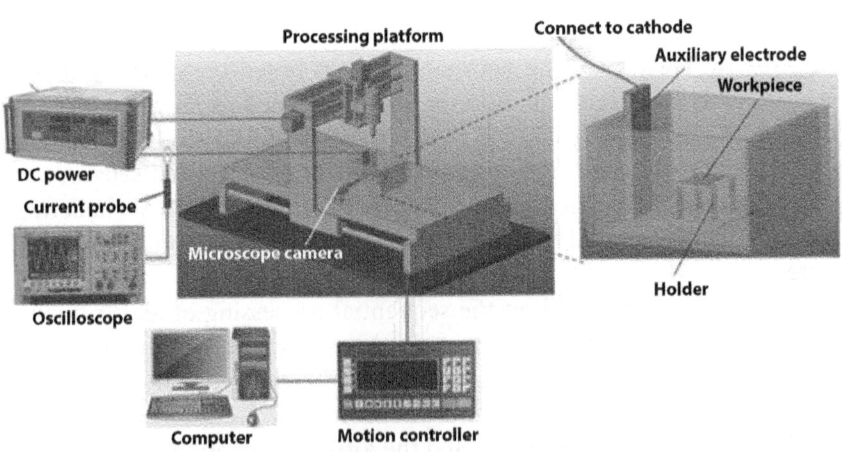

Figure 6.1 Experimental apparatus of ECDM [26].

other auxiliary equipment are all displayed in Figure 6.2. In this investigation, a Nd:YVO4 laser is used, working using 1064 nm wavelength, 12-ps pulse duration, and good beam quality. While the frequency varied from 0.2 to 1 MHz, subsequent highest pulse energy up to 260 J and the highest output power was approximately 70 W. In this laser system, the output capacity was changed by varying the high voltage (HV) level, which is advanced to adjusting the pump current because the HV level has no impact on the laser spot size. Additionally, a higher HV level causes a greater energy emission, which increases laser power. The laser beam was directed from the generator ultimately contacted the target specimen surface. For further security, the light route also has a mechanical beam blocker. The filtered water was looped within the laser machine to eliminate the produced heat and so retain a consistent temperature to maintain the laser in a steady working condition. This optical system produced a 20 μm-diameter focused laser point in the focal plane. The laser parameters are scanning speed, frequency, power, etc.

In this experimental work, ECDM and laser micromachining were initially done separately. The workpiece for the ECDM experiment was quartz that was bracketed to the bottom side of the solution container. The solution level was set to 2 mm above the job and was kept in NaOH solution. The ECDM settings utilized in this experiment are peak current, voltage, frequency, and feed speed. The layer was served three times at a rate of 0.2 mm every time. The laser used to heat the material in the laser micromachining procedure may result in vaporization, melting, direct phase explosion, and plasma production. The workpiece was mounted to the laser system's worktable and using the laser's included CCD camera for precise focusing, the laser was directed at the workpiece's upper surface to create microgrooves. The scanning space between two consecutive

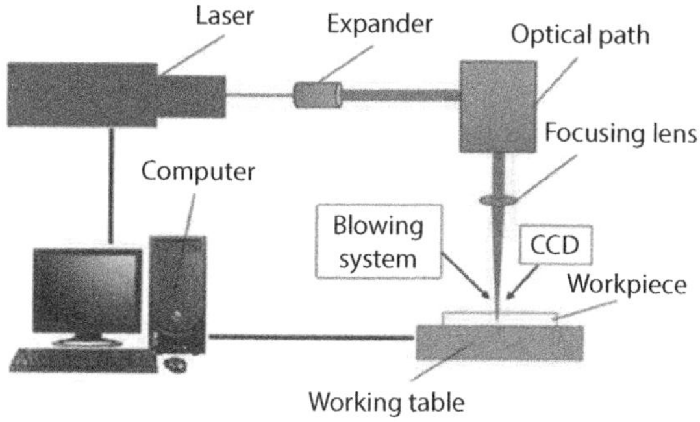

Figure 6.2 Picosecond laser machining system [26].

corresponding lines in the experiment was 10 μm, there were 20 scanning lines having scanning length of 2 mm. The scanning path was created by the computer laser system. The laser focus supplied down by 20 μm to scan a layer again after every scan, for a total of 10 feeds. This technique processed microgrooves with a width of about 200 μm due to the 20 μm laser focus diameter, 10 μm scanning space between nearby parallel lines, and 20 scanning lines. Following the completion of the laser machining, the workpiece was submerged in an alcohol solution for 10 minutes of ultrasonic cleaning before being dried to eliminate products.

Figure 6.3 depicts the cross-sectional area of microgrooves after being subjected to laser and ECDM processing, respectively. The findings demonstrate that while the cross-sectional taper produced by laser exhibits a distinct V-shape, the edge profile is comparatively wide, flat, and clean. The microgrooves produced by ECDM, on the other hand, have essentially no taper but pronounced wave-shaped HAZ along the margins.

Fixing the job to the laser fixturing system is the next step. After that, use the CCD camera to see where the structure is located and the computer's software to concentrate on the microgroove's centre. This allows the microgrooves produced by the laser and the ECDM in the first stage to entirely overlap. These two steps used the same settings as those previously mentioned. However, in this ECDM experimentation, there was only one layer feed, and the spinning rate was 0 RPM. A laser was utilized in the first phase to machine the workpiece's microgrooves. The job is then dried after being submerged in alcohol for 10 minutes to perform ultrasonic cleaning. The ECDM platform must be fixed as the next stage. Then, using a microscopic camera to determine the location of the microgroove, align the tool with the microgroove's center, and carry out the ECDM. These two steps used the same settings as those previously mentioned except that

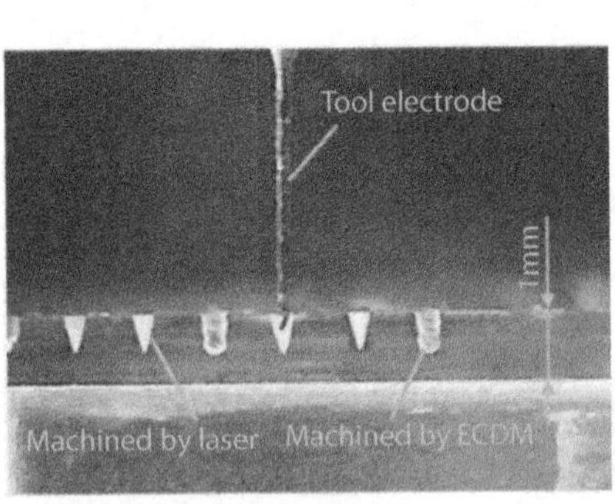

Figure 6.3 Cross-sectional structures of microgrooves correspondingly produced by laser and ECDM [26].

there is no rotational speed. The glass underwent processing before being cleaned and dried. The surface structures were analyzed using a scanning electron microscope (SEM, Hitachi S-3400N). Because glass is an insulator by nature, it must first be sprayed with a metal or carbon film to create a conductive layer before being seen.

6.3 Results and Discussion

6.3.1 ECDM Pre-Process

Figure 6.4 shows the SEM picture of an ECDM-processed microgroove. The figure's left side displays the structure of a microgroove formed just by ECDM, whereas the right side corresponds to the outcome of postprocessing by laser machining followed by ECDM. In Figure 6.4a, at the rear of the microgroove, protrusions are frequently observed; they are cylindrical and have steep sidewalls. Furthermore, the bottoms of two nearby protrusions are not joined, and the space between their peaks is around 70 μm.

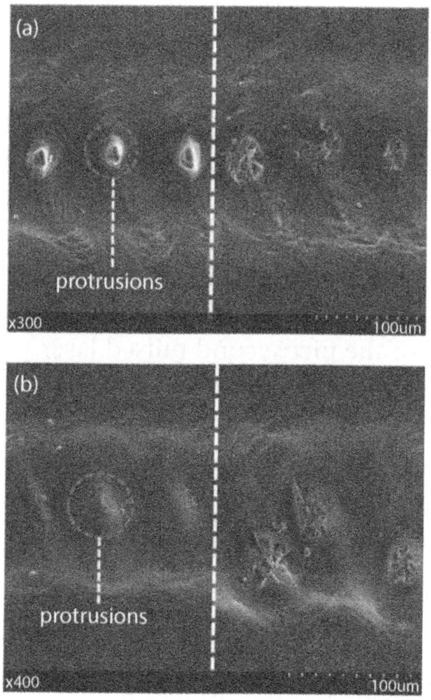

Figure 6.4 Protrusions difference of microgrooves processes by ECDM and hybrid technique with laser machining. (a) 1000 Hz and (b) 800 Hz [26].

Figure 6.4b displays the microgrooves that were treated using various ECDM parameters (800 Hz and 0 RPM). The protrusions have a soft sidewall and are almost cone-shaped. The bottoms of the two adjacent protrusions are joined, and there is a 30- to 80-μm gap between their tops. Two factors could account for how this protrusion structure formed. On the one hand, the tool electrode's flat-bottomed centre makes it harder to generate the discharge. The current density of the tool electrode with this form was specifically the subject of a finite element analysis [27, 28]. The current density is higher close to the contour's edge than it is in the tool electrode's centre. Given that sparks prefer to develop along the tool electrode's edge and have a harder time doing so towards the geometric centre of the bottom, the material in this portion is more challenging to eliminate if geometric imperfections are not considered. A crater is made by the ECDM method employing a tubular tool with a diameter of 0.5 mm and a 2-second discharge time. This crater shows that material at the tool's rim was removed, suggesting the release of sparks that were dispersed across the rim. Alternatively, bubbles will form at the tool electrode's bottom during the ECDM. Due to the higher viscosity of the solution and the fact that the bubbles are mostly strained in the perpendicular direction, they tend to accrue at the bottom side of the electrode and are challenging to sparking, which prevents the electrolyte from entering the opening between the electrodes, making it challenging to generate electrochemical discharge [29] and making the material challenging to eliminate. Consequently, the bottom-protruding structures are produced.

The structure makes it abundantly evident that the protrusions were eliminated during laser machining, leaving a quieter surface. Because ECDM preprocessing can encourage the high-energy pulsed laser's nonlinear absorption and generate plasma formation. It can delete specific protrusion structures of various forms created in the ECDM. Additionally, the high peak power of the picosecond pulsed laser when it interacts with a transparent medium can cause a robust curved effect within the translucent medium and efficiently limit the HAZ, which significantly enhances the quality of the morphology. However, this technique has clear disadvantages. Although protrusion formations can be removed by laser scanning, the surface finish was not excellent, and the eliminated products was still present at the bottom side. The order of LBM and ECDM was adapted to generate a laser aided ECDM structure machining technique, as described in the facts of the following section, to more enhance the structure superiority and merged consequence and to investigate the impact of laser preprocessing on the gas film creation and material removal in ECDM.

6.3.2 Laser Pre-Process

In the laser preprocess studies, the structure was first laser-machined, and then ECDM was applied as a postprocess. The diameter of the tool electrode was specifically chosen based on the width of the microgroove produced by the laser premachining. To ensure that ECDM and LBM outline overlapped precisely and absolutely, the relative location of the tool tip and the pregroove throughout the machining process was tracked by attaching the microcamera to the computer.

Figure 6.5 shows a comparison of the 200 μm-wide microgrooves created using the combined approach and just the ECDM. In order to avoid making direct contact with the sidewall, the tool was initially lowered until its tip was just above the microgroove inlet level. To get rid of burrs without creating HAZ on the microgroove's edge, ECDM was used. The tool kept diving until it reached to a particular bottom position of microgroove. The tool and sidewall of the microgroove should be in contact at this point. However, the discharge caused this section to form a melting zone, which the electrolyte's activity will cause to further melt and erode. The tool electrode eventually descended to a particular position, the microgroove's bottom, as it kept going down. As ECDM continued, the cross-section steadily changed from V to U shape, and the taper dramatically shrank. It is evident that the microgroove treated by ECDM alone has sinuous edges. Additionally, the processing quality is often subpar, and the edge has a wave-shaped HAZ. This is since during processing, molten material overflowed due to thermal shock and gas film agitation, cooled, hardened, and gathered on the edge of the microgroove, finally producing a wave pattern that was marginally greater than the material's surface.

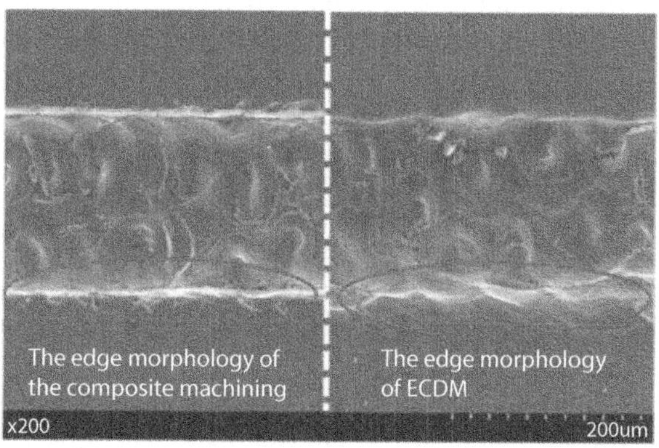

Figure 6.5 Comparison of microgrooves processed by combination method and ECDM only [26].

The combined method produces a microgroove with a flat edge, and regular side walls, and no HAZ effect on the structure, which only reveals the structure of laser machining. An essential component that forms across the tool and triggers discharge actions in the gas film. The amount of material removed, and the precision of the microgrooves were greatly influenced by the size and stability of the gas film. A more stable gas film creates the ideal environment for consistent discharges. The microgrooves' edges demonstrate the morphology of laser machining as opposed to ECDM, demonstrating how the microgrooves produced by laser preprocessing can improve quality and prevent overcutting while also reducing gas film thickness and electrical discharge erosion distance. Due to the paucity of material created by laser processing, the liquid layer's thickness between the edge of microgroove and tool discharge area increased, which improved the gas film's stability. The gas film on the tool electrode's tip was squeezed by the processed microgrooves at the same time, which had a standard and limiting impact on the ECDM energy. The combined effect significantly increased machining accuracy, decreased overcutting, and improved ECDM localization. As a result, the discharge had no effect on the laser preprocessed microgroove's edge, which continues to show a clean laser edge.

While the bottom surface displays ECDM's morphological traits rather than those of laser machining. The width of microgrooves narrows as the depth rises because the microgrooves processed by the laser exhibit a particular taper. The material at that location will be degraded by an electric discharge when the gap between the tool and material is less than the significant distance for ECDM. Due to the ultimate location of the tool electrode in the groove approaching the efficient processing range and ECDM's deeper cutting depth than laser, the base surface exhibits the morphological traits of ECDM rather than laser machining. ECDM only affects the middle and lower sections, but laser-assisted ECDM can preserve the laser processing on the upper border of the microgroove. As a result, the experiment can produce high-quality microgrooves with minimal tapers and flat edges. In general, the combined machining produces microgrooves that are narrower than those produced by ECDM. The overcut phenomenon brought on by ECDM was improved by the combined technique in this study using various voltage parameters, which also improved the machining quality.

In inference, the groove was initially laser-processed before a tool electrode with the proper diameter was chosen for ECDM in accordance with the groove's width. Combining the two techniques can successfully raise the quality of the microgrooves' edges, resulting in less overcutting and a more straight and tidy edge morphology. At the similar time, the cross section is quicker to a U-shape and the microgroove's taper can be efficiently minimized.

6.4 Conclusions

This study investigated the morphological variations between single machining technique and combined approach, comparing ECDM and laser micromachining of quartz. These are the findings of the experiment:

- Laser micromachining can efficiently and arbitrarily eliminate the protrusions at the bottom side as a postprocessing after ECDM. There is not a noticeable heat-affected zone due to the picosecond laser's "cold processing" capability. As a result, the experiment revealed a microgroove's bottom structure to be more regular.
- In the laser preprocessing, the pre-etched structure can improve quality by reducing gas film thickness and electrical discharge erosion distance. According to the results, the upper surface had a laser structure, while the bottom side of the microgroove displayed an ECDM structure, and there was no discernible wave-shaped HAZ. The overcut phenomenon brought on by ECDM can be improved by the combined processing approach.
- The V-shaped microgrooves steadily change to the U-shape after the postprocessing by ECDM, and the taper of the microgrooves is decreased.

References

1. Lucas, H. and Jana, A.Z., Micro-hole drilling on glass substrates—A review. *Micromachines*, 8, 53, 2017.
2. Sabahi, N., Hajian, M., Razfar, M.R., Experimental study on the heat-affected zone of glass substrate machined by electrochemical discharge machining (ECDM) process. *Int. J. Adv. Manuf. Technol.*, 97, 1–8, 2018.
3. Schwartzentruber, J. and Papini, M., Abrasive waterjet micro-piercing of borosilicate glass. *J. Mater. Proc. Technol.*, 219, 143–154, 2015.
4. Zhu, H., Wang, J., Yao, P., Huang, C., Heat transfer and material ablation in hybrid laser-waterjet microgrooving of single crystalline germanium. *Int. J. Mach. Tools Manuf.*, 116, 25–39, 2017.

5. Zhu, H., Wang, J., Yao, P., Huang, C., Heat transfer and material ablation in hybrid laser-waterjet microgrooving of single crystalline germanium. *Int. J. Mach. Tools Manuf.*, 116, 25–39, 2017.
6. Xie, X., Zhou, C., Wei, X., Hu, W., Ren, Q., Laser machining of transparent brittle materials: From machining strategies to applications. *Opto-Electron. Adv.*, 2, 11–23, 2019.
7. Zhu, H., Zhang, Z.Y., Xu, J.L., Xu, K., Ren, Y., An experimental study of micro-machining of hydroxyapatite using an ultrashort picosecond laser. *Precis. Eng.*, 54, 154–162, 2018.
8. Chichkov, B.N., Momma, C., Nolte, S., Von Alvensleben, F., Tunnermann, A., Femtosecond, picosecond and nanosecond laser ablation of solids. *Appl. Phys. A*, 63, 109–115, 1996.
9. Ghosh, A., Electrochemical discharge machining: Principle and possibilities. *Sadhana*, 22, 435–447, 1997.
10. Jain, V.K., Dixit, P.M., Pandey, P.M., On the analysis of the electrochemical spark machining process. *Int. J. Mach. Tools Manuf.*, 39, 165–186, 1999.
11. Bhattacharyya, B., Doloi, B.N., Sorkhel, S.K., Experimental investigations into electrochemical discharge machining (ECDM) of non-conductive ceramic materials. *J. Mater. Proc. Technol.*, 95, 145–154, 1999.
12. Lee, J.Y., Lee, S.W., Lee, S.K., Park, J.H., Through-glass copper via using the glass reflow and seedless electroplating processes for wafer-level RF MEMS packaging. *J. Micromech. Microeng.*, 23, 085012, 2013.
13. Zhang, Z.Y., Huang, L., Jiang, Y., Liu, G., Nie, X., Lu, H., Zhuang, H., A study to explore the properties of electrochemical discharge effect based on pulse power supply. *Int. J. Adv. Manuf. Technol.*, 85, 2107–2114, 2016.
14. Ahmmed, K., Colin, G., Anne-Marie, K., Fabrication of Micro/nano structures on metals by femtosecond laser micromachining. *Micromachines*, 5, 1219–1253, 2014.
15. Shin, J., Investigation of the surface morphology in glass scribing with a UV picosecond laser. *Opt. Laser Technol.*, 111, 307–314, 2019.
16. Jawalkar, C.S., Sharma, A.K., Kumar, P., Investigations on performance of ECDM process using NaOH and $NaNO_3$ electrolytes while micro machining soda lime glass. *Int. J. Manuf. Technol. Manage.*, 28, 80–93, 2014.
17. Zheng, Z.P., Cheng, W.H., Huang, F.Y., Yan, B.H., 3D microstructuring of Pyrex glass using the electrochemical discharge machining process. *J. Micromech. Microeng.*, 17, 960–966, 2007.
18. Pawar, P., Ballav, R., Kumar, A., Revolutionary developments in ECDM process: An overview. *Mater. Today Proc.*, 2, 3188–3195, 2015.
19. Xuan, D.C., Bo, H.K., Chong, N.C., Hybrid micromachining of glass using ECDM and micro grinding. *Int. J. Precis. Eng. Manuf.*, 14, 5–10, 2013.
20. Nath, C., Lim, G.C., Zheng, H.Y., Influence of the material removal mechanisms on hole integrity in ultrasonic machining of structural ceramics. *Ultrasonics*, 52, 605–613, 2012.

21. Ho, C.C., Wu, D.S., Chen, J.C., Flow-jet-assisted electrochemical discharge machining for quartz glass based on machine vision. *Measurement*, 128, 71–83, 2018.
22. Xu, Y., Chen, J., Jiang, B., Liu, Y., Ni, J., Experimental investigation of magnetohydrodynamic effect in electrochemical discharge machining. *Int. J. Mech. Sci.*, 142, 86–96, 2018.
23. Jain, V.K. and Adhikary, S., On the mechanism of material removal in electrochemical spark machining of quartz under different polarity conditions. *J. Mater. Process. Technol.*, 200, 460–470, 2008.
24. Gaurav, S., Alakesh, M., Sethi, A.S., Investigations on performance of ECDM process using different tool electrode while machining e-glass fibre reinforced polymer composite. *Mater. Today Proc.*, 28, 1622–1628, 2020.
25. Kim, D.J., Ahn, Y., Lee, S.H., Kim, Y., Voltage pulse frequency and duty ratio effects in an electrochemical discharge microdrilling process of Pyrex glass. *Int. J. Mach. Tools Manuf.*, 46, 1064–1067, 2006.
26. Zhao, D., Zhang, Z., Zhu, H., Cao, Z., Xu, K., An investigation into laser-assisted electrochemical discharge machining of transparent insulating hard-brittle material. *Micromachines*, 12, 22, 2021.
27. Jiang, B., Lan, S., Ni, J., Zhang, Z., Experimental investigation of spark generation in electrochemical discharge machining of non-conducting materials. *J. Mater. Process. Technol.*, 214, 892–898, 2014.
28. Jiang, B., Lan, S., Wilt, K., Modeling and experimental investigation of gas film in micro electrochemical discharge machining process. *Int. J. Mach. Tools Manuf.*, 90, 8–15, 2015.
29. Yang, C.K., Wu, K.L., Hung, J.C., Lee, S., Lin, J., Yan, B., Enhancement of ECDM efficiency and accuracy by spherical tool electrode. *Int. J. Mach. Tools Manuf.*, 51, 528–535, 2011.

7

Laser-Assisted Hybrid Micromachining Processes and Its Applications

Ravindra Nath Yadav

BBD National Institute of Technology and Management, Lucknow, India

Abstract

Multifunctional properties of miniatures within limited space attract the attention of advanced industries to meet the current and futuristic demands. The fabrication of miniatures at micro/nanoscale is challenging for traditional processes because of size constraint and tool hardness more than workpiece. In existing nontraditional processes, laser is one of versatile shaping process that uses for fabrication of miniatures at micron and submicron scales with precise surface quality due to highly flexible nature with application of short and ultrafast lasers. Such lasers are broadly used in microfabrication of electrical, mechanical, and photonic devices for surface structuring. Even though, laser machining suffers with several drawbacks like spattering, burr formation, recast layer, and thermal stresses. Instead of this, dimensional control and straight cut profiles become challenging. To overcome these problems, laser is hybridized with other processes (traditional/nontraditional) referred as laser-assisted hybrid processes. Generally, performances of hybrid processes are better than the individual one at similar conditions. In present chapter, the various laser-based hybridized processes are discussed with their applicability for shaping of materials at micron/submicron level. Here, laser is considered as energy source and their hybridization with other processes is presented with name of laser-assisted hybrid micromachining processes (LA-HMMPs). Instead of this, the present chapter covers the different types of LA-HMMPs and their basic mechanism of material removal with suitability for microfabrication in such a way that it becomes easier for readers.

***Keywords*:** Arc, machining, laser, hybrid, EDM, ECM, ECSM, grinding

Email: mechrny@gmail.com

Sandip Kunar, Golam Kibria, Prasenjit Chatterjee and Asma Perveen (eds.) Hybrid Micromachining and Microfabrication Technologies: Principles, Varieties and Applications, (151–178) © 2023 Scrivener Publishing LLC

7.1 Introduction

Laser (light amplification by stimulated emission of radiation) is an electromagnetic radiation that uses in various fields of manufacturing, like cutting, welding, forming, sintering, heat treatment, and surface coating. It has several unique features like monochromic (same wavelength), coherent (same phase), focus in small wavelength ranges (0.50–70 μm), and high-power intensity (in order to 1 MW/cm^2) [1–3]. It shows their ability to travel long distance without scattering and also focuses in very small and narrow area, as well as inaccessible location with application of focusing lens [3, 4]. The key element of laser is photons that responsible for heat generation to make laser for practice applications. Instead of this, the flexible nature of laser with low heat input and high lateral resolution makes it a superior process for fabrication of objects at micron and submicron scales.

In laser beam machining (LBM), the beam of laser is used as heat source, which is generated by stimulated emission of photons. Mostly, flash lamp, reflective mirrors, and laser source are required for generation of laser (Figure 7.1). In LBM, the laser source/medium becomes important parameter for laser light generation. Generally, solid, liquid, gas, or fiber lasers are used as laser medium. In solid laser, the ruby is the first solid laser that used in different industrial applications. Instead of this, neodymium doped yttrium-aluminium-garnet (Nd-YAG), neodymium doped glass (Nd-glass) and neodymium doped yttrium lithium fluoride (Nd-YLF) are also used as solid laser for industrial applications [2, 3, 5]. Fiber lasers are also one type of solid laser where optics fiber especially fiber glass, which is used as laser medium. The carbon dioxide (CO_2), carbon monoxide (CO),

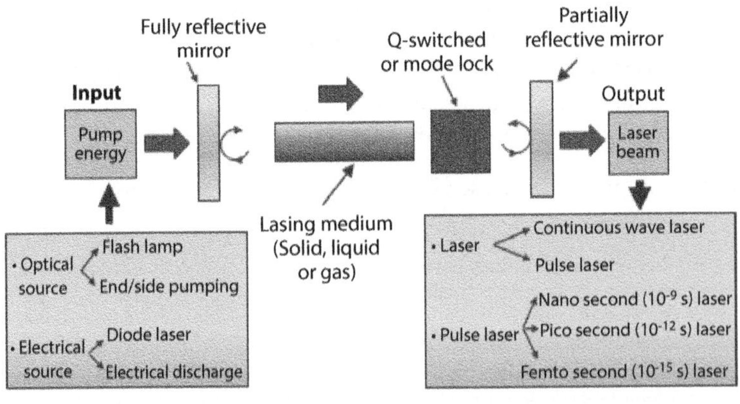

Figure 7.1 Laser generation system.

nitrogen, hydrogen, argon, and excimer lasers are common example of gaseous lasers [1–3], while liquid die is considered as laser medium for liquid laser [2]. Based on the pulse, the laser may be millisecond (ms), microsecond (μs), nanosecond (ns), picosecond (ps), and femtosecond (fs). Generally, long pulse (ms and μs) lasers are used for macromanufacturing, like cutting, drilling, milling, and welding applications, while short pulse (ns) laser and ultrashort pulse (ps/fs) lasers are applied for micromachining and microfabrication.

In simple, the melting and vaporization play important roles in material removal when laser beam has been focused on workpiece surface. In case of micromachining, the laser ablation is responsible for material removal, which can be defined as removal of material from workpiece by pulse laser [3, 5]. In Figure 7.2, a typical laser ablation is presented for a pulse laser [5]. Due to ablation, high heat is generated because of conversion of laser energy into heat energy causes melting and subsequently vaporization of metal from work surface. Such phenomena are completed in different phases, such as heating, melting, vaporization, and ablation (Figure 7.3).

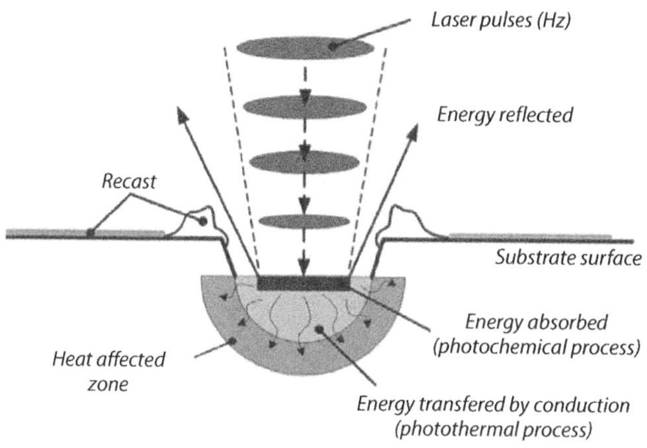

Figure 7.2 Laser ablation phenomena [5].

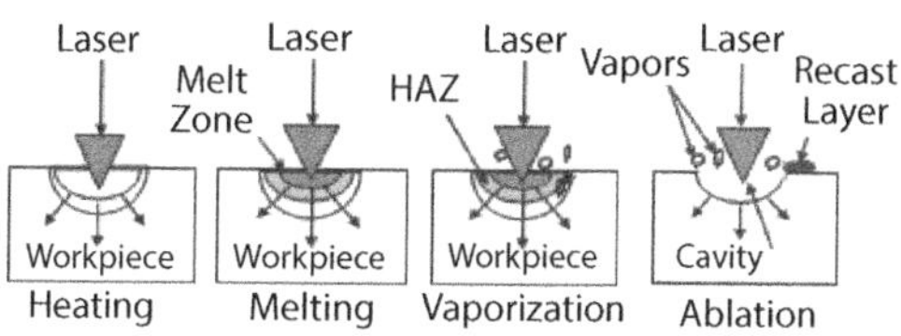

Figure 7.3 Various phases of laser ablation.

In laser beam micromachining (LBMM), the rate of energy deposition is much smaller than the energy transfer time and electron-photon coupling time as which no collateral damages occur [3]. Due to this, negligible heat-affected zone (HAZ) during micromachining by laser radiation, which makes it suitable for processing of materials at micron and submicron scales.

In LBMM, a variety of parameters (controllable/uncontrollable) are significantly affected the performances of material processing at micro and nanoscales. The controllable parameters are covered as laser parameters (wavelength, pulse frequency, and width,), material parameters (thickness and properties), beam parameters (power source, beam diameter, focal length, standoff distance), and other parameters (interaction time, spot diameter, focal spot length). On other hand, the humidity, temperature, work environment, operator skill and machine efficiency are also affected the performances of process. These parameters are come in category of uncontrollable because of challenges in controlling. The various controlled parameters related the LBMM are graphically presented in Figure 7.4, while Table 7.1 presents the effect of control factors on responses.

Off course, laser is wonderful tool for microfabrication with application of short and ultrashort pulse lasers [3]. Due to their unique characteristics, laser can be employed to machine all materials including extremely hard like diamond, ceramic, sapphire to soft like polymers [3, 6–8]. It is

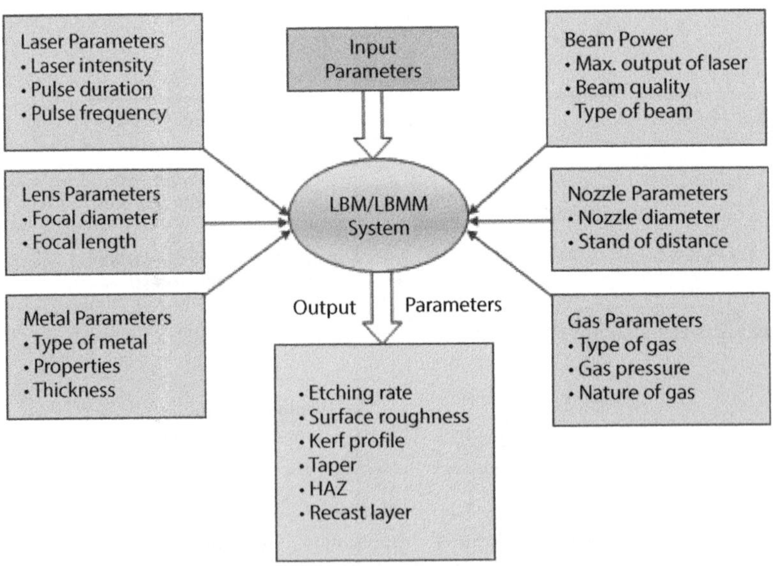

Figure 7.4 Process and performance parameters of LBMM.

Table 7.1 Process parameters and their effects.

Sl. no.	Process parameters	Effects
1	Pulse intensity	Penetration depth, crater profile
2	Pulse width	Ablation rate, crater profile, HAZ
3	Pulse interval	Ejection rate, SQ, recast layer
4	Beam power	Ablation, HAZ, crater geometry
5	Pulse frequency	Penetration depth, SQ, ablation, feature size
6	Gas pressure	Ejection, recast layer, SQ
7	Focal length	Ablation rate, feature size
8	Nozzle distance	Ejection rate, profile of cut
9	Beam profile	Feature shape, informality in micromachining
10	Material thickness	Taper, kerf profile, feature at exit and entry

effectively applied for manufacturing of microelectromechanical (MEM) devices for biomedical, robotics, sensor, actuators, electronics, and computer applications. It also shows their potential in fabrication of microfluidic channels and tubes for drugs supply system. It is significantly used for manufacturing of biomedical implantable items especially cardiovascular stent for medical applications [9].

The applications of laser in micromachining significantly meet the demand of advanced industries but it suffers with several inherent problems like taper profile and uneven diameter at entry and exit of holes. The basic reason behind this is to divergence and convergence nature of beam. Due to this, the straight profiles of laser-drilled holes are difficult to achieve. Such phenomenon needs secondary operation to get the straight and parallel profile of drilled holes as which enhancement in cost and time. Instead of this, high-aspect ratio holes drilling and recast layer formation also limits the applicability of laser for microprocessing of materials. To overcome these problems, laser has been combined with other processes to enhance the machinability and reduce the deficiency of LBMM process. In case of combined processes, the laser mostly assists the performance of combined processes by thermal heating during machining. Such combined process is known as laser-assisted hybrid micromachining processes (LA-HMMPs).

7.2 Laser-Assisted Hybrid Micromachining

Hybrid micromachining processes (HMMPs) can be defined as combination of two or more machining processes in such a way that they utilize their merits and simultaneously minimize/eliminate the demerits of constituent processes to achieve micromachining by removal of material at micron level. On other hands, HMMPs can be defined as an enhanced micromachining processes that aim to achieve the significantly better machining effect by combining at least two different forms of energies like mechanical, thermal, chemical, electrochemical, magnetic, or electromagnetic. Based on involvement, the HMPs (hybrid machining processes) are categorized in two categories as associated HMPs and assisted HMPs [10, 11]. In associated-HMPs, all integrated processes are involved in machining to enhance the performance of process. On other hand, one process involves in machining while other either assists/facilitates the machining or material removal in case of assisted HMPs. The main objectives of HMPs are to exploit the merits and simultaneous eliminate/minimize the demerits of constituent processes. The HMPs significantly enhance the material removal and SQ with reduction in tool wears and specific energy consumption [10–14].

In simple, LA-HMMPs can be defined as combination of LBMM with other (traditional/nontraditional) processes to enhance machinability with combined effort of integrated processes. On other way, LA-HMMPs can be defined as integration of various shaping processes with LBM/LBMM that aim to achieve significantly better machining effects by utilizing at least one or more forms of energies like chemical, electrochemical, thermal, magnetic or mechanical. The main objectives of LA-HMMPs are to utilize the merits with simultaneous eliminate/minimize the demerits of individual process during machining. The major benefits of LA-HMMPs are as higher productivity and machining rate with better SQ, negligible recast layer and lesser cracks formation.

Generally, laser is assisting in nature with integrated machining processes and referred as LA-HMMPs, which are developed with combination of LBM/LBMM with either traditional or nontraditional processes. Hence, LA-HMMPs can be categorized as laser-assisted traditional-HMMPs and laser-assisted nontraditional-HMMPs as presented in Figure 7.5. In laser-assisted traditional-HMMPs, mostly cutting processes (turning, drilling, milling and grinding) are combined with laser while electrodischarge machining (EDM), electrochemical machining (ECM), electrochemical discharge machining (ECSM) or water jet machining (WJM) are combined with laser to achieve laser-assisted nontraditional-HMMPs.

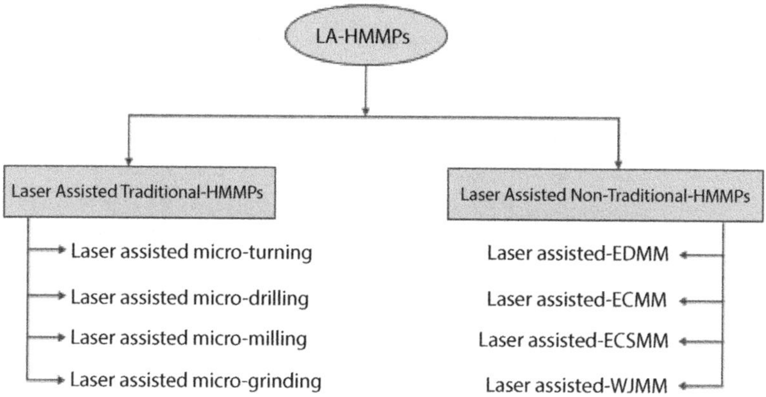

Figure 7.5 General classification of LA-HMMPs.

7.3 Laser-Assisted Traditional-HMMPs

Generally, laser light is employed for removal of material through melting and vaporization in LBM/LBMM process. However, it can be partially applied for heating and softening of materials when combined with traditional metal cutting (turning, drilling or milling) processes [15]. Here, laser energy is utilized for softening of objects while traditional processes are responsible for machining. In traditional machining, tool hardness is always greater (1.35–1.50 times) than workpiece hardness [16]. Hence, machining as well as micromachining of extremely hard materials becomes difficult for traditional processes due to excessive tool wears, poor SQ, chattering and high vibration. The application of laser in traditional machining/micromachining leads in material softening as a result reduction in tool wears, better SQ, lower vibration and chattering at low specific energy consumption. The common laser-assisted traditional-HMMPs are summarized as:

- Laser-assisted microturning process
- Laser-assisted microdrilling process
- Laser-assisted micromilling process
- Laser-assisted microgrinding process

7.3.1 Laser-Assisted Microturning Process

Laser-assisted microturning (LAMT) is utilized laser as heat source for softening of workpiece for turn cutting at macro/microscales as shown

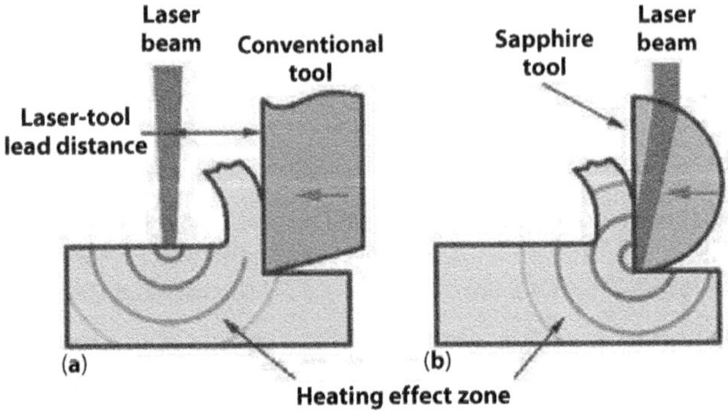

Figure 7.6 Various modes of LAMT process; (a) application of laser infront of tool; (b) application of laser through tool [17].

in Figure 7.6 (a) [17]. In simple, it can be defined as hybrid cutting process where hardness of material decreases locally and instantaneously with application of laser energy [17]. Here, rotating object is preheated infront of cutting tool by means of laser energy and soften material prior to removal of metal in form of chips without phase transformation [15, 17]. It can be employed for turn cutting of extremely hard materials like high hardness steels, tungsten carbide (WC), silicon, ceramics, composites, glass, sapphire, mold and die steels [17–20]. It effectively improves the efficiency of integrated process and quality of products. It significantly decreases in cutting forces and energy consumption than normal turn cutting [17].

Generally, laser is employed for heating of material infront of tool causes loss in heat energy due to heat transfer at actual cutting point. Hence, excessive heat is required to achieve the desired cutting efficiency. Due to extra heating means local bucking, phase change, physicochemical interactions, and thermal cracking [21]. Hence, direct application of microlaser is tested by researchers to overcome such limitations of LAMT process. Here, laser beam is directly passed through cutting tools during turning as shown in Figure 7.6 (b). For this, infrared (IR) diode laser has been integrated with diamond tool to focus at desired point on workpiece passing through diamond cutting tool.

Off course, direct application of laser enhances the machinability of turn cutting process. However, it requires microcutting tools (5μm~1mm) made of diamond [19, 22]. The high cost and difficulty in manufacturing limits the applicability of such integrated process. In this situation, sapphire becomes an alternative of diamond tool because of their unique properties as high hardness and transmissivity. It can sustain hardness with small variations with increase in temperature and also allow laser to easily pass through it.

The performances of LAMT are always better than normal microturning in term of better SQ at low specific energy consumptions. The SQ of LAMT turned surface was found about 80% better than normal turning of silicon (Figure 7.7) [19]. The SEM images of unturned and turned surfaces (normal/laser-assisted) are presented in Figures 7.8 (a)–(c) respectively [19]. The application of laser is not always beneficial and excessive heat transfer means reduction in quality of turned surfaces. Hence, desired quantity of laser heat input is always preferred for quality turning of materials.

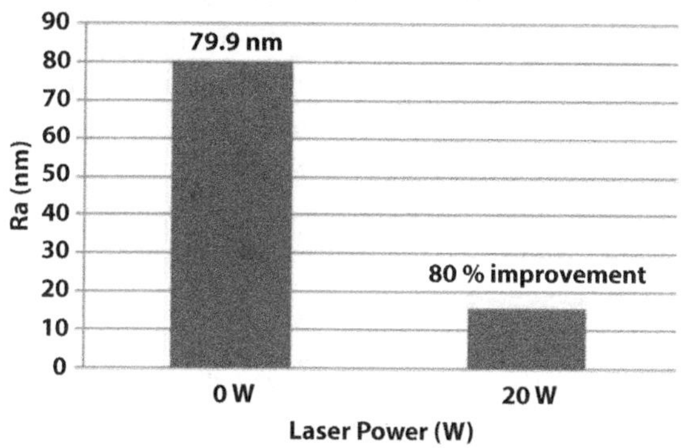

Figure 7.7 Effect of laser power on surface roughness [19].

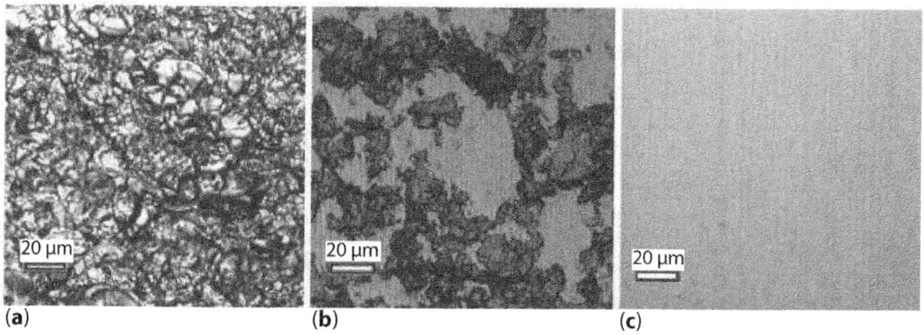

Figure 7.8 SEM images (a) unmachined surface; (b) turned surface without laser; (c) laser-assisted turned surface [19].

7.3.2 Laser-Assisted Microdrilling Process

Drilling is an important method for making holes into transmission systems of automotive at macro/microscales. Even though, it faces challenges and becomes time consuming for difficult to cut metals that limits the applicability of drilling. The process becomes more difficult for drilling of holes at micron and submicron scales with high aspect ratio into such difficult to cut metal and alloys. To overcome the problems, laser energy is employed with conventional microdrilling for softening of material. Such integrated process is known as laser-assisted microdrilling (LAMD) as schematically presented in Figure 7.9.

LAMD is one of powerful tool for making of holes at microscale with application of laser during drilling process. In this process, a tiny area is heated by laser just under tip of drill. Due to this, temperature of workpiece increases, and material becomes softer [23, 24]. Hence, drilling of softer material becomes easier because of reduction in cutting forces and resistance. It gives better performances than conventional drilling in term of higher productivity, reduction in tool wears, precise geometrical controls with significant reduction in cycle time related to machine and manpower [23–25]. The pre-heating by laser significantly reduces drilling time as 16% to 40% in different conditions for drilling of cast iron, stainless steel and alloy (C45) steel [25].

LAMD process is suitable for drilling of microholes into high hardness materials, shafts, and other similar objects. It is mostly preferred in automobile sectors for macro/microholes drilling into crank shafts, connecting rods and other transmission systems [23, 25]. Even though it suffers with several demerits, such as spike formation and softening of drill bit for drilling of high aspect ratio holes into difficult and hard metals.

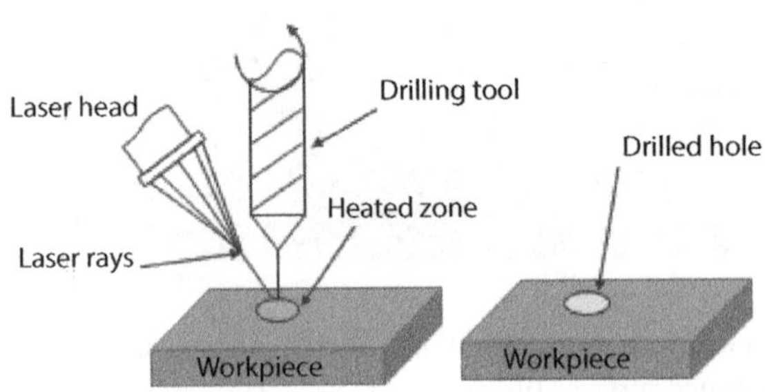

Figure 7.9 Schematic of LAMD process.

7.3.3 Laser-Assisted Micromilling Process

In traditional cutting processes, micromilling is more desirable and flexible process for manufacturing of 3D complex profiles especially for dies and mold steels with microfeatures. However, the productivity of micromilling is limited due to rapid tool wears, relatively low stiffness and bending as compared to traditional milling cutter [26–30]. Instead of this, long length to width ratio leads in probability of tool deflection and breakage in case of micromilling [29]. To handle such limitations, the laser-assisted micromilling (LAMM) becomes a promising method for 3D microstructuring.

LAMM process can be developed by combining the features of laser with micromilling process. Here, workpiece is preheated ahead milling tool by laser radiation causes of softening of material as presented in Figures 7.10 (a)–(b) positively. Due to this, enhancement in machinability of milling in term of reduction in cutting forces and higher tool life with better SQ [28, 29]. The LAMM process is suitable for creation of 3D microstructuring, microgrooves and complex geometries. It is a suitable method to manufacture complex profiles made of heat-treated alloys, nickel-based super alloys, titanium alloys, tool steel, hot and cold rolled steels especially for dies and molds with microfeatures [28–30]. It is significantly used in microgrooving, microslotting, and microchannelling of difficult to cut materials.

Off course, LAMM is a novel process for micromanufacturing but the controlling of laser radiation with tool path accordance to profile of object becomes difficult for manufacturers. The complex geometry and path programming become also necessary for laser source that makes it costly process. The smaller size of milling cutter and it manufacturing at micron level

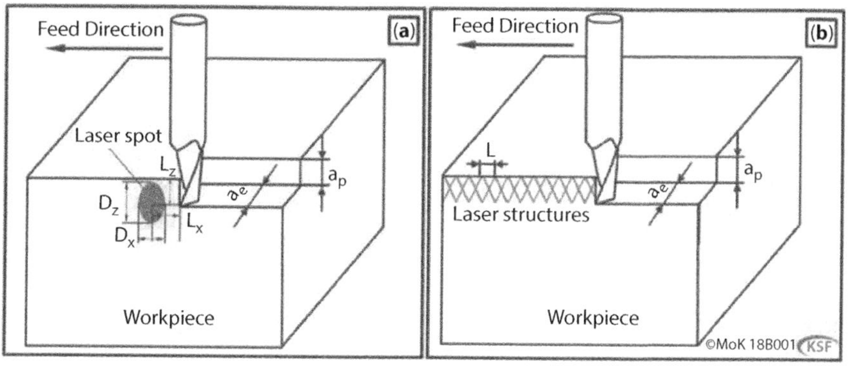

Figure 7.10 Schematic view of LAMM process; (a) softened area; (b) structures area [30].

becomes challenges for industries. Instead of this, deformation of tool due to thermal softening also limits the applicability of LAMM process.

7.3.4 Laser-Assisted Microgrinding Process

Laser-assisted microgrinding (LAMG) comprises laser features with mechanical grinding where ultra pulse laser is employed for pre-heating of object and material removal occurs by microgrinding phenomena [31, 32]. A typical schematic diagram of LAMG process and experimental setup is shown in Figure 7.11 (a)–(b), respectively [33]. The basic reason behind the development of LAMG process is to difficulties in micromachining of difficult to hard materials with high SQ by mechanical grinding process. Off course, conventional microgrinding (CMG) is the most common method for super finishing of objects at micro/nano scales, but high grinding forces causes of tool deflection and rapid wears [32]. In such conditions, the application of laser energy in CMG becomes an effective method to remove such limitations of CMG process.

In LAMG process, the thermal heating by laser means softening of material prior to grinding that assists the grinding process without melting of material and altering of microstructures. Due to this, the LAMG becomes an advantageous for microfabrication of hard and difficult to shape materials such as carbide alloy, quartz, glass, silicon nitride (Si_3N_4), aluminium oxide (Al_2O_3) and silicon carbide (SiC) [32–34]. Mostly, performance of LAMG is always better than CMG in term of lower in cutting forces and tool wears with better SQ. Instead of this, the microtip of wheel may significantly sharpen by laser during microgrinding without cracks and burr

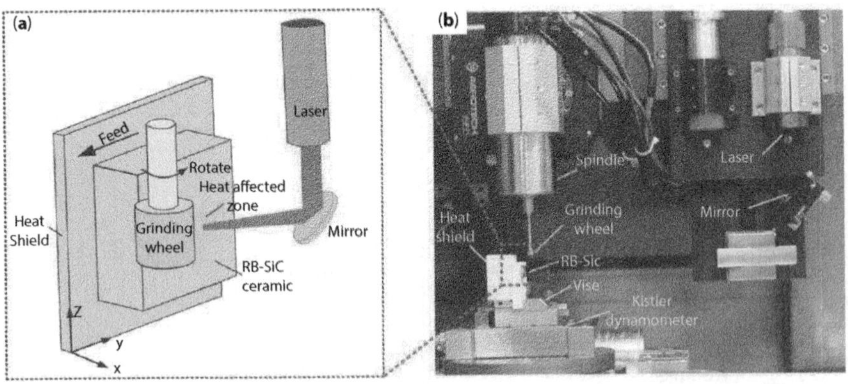

Figure 7.11 Detail of LAGM process, (a) schematic view of setup (b) experimental setup [33].

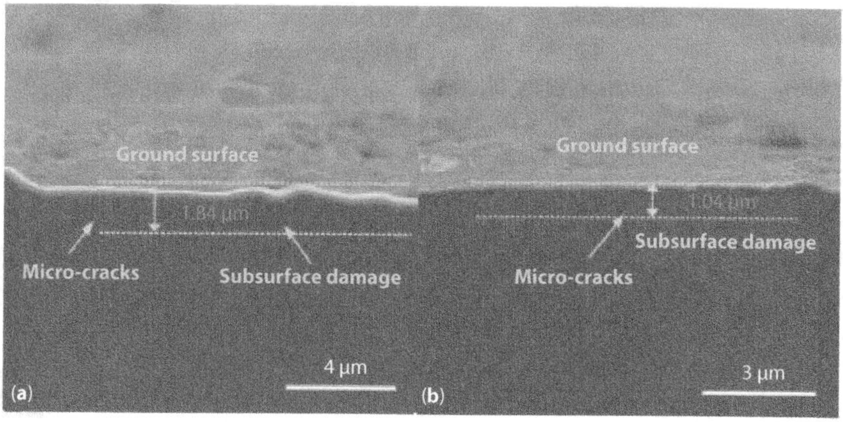

Figure 7.12 SEM of surface damages, (a) Conventional grinding (b) LAMG process [33].

formation [34]. The quality of machined surface by LAMG is much better than CMG without any subsurface damages [31].

With properly controlled parameters means significantly enhancement in the SQ and reduction in cutting forces for LAMG process. It effectively reduces the cutting forces about 40% [32] while surface roughness and subsurface damages are by 37% to 40% and 22% to 50% respectively [33]. The subsurface damage for grinding and laser-based grinding is presented in Figures 7.12 (a)–(b) [33]. It also provides deeper groves than CMG process without surface damage because of thermal softening of material [31].

The application of laser in microgrinding significantly assists the performance of grinding but simultaneously dominates the melting point of materials and bonding strength of wheel [34]. Due to thermal softening, losses of abrasives particles without wears as a result reduction in wheel life. It may also be possible that microchips easily entrapped/bonded with bond materials causes wheel loading. Even though application of laser is limited only for flat surfaces and challenging for complex and deep holes grinding. Instead of this, the initial setup cost and skill operator also limits the applicability of laser in microgrinding.

7.4 Laser-Assisted Nontraditional HMMPs

In laser-assisted nontraditional HMMPs, the laser energy is utilized to assist the advanced micromachining processes (EDM, ECM, or WJM). The common laser-assisted nontraditional HMMPs are listed as follows:

- Laser-assisted electrodischarge micromachining
- Laser-assisted electrochemical micromachining
- Laser-assisted electrochemical spark micromachining
- Laser-assisted water jet micromachining

7.4.1 Laser-Assisted Electrodischarge Micromachining

In various noncontact machining processes, EDMM (electrical discharge micromachining) broadly accepted for machining of hard and difficult to cut metals at micro/nano scales. In this process, material removal occurs due to melting and vaporization when spark is created between electrodes in presence of electrolyte [35]. It provides high quality products free from mechanical stresses because of noncontact nature of process. Instead of this, it gives less HAZ as compared to laser microprocessing of metals [36]. Even though, low machining rate and high tool wear are major limitations of EDMM process [35]. On other hand, thermal energy-based LBMM is used to create desired profiles on workpiece surface through various configurations (drilling, milling, grooving or turning) of laser. Off course, LBMM is a quick material removal process but suffers with poor SQ, tapered profiles and larger HAZ [36]. Hence, laser combined with EDMM in sequentially and referred as laser-assisted electrodischarge micromachining (LA-EDMM) process that able to overcome the limitations of constituent processes.

LA-EDMM comprises the features of LBMM and EDMM processes in a sequential manner as shown in Figure 7.13. In this process, laser (short/ultra short pulse) is utilized for pre-machining (drilling, grooving, cutting etc.) or simply say rough cutting on base material. Subsequently, EDMM process is applied to remove recast layer and HAZ formed by laser processing [36–38]. It is an effective way to finish the objects machined by

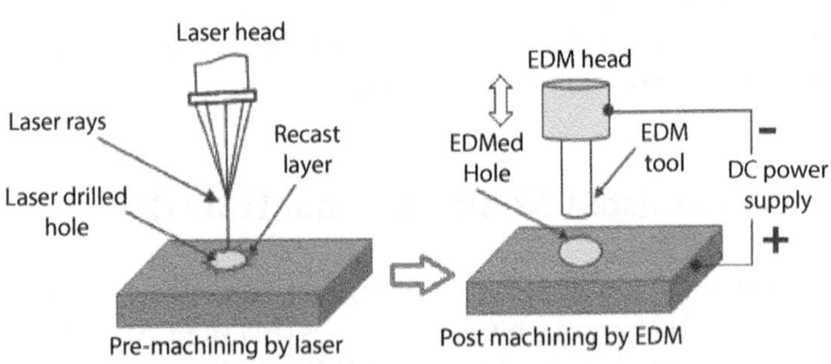

Figure 7.13 Sequential application of LBM and EDMM in LA-EDMM process.

LBM/LBMM to make them defects free with effective elimination of recast layer and HAZ. It also reduces the material processing time with quality products. Hence, low material removal limitation of EDMM process can be eliminated effectively [36, 38].

LA-EDMM process is significantly reduced the machining time about 70% [38] and 50% to 65% [37] with better quality than microholes drilled by EDMM process. The SEM images of drilled holes by EDMM and LA-EDMM processes are presented in Figures 7.14 (a)–(b) [38]. Furthermore, Figures 7.15 (a)–(c) [38] show that accuracy of LA-EDMM machined hole (approx. 140μm) is better than laser drilled hole (100μm) and EDMM drilled hole (approx. 146μm) [38]. It is applied for microdrilling into aeronautical and turbines parts like combustion casing, vanes, and blades for effective cooling [39]. It also becomes a unique method for drilling of holes of fuel injector nozzle for diesel fuel engines [38].

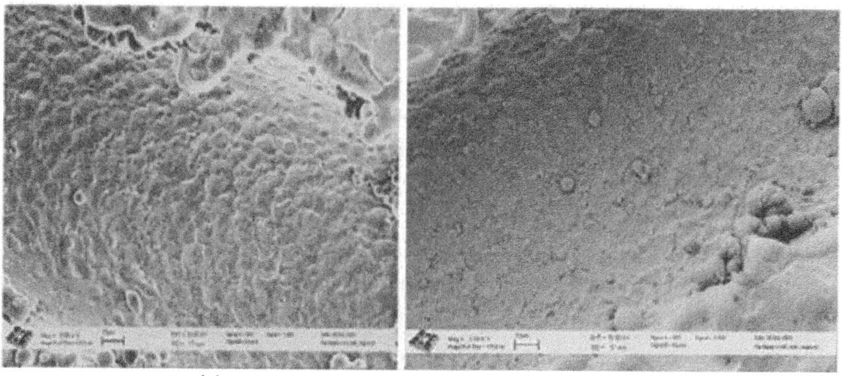

Figure 7.14 SEM images drilled holes; (a) EDMM process; (b) LA-EDMM process [38].

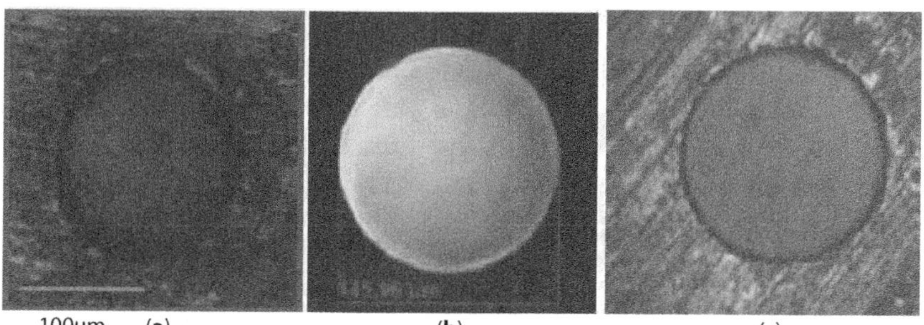

Figure 7.15 Drilled holes by different processes; (Laser pilot hole); (b) EDM drilled hole; (c) hole drilled by laser-EDMM processes [39].

LA-EDMM process has been found an appropriate method for micro-drilling and microfabrication, but it requires two separate platforms for LBMM and EDMM operations. Due to this, increase in the changeover time from one platform to another. Instead of this, the initial cost for LA-EDMM process is higher and it requires well skilled operator for laser as well as EDM process.

7.4.2 Laser-Assisted Electrochemical Micromachining

Electrochemical micromachining (ECMM) is one the most prominent method for shaping of electrically conductive materials at micron and sub-micron levels where material removal occurs by metallic dissolution in presence of electrolyte. The machined parts are highly precise and stress free [40, 41]. The high material removal capability without tool wear makes it wonderful technique for processing of difficult to shape electrically conductive materials in different profiles including complex geometry and 3D structures [40–42]. It is applied in different manufacturing areas as aerospace, automobiles, defence and medicals [40, 42]. The performance of ECMM depends on chemical dissolution and high dissolution means high machining rate. Hence, the controlling of dissolution becomes necessary for precise and efficient machining which is a difficult task. Instead of this, the stray losses surrounding tool also limits the applicability of ECMM process.

To enhance the machinability of ECMM, various approaches are tested such as minimize the IEG (inter electro gap), application of insulated electrodes, rotating tool and low voltage for machining [41]. Even though, the ionic dissolution is not shown significant improvement in most of cases. Hence, another methodology refers as hybridization of ECMM with laser that shows potential in fast and rapid dissolution of metallic workpiece. As a result higher productivity in term of material removal. Such process is known as laser-assisted-ECMM (LA-ECMM) process.

In hybridization of laser with ECMM, low-power (375–570 mw) laser energy is utilized for localized heating of electrolyte with effective application of solid (Nd:YAG or fiber) laser into IEG [40, 43]. The localised heating of electrolyte by laser leads to increase the temperature resulting higher dissolution. Hence, higher material removal can possible with precision and accuracy [40]. Here, laser radiation also helpful for removal of oxide layers formed on workpiece surface in presence of certain electrolyte. Furthermore, another combination of laser with electrochemical jet micromachining (ECJMM) is developed by researchers to enhance the performance of exiting LA-ECMM process which is known as LA-ECJMM process.

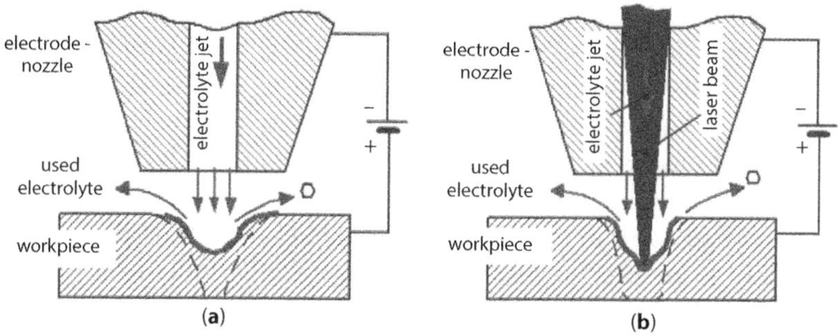

Figure 7.16 Hybridization of ECJMM and laser beam; (a) ECJMM without laser process; (b) LA-ECJM/ECJMM process [44].

LA-ECJMM comprises the laser with electrolyte jet where material removal occurs due to electrochemical dissolution [44]. The typical diagram of ECJM/ECJMM and hybridized ECJMM with laser are presented in Figures 7.16 (a)–(b). Laser energy significantly enhances the precision machining with high material removal by localized thermal heating of electrolyte resulting enhancement in electrochemical dissolution at target area. It also prevents the unwanted machining especially stray losses [40, 43–45]. The key advantage of LA-ECJMM is to accelerate the dissolution along with flushing in any desired location. On another way, the application of laser energy in ECMM/ECJMM is significantly raised the boiling nature of electrolyte and electrical discharge that adverse effects on performance of process. Generally, laser beam and electrolyte jet must be coaxially in LA-ECJMM, but the hydrodynamic behaviour of electrolyte jet disturbs the alignment in several cases. The evaluation of gases at cathode also crates disturbance and turbulence for jet of electrolyte near target area during machining of objects. However, machining of miniatures (5–500 μm) made of ceramics, composites, or alloys with high accuracy (1–10 μm) is still challenges for it [40].

7.4.3 Laser-Assisted Electrochemical Spark Micromachining

Electrochemical spark micromachining (ECSMM) is simply known as ECSM (electrochemical spark machining) that shows their ability in micromachining of ceramics materials including glass, quartz, and Pyrex [46]. In general, it is hybridization of EDM and ECM to eliminate the limitation of such processes as needs of electrically conductive metal and alloys for processing. On other hand, ECSM can be defined as a noncontact thermal energy-based method that comprises features of EDM and ECM.

Here, material removal occurs due to melting and vaporization when spark is created between tool electrode and gaseous (H_2 gas) layer with application of suitable discharge voltage in presence of electrolyte [46, 47]. It is applicable for micro/nano machining of variety of materials including ceramics and electrically nonconductive as well as conductive materials [46]. Even though, it suffers with recast layer, HAZ and microcracks due to rapid cooling [46]. In such situations, hybridization of ECSM/ECSMM with laser in sequential manner is become an effective tool to minimize the associated problems. Such combined process is referred as laser-assisted-ECSMM (LA-ECSMM) process.

Like LA-EDM, laser and ECSM processes are applied sequentially for processing of materials through LA-ECSMM. In this process, laser is applied either pre-processing or post-processing [47]. Mostly, the bottom surface microgrooves in glass ceramic consists with uneven surface, protrusion like structures, over cutting, HAZ with taper profile when processed by ECSMM. On other hand, microgrooves straight edge profile is not possible when laser is used for processing individually. Furthermore, the laser machining of microgrooves shows V-profile with larger taper and HAZ [47]. The combined effort of laser and ECSM effectively reduces the HAZ, recast layer and taper formed by either LBMM or ECSM/ECSMM processes.

In LA-ECSMM process, either laser or ECSMM are applied sequentially to improve the machinability of process. When laser is applied as post processing for shaping of bottom of microgrooves formed by ECSMM then the protrusions are significantly removed by laser, but edge quality of grooves is still poor. On other way, when ECSMM process is applied as post processing of microgrooves formed by laser machining then the grooves obtained with better edge quality with improvement in taper and V-shape profile converted into U-shape profile of microgrooves [47]. Hence, application of ECSMM for post operation after laser is better for machining of microprofiles like grooves, chute, holes or other microprocessing in/on ceramic and nonconductive materials. Even though, it requires separate platform for each process as which enhancement in processing cost.

7.4.4 Laser-Assisted Water Jet Micromachining

Laser-assisted-water jet micromachining (LA-WJMM) is a better technique for microprocessing of brittle materials (ceramics, glass and composites) without thermal damages of machined surfaces. The extremely hard metals like diamond, cubic boron nitride (CBN) and sapphire can be cut with better SQ by LA-WJMM process [48–50]. The basic objective for combining laser and WJM (water jet machining) is to minimize the thermal damages occurs due to laser machining. Mostly, high-power beam intensity (1 MW/cm^2) is utilized for laser machining where material removal occurs by melting and vaporization causes of thermal damages [2]. On other way, less thermal energy is required for LA-WJMM process that utilizes for softening of material resulting reduction in thermal damages of test material. Such softened material is removed by expulsion of high-pressure water jet.

The typical diagram of LA-WJMM is presented in Figure 7.17 [49]. In this process, the laser and water jet moves simultaneously during material removal process. Here, low-power laser (max. 1.5 kW) is used for precise heating in a tiny area on workpiece surface as which temperature gradient zone is created equal to depth of thermal diffusion [48]. Hence, heated zone tends to expand that restrained by surroundings resulting high compressive stresses in heated zone. Subsequently, compressive nature of heated zone is changed into tensile nature due to rapid cooling by low pressure water jet resulting microcracks are formed on workpiece material [50]. Such cracks may extend and run along thickness of material depending upon fracture nature of material and tensile force [48].

In general, cracks follow the laser-water jet path along length of workpiece because the tensile stresses only generated in such location. Subsequently, the separation of material will be accomplished along

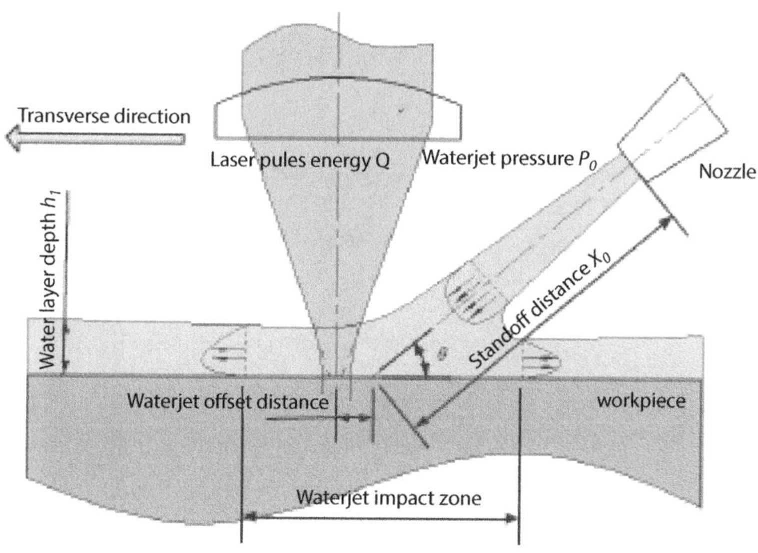

Figure 7.17 LA-WJMM process and process parameters [49].

length of workpiece through controlled propagation of cracks [48]. Due to this, material removal occurs and machined zone expected to free from residual stresses, HAZ and thermal damages. The machined surfaces are also free from visual defects because tiny debris particles are washed out by water jet.

Instead of LA-WJMM process, another combination of laser and water jet is also come in existence for microprocessing of materials where water jet is used to guide the laser beam. Such process is known as water jet guided laser (WJGL) cutting process. It is an effective way for precision cutting without burrs with less heat with pollution free cutting than tradition processes [51–55]. The basic concept of water jet in laser machining is created to replace the assist (inert) gas by water stream that broadly used in conventional laser cutting process. The WJGL process shows superior machinability than laser machining and varieties of material can be machined such as brittle and ultra-hard, heat treated and mechanically sensitive, electrically conductive/isolating materials [52–54]. The basic advantages of WJGL process are to transfer of energy into water than material and cooling of cut edges by water jet resulting less HAZ, thermal damages and thermal residual stresses [52].

Figure 7.18 shows the typical diagram of WJGL cutting process and their main components that affects the performance of process [52]. Here, laser is guided through high pressure (50–80 MPa) water jet and employed for cutting [52]. In WJGL process, the water jet is worked as optic surface

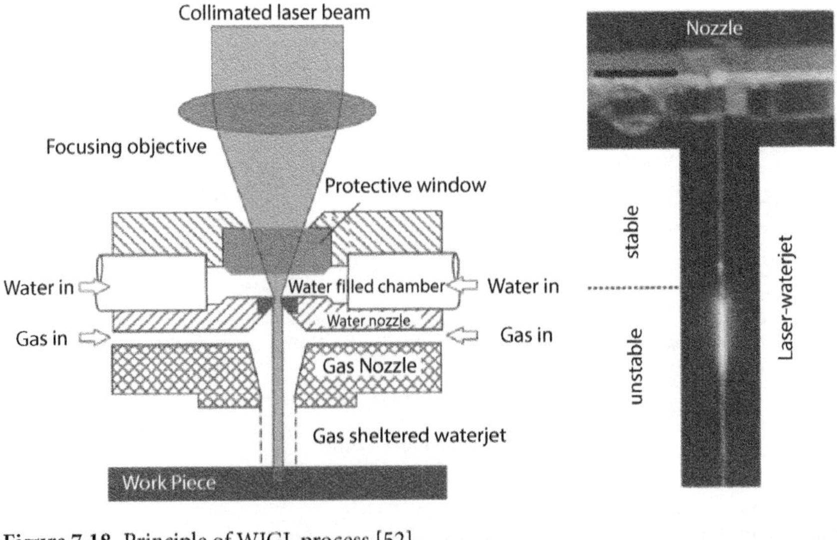

Figure 7.18 Principle of WJGL process [52].

to reflect laser at work material. A focusing lens is utilized to focus the laser into small diameter (diameter = 20–150 μm) of nozzle [51]. The nozzle is fitted beneath the chamber of de-ionized water. The laser gets reflected at interface of water and air when comes in contact of pressurized water into nozzle. Hence, ablation of laser beam occurs causes of melting and vaporization of material. On other hand, the water jet cools the target material resulting reduction in generation of harmful gases and smokes.

In WJGL cutting becomes beneficial in many ways as not requirement of focal optics because water works as optics. Here, a very small amount of mechanical force (<0.10 N) delivered by jet unlike to conventional jet machining (1–5 N) causes of scathing of material from the surface of workpiece [55]. Instead of this, the controlling of laser within the guided water is difficult that affects the precise machining and efficiency.

7.5 Capabilities and Shortfalls of LA-HMMPs

Off course, LA-HMMPs show there potential for microprocessing of difficult to cut metals. It is widely utilized in precision manufacturing and advanced industries especially aeronautical, aerospace, automobile, military, dies and molds because of high hardness and wear resistance characteristics in nature. In conventional LA-HMMPs, the localized heating of material leads in thermal softening of workpiece as which significant reduction in hardness, yield strength and strain hardening. Subsequently, thermal softening causes change the deformation nature of materials from brittle to ductile. Hence, cutting becomes easier with better SQ, reduction in cutting forces and tool wears.

In conventional LA-HMMPs, short/ultrashort lasers are applied infront of cutting tool for thermal softening of work. Here, to maintain a uniform temperature throughout cutting is difficult for turning as well as other processes. Hence, application of two-lasers becomes a novel approach but raises the initial cost and complexity. Instead of this, other issues may also be raised (local buckling, thermal cracks and phase changes) due to excessive heating [21]. To overcome the problem, the direct heating-based LAMT is a unique technique where microlaser directly applied at desired location throughout cutting tool [17]. Off course, direct heated LAMT significantly reduces the heat loss with reduction in cutting force by 9% to 15% and also reduction in surface roughness by 0.34 to 10 μm for turning of aluminium alloy [17]. Even though, difficulties in manufacturing of microtool made of diamond also limits the applicability of direct heating-based LAMT process.

LAMD is one of popular method for making through, holes, blind holes or partial holes into difficult to cut and hard materials with precise geometry in minimum possible time than drilling process. It significantly reduces the drilling time by 16% to 40% with reduction in holes diameter by 85% to 140% in different conditions of machining for chrome steel, high carbon steel and stainless steel [25]. The spike like formation and application of laser beam for high aspect hole drilling are major issues. The thermal softening of drill affects the cutting ability of tool and leads in thermal deformation of drill tools. Similarly, micromilling cutter and their manufacturing for application of laser beam in desired location is always challenges in case of LAMM process. Even though, the LAMM is applicable for 3D structuring, grooves formation, complex geometry made of heat-treated alloys, super alloys, titanium alloys, hot and cold rolled steels [29, 30].

In category of finishing processes, CMG is the broadly accepted process that applicable for hard and brittle materials. The application of laser with CMG significantly enhances the machinability with better SQ and low specific energy consumption [33]. The thermal softening of workpiece infront of grinding wheel assists in deeper grinding with high feed rate and better SQ without subsurface damages. The effective application of laser with CMG improves the surface roughness by 37% to 40% while reduction in subsurface damages by 22% to 50% than CMG process with lesser (1.5 times) processing duration for reaction bonded SiC workpiece [33]. The thermal softening also affects the bonding strength and finally loss of abrasive particles without wear at low cutting forces resulting poor wheel life. Such phenomenon increases the cost of grinding wheel. In some cases, chips particles easily entrap with softened bond material and creates wheel loading problem. Instead of this, the application of laser with CMG only limited to flat surface and not suitable for complex geometry and deep holes grinding.

The sequential application of EDMM after LBM provides better responses with better quality of machined object in term of SQ, HAZ, taper and recast layer formation. The application of LA-EDMM in microholes drilling significantly reduces the processing time about 50% to 65% without tempering of quality as compared to EDMM process [37]. In some cases, reduction in processing time upto 70% can possible with selection of appropriate parameters [38]. It is an effective method for drilling of microholes into turbine casing and aeronautical parts for cooling purposes and also drilling of fuel injector nozzle of internal combustion engines [38, 39]. Though, it requires separate platform for LBM/LBMM and EDMM that affects the overall machining time and process becomes costly.

Similar to LA-EDMM, other laser-assisted unconventional process as LA-ECMM is only applicable for micromachining of electrically conductive hard and brittle materials in presence of electrolyte. In ECMM, laser energy enhances the electrolysis process as which improvement in material removal capacity of process. Here, laser only assists the electrolysis process while machining occurs by metallic dissolution due to electrochemical reaction. It is more suitable process for 3D structuring, machining of complex profiles at micro/nano scales and applicable in aerospace, automobiles, defence, nuclear and medical industries [40, 41]. Here, controlling of electrolysis process becomes necessary for precise machining which is a difficult for manufacturing unit. In same way, the stray loss surrounding the tool also affects the quality of machined objects. To overcome the above limitations, LA-ECJMM becomes a powerful tool where laser is applied with electrolyte jet at desired location only. Hence, enhancement in electrochemical reaction of electrolysis and reduction in stray losses that prevents the unwanted stray machining [43, 44]. In some cases, hydrodynamic nature of electrolyte jet disturbs the focusing direction of laser beam as which adverse effect on machining performances. The formation of hydrogen gases also creates turbulence in jet of electrolyte near target area. Such phenomenon may also be responsible for declination of machinability performances.

Noncontact based another process refers as ECSM significantly eliminates the limitations of EDM, and ECM as needs of electrically conducive metals. The application of laser with ECSM significantly enhances the performance of LA-ECSMM process but boiling of electrolyte and uncontrolled generation of hydrogen create issues during machining. Off course, ECSM is a better machining method for microfabrication of objects made of either electrically conductive or nonconductive materials. Here, laser and ECSM/ECSMM processes are applied in sequential modes and post application of ECSM significantly provides better responses with less HAZ and recast layer.

LA-WJMM shows their potential for microprocessing of ceramic and nonconductive materials. It is an effective process for shaping of hard and brittle materials like diamond, cubic boron nitride (CBN) and sapphire without HAZ, thermal damages along with residual stress free surfaces [47–49]. In this process, the high pressure water jet necessary for removal of material from target surface but simultaneously creates disturbance for laser beam as which beam may divert from desired location. To minimize the turbulence, WJGL process shows better responses than LA-WJMM because laser heat is transmitted to water not to workpiece material. Due to this, WJGL becomes more suitable for precision machining of brittle, ultra-hard, heat treated, electrically conductive or isolating materials [52–54]. Even though, the controlling of laser with guided water is difficult that affects the precise machining and efficiency of process.

7.6 Conclusion

Laser is a powerful source of energy that utilizes by advanced industries to meet the current demands for micromanufacturing and nanomanufacturing. In a simple way, LBMM has been suffered to fulfil the demands of industries in present scenarios with high quality miniatures. To overcome the situations, hybridization of laser energy with mechanical, chemical, or electrical energies provides a better platform for manufacturing industries. The present chapter covers the various types of HMMPs that can be developed by combining of laser energy. Here, various traditional and non-traditional processes and their hybridization with laser are discussed and explained in brief with merits, demerits, and applications. Instead of this, the capabilities and shortfalls of LA-HMMPs are discussed briefly which makes easier for readers. Overall, the present chapter provides a platform for manufacturing industries to choose a better machining process for fabrication of miniatures with application of laser-based hybrid processes. The present chapter also provides a futuristic direction for researchers to find an appropriate research field for their works related to laser-based hybrid micromachining of extremely hard and difficult to machine materials.

Acknowledgment

The authors gratefully acknowledge Elsevier publishers and Journals for their kind reprint/reuse permission of Figures and Tables form already published papers. The authors are also thankful to the copyright center for their assistance to get reprint/reuse permission quickly from desired location.

References

1. Jain, V.K., *Introduction to Micromachining*, Narosa Publishing House, New Delhi, India, 2010.
2. Dubey, A.K. and Yadava, V., Laser beam machining-a review. *Int. J. Mach. Tools Manuf.*, 48, 609–628, 2008.

3. Mishra, S. and Yadava, V., Laser beam micromachining (LBMM)-a review. *Opt. Lasers Eng.*, 73, 89–112, 2015.
4. Sugioka, K., Meunier, M., Pique, A., *Laser Precision Microfabrication*, Springer, Berlin Heidelberg, 2010.
5. Desbiens, J.P. and Masson, P., ArF excimer laser micromachining of Pyrex, SiC and PZT for rapid prototyping of MEMS components. *Sens. Actuators A Phys.*, 136, 554–563, 2007.
6. Pan, C.T., Hwang, Y.M., Hsieh, C.W., Dynamic characterization of silicon-based microstructure of high aspect ratio by dual-prism UV laser system. *Sens. Actuators A Phys.*, 122, 45–54, 2005.
7. Ali, B., Litvinyuk, I.V., Rybachuk, M., Femtosecond laser micromachining of diamond: Current research status, applications and challenges. *Carbon*, 179, 209–226, 2021.
8. Gower, M.C., Industrial applications of laser micromachining. *Optics Express*, 7, 56–67, 2000.
9. Meng, H., Liao, J., Zhou, Y., Zhang, Q., Laser micro-processing of cardiovascular stent with fiber laser cutting system. *Opt. Laser Technol.*, 41, 300–302, 2009.
10. Yadav, R.N., Yadava, V., Singh, G.K., Multi-objective optimization of process parameters in electro-discharge diamond face grinding based on ANN-NSGA-II hybrid technique. *Front. Mech. Eng.*, 8, 319–332, 2013.
11. Rajurkar, K.P. and Gu, L., Resent research and developments in hybrid machining processes. *Proc. 3rd Int. 24th All India Manuf. Technol. Des. Res. Conf*, Vishakhapatnam, India, pp. 39–44, 2010.
12. Srivastava, P. and Dubey, A.K., Electrical discharge machining-based hybrid machining processes: A review. *Proc. Inst. Mech. Eng. B J. Eng. Manuf.*, 228, 799–825, 2014.
13. Chavoshi, S.Z. and Luo, X., Hybrid micro-machining processes: A review. *Precis. Eng.*, 41, 1–23, 2015.
14. Zhu, Z., Dhokia, V.G., Nassehi, A., Newman, S.T., A review of hybrid manufacturing processes-state of the art and future perspectives. *Int. J. Comput. Integr. Manuf.*, 26, 596–615, 2013.
15. Lee, C.M., Woo, W.S., Kim, D.H., Oh, W.J., Oh, N.S., Laser-assisted hybrid processes: A review. *Int. J. Precis. Eng. Manuf.*, 17, 257–267, 2016.
16. Ghosh, A. and Mallik, A.K., *Manufacturing Science*, East-West Press, New Delhi, 1999.
17. Wei, Y., Park, C., Park, S.S., Experimental evaluation of direct laser assisted turning through a sapphire tool. *Proc. Manuf.*, 10, 546–556, 2017.
18. You, K., Fang, F., Yan, G., Zhang,Y., Experimental investigation on laser assisted diamond turning of binderless tungsten carbide by in-process heating. *Micromachines*, 11, 1104–15, 2020.
19. Mohammadi, H., Ravindra, D., Kode, S.K., Patten, J.A., Experimental work on micro laser-assisted diamond turning of silicon (111). *J. Manuf. Process.*, 19, 125–128, 2015.

20. Kim, J. and Kang, B., Machining characteristics of micro lens mold in laser-assisted micro-turning. *J. Mech. Sci. Technol.*, 32, 1769–1774, 2018.
21. Chryssolouris, G., Anifantis, N., Karagiannis, S., Laser assisted machining: an overview. *J. Manuf. Sci. Eng.*, 119, 766–769, 1997.
22. Mohammadi, H., Poyraz, H.B., Ravindra, D., Patten, J.A., Surface finish improvement of an unpolished silicon wafer using micro-laser assisted machining. *Int. J. Abras. Technol.*, 7, 107–121, 2015.
23. Wang, Y., Zhang, G., Zheng, L., Ren, Y., Eichstaedt, O., Chen, X., Experimental study of laser assisted drilling and parameters analysis. *Proc. Int. Conf. Indus. Technol. Manag. Sci. (ITMS 2015)*, pp. 145–149, 2015.
24. Jen, T.C., Chen, Y.M., Tuchowski, F., Experimental and numerical studies of laser-assisted drilling processes. *Proc. Heat Trans./Fluids Eng. Summer Conf*, Charlotte, North Carolina, USA, pp. 1–9, July 11-15, 2004.
25. Zhang, G., Jiang, C., Zhang, S., Wang, Y., Chen, X., Yu, L., An experimental investigation of laser assisted drilling process. *Mater. Res. Innov.*, 19, 889–894, 2015.
26. Bissacco, G., Hansen, H.N., Chiffre, D.L., Micromilling of hardened tool steel for mould making applications. *J. Mater. Process. Technol.*, 167, 201–207, 2005.
27. Bissacco, G., Hansen, H.N., Chiffre, D.L., Size effects on surface generation in micro milling of hardened tool steel. *CIRP Ann.-Manuf. Technol.*, 55, 593–596, 2006.
28. Melkote, S., Kumar, M., Hashimoto, F., Lahoti, G., Laser assisted micro-milling of hard-to-machine materials. *CIRP Ann.-Manuf. Technol.*, 58, 45–48, 2009.
29. Hojati, F., Azarhoushang, B., Daneshi, A., Biermann, D., Laser prestructure-assisted micromilling of Ti_6Al_4V titanium alloy. *Int. J. Adv. Manuf. Technol.*, 120, 1765–1776, 2022.
30. Kadivar, M., Azrhoushang, B., Zahedi, A., Muller, C., Laser-assisted micro-milling of austenitic stainless steel X5CrNi18-10. *J. Manuf. Process.*, 48, 174–184, 2019.
31. Chang, W., Luo, X., Zhao, Q., Sun, J., Zhao, Y., Laser assisted micro grinding of high strength materials. *Key Eng. Mater.*, 496, 44–49, 2012.
32. Kadivar, M., Shamray, S., Soltani, B., Daneshi, A., Azarhoushang, B., Laser-assisted micro-grinding of Si3N4. *Precis. Eng.*, 60, 394–404, 2019.
33. Luo, X., Li, Z., Chang, W., Cai, Y., Sun, J., Ding, F., Zhang, F., Liu, H., Sun, Y., Laser-assisted grinding of reaction-bonded SiC. *J. Micromanuf.*, 3, 93–98, 2020.
34. Hu, M.F., Xie, J., Su, H.H., Liu, J.N., Study on laser-assisted dry micro-ground surface of difficult-to-cut materials. *Int. J. Adv. Manuf. Technol.*, 94, 2919–2928, 2018.
35. Yadav, R.N., Mishra, S., Yadav, S.K.S., Electro spark process for microfabrication, in: *Micro Electro-Fabrication*, T. Saleh, K. Takahata, M.S.M. Ali, (Eds.), pp. 1–31, Elsevier, Netherlands, 2021.

36. Rashid, M.A.N., Saleh, T., Noor, W.I., Ali, M.S.M., Effect of Laser parameters on sequential laser beam micromachining and micro electro-discharge machining. *Int. J. Adv. Manuf. Technol.*, 114, 709–723, 2021.
37. Ahmaria, A., Rasheed, M.S., Mohammed, M.K., Saleh, T., A hybrid machining process combining micro-EDM and laser beam machining of nickel-titanium based shape memory alloy. *Mater. Manuf. Process.*, 31, 447–455, 2016.
38. Li, L., Diver, C., Atkinson, J., Wagner, R.G., Helml, H.J., Sequential laser and EDM micro-drilling for next generation fuel injection nozzle manufacture. *CIRP Ann.*, 55, 179–182, 2006.
39. Antar, M., Chantzis, D., Marimuthu, S., Hayward, P., High speed EDM and laser drilling of aerospace alloys. *Proc. CIRP*, 42, 526–531, 2016.
40. Mandal, I. and Doloi, B., State of the art on laser assisted electrochemical machining. *Mater. Sci. Eng.*, 653, 012030-6, 2019.
41. Skoczypiec, S., Application of laser and electrochemical interaction in sequential and hybrid micromachining processes. *B. Pol. Acad. Sci. Tech.*, 63, 305–314, 2015.
42. Bhattacharyya, B., Munda, J., Malapati, M., Advancement in electrochemical micro-machining. *Int. J. Mach. Tools Manuf.*, 44, 1577–1589, 2004.
43. Wyszynski, D., Skoczypiec, S., Grabowski, M., Ruszaj, A., Lipiec, P., Electrochemical microprocessing assisted by diod pumped solid state Nd:YAG pulse laser. *J. Mach. Eng.*, 12, 131–142, 2012.
44. Pajak, P.T., Desilva, A.K.M., Harrison, D.K., Mcgeough, J.A., Precision and efficiency of laser assisted jet electrochemical machining. *Precis. Eng.*, 30, 288–298, 2006.
45. Zhang, H., Gao, P., Xu, J., Yuan, L., Effect of electrochemical dissolving in laser drilling assisted with jet electrochemical machining. *Int. J. Electrochem. Sci.*, 16, 1–16, 2021.
46. Yadav, R.N., Electro-chemical spark machining-based hybrid machining processes: Research trends and opportunities. *Proc. Inst. Mech. Eng. B J. Eng. Manuf.*, 233, 1037–1061, 2019.
47. Zhao, D., Zhang, Z., Zhu, H., Cao, Z., Xu, K., An investigation into laser-assisted electrochemical discharge machining of transparent insulating hard-brittle material. *Micromachines*, 12, 1, 1–22, 2020.
48. Kalyanasundaram, D., Shehata, G., Neumann, C., Shrotriya, P., Molian, P., Design and validation of a hybrid laser/water-jet machining system for brittle materials. *J. Laser Appl.*, 20, 127–134, 2008.
49. Wang, L., Huang, C., Wang, J., Zhu, H., Liang, X., An experimental investigation on laser assisted water jet micro-milling of silicon nitride ceramics. *Ceram. Int.*, 44, 5636–5645, 2018.
50. Elperin, T. and Kornilov, A., Formation of surface micro crack for separation of nonmetallic wafers into chips. *J. Electron. Packag.*, 122, 317–322, 2000.

51. Tabie, V.M., Koranteng, M.O., Yunus, A., Kuuyine, F., Water-jet guided laser cutting technology-an overview. *Lasers Manuf. Mater. Process.*, 6, 189–203, 2019.
52. Liu, Y., Wei, M., Zhang, T., Qiao, H., Li, H., Overview on the development and critical issues of water jet guided laser machining technology. *Opt. Laser Technol.*, 137, 2021.
53. Sokołowski, Z. and Malinowski, I., Perspectives of applications of micromachining utilizing water jet guided laser, pp. 365–369, Springer, Singapore, 2007.
54. Levesque, T., Perrottet, B., Richerzhagen, B., Damage-free cutting of medical devices using the water-jet-guided laser. *Proc. Mater. Process. Med. Dev. Conf*, Boston, USA, pp. 73–76, November 14-16, 2005.
55. Richmann, A. and Richerzhagen, B., Comparison study: Cutting with the laser microjet© vs. well-established and new micro-machining technologies for applications of the watch industry. *International Congress on Application of Lasers and Electro-Optics*, vol. 1, pp. 269–277, 2014.

8

Hybrid Laser-Assisted Jet Electrochemical Micromachining Process

Sivakumar M.*, **J. Jerald, Shriram S., Jayanth S. and N. S. Balaji**

Department of Production Engineering, National Institute of Technology, Tiruchirappalli, India

Abstract

Laser-assisted jet electrochemical micromachining (LAJECM) is a process in which laser beam is combined with a jet of electrolyte to create a hybrid cutting process. This technique is also known as hybrid laser-assisted jet electrochemical micromachining. The jet electrolyte and laser beam produce a noncontact tool electrode that dissolves metal by electrochemical reactions. The laser beam can direct the dissolution to particularly focused areas, which results in an efficient improvement in the accuracy of the process. The material removal mechanism and process energy distribution mechanism of hybrid laser-assisted jet electrochemical micromachining is discussed in this chapter. The material removal depends on electrochemical dissolution and with assisted laser beam offers an increase in both the precision and productivity of the machining process. The workpiece surface is thermally activated by the source of the laser. Hence, the geometry of the holes and cavities is improved after the LAJECM process. The material removal rate and reduction in tapper are enhanced with the assistance of the laser in LAJECM for Hastelloy, Aluminium alloy, Titanium, and Stainless steel.

Keywords: Hybrid laser-assisted jet electrochemical micromachining, electrochemical machining, jet electrochemical machining, reduction in taper, surface roughness, material removal rate

**Corresponding author*: siva.sky06@gmail.com

Sandip Kunar, Golam Kibria, Prasenjit Chatterjee and Asma Perveen (eds.) Hybrid Micromachining and Microfabrication Technologies: Principles, Varieties and Applications, (179–204) © 2023 Scrivener Publishing LLC

8.1 Introduction

Machining processes have a high degree of accuracy and surface quality in finished products. The tools which are harder than the workpiece material is used in conventional machining. A relative motion between the workpiece and tool is responsible for forming or generating the required shape. But as the hardness of the workpiece material increases, it became difficult to provide a proper tool for machining. The demand for greater machining productivity with high accuracy and surface quality paved the way for the hybrid machining process. This hybrid machining process sometimes includes a primary process that acts as a major material removal mechanism and a secondary process that assists the primary in material removal.

Micromachining is a term used to describe the minimal requirements for microproducts or microcomponents, which typically range from a few μm to 1000 μm [1, 2]. There are various types of nontraditional machining processes, among those electrochemical machining (ECM) seems to be promising, effective, and used commercially in several industrial areas [4, 5]. A combination of traditional and non-traditional processes are employed in the hybrid micro-machining process in order to make machining energy from several processes possible, either in an aided or sequential fashion [3]. Large metallic items and intricately shaped parts are both machined using the ECM technique. The small tool wear, burr-free machining, high MRR, better surface finish, and the ability to manufacture complicated forms regardless of their hardness are some of the benefits of ECM over conventional machining methods.

A free jet of electrolyte that is used in jet-ECM sometimes referred to as electro jet drilling, electro jet machining, or electrochemical jet machining, dissolves material from an electrically conductive workpiece on-site by anodic dissolution. In Jet-ECM, an electrolyte jet is ejected via a nozzle under pressure to dissolve material just where the jet makes contact with the workpiece [6]. Generally, the jet-ECM is used to create microsize features independent of material hardness without creating the heat-affected zone, surface tension, microcracks, and burrs [7].

Increasing the rate of material dissolution from certain places in an axial direction is the primary goal of using laser beams. By thermally heating the surface layer, laser beams to aid in the breakdown of metal. It slows down the rate of lateral dissolution, which boosts the machining rate while improving surface quality and dimensional accuracy.

LAJECM is a combination of electrochemical dissolution and localized heating by laser energy [8]. The merits of LAJECM is that the laser beam may be conveniently directed on the material surface, allowing dissolution to be accelerated in any required direction in conjunction with the cleansing electrolyte jet [9]. The pace of electrochemical processes is accelerated by the thermal energy from laser beams, which accelerates material removal [10]. LAJECM has been widely expanded in the area of micromachining applications in recent years. LAJECMM goods can be made from a variety of materials, such as ceramics, steels, specialty alloys, etc., and can have complicated 3D geometries with surface roughness (Ra) values higher than 0.5 μm and high tolerances generally below 1 μm for various applications in automobile, aerospace and electronics industries. The book chapter describes the hybrid laser-assisted jet electrochemical micromachining working principle, material removal mechanism, temperature distribution mechanism, and its applications.

8.2 Overview of Electrochemical Machining

ECM was parented by Gusseff in 1929. ECM has been used to shape, mill, and finish large parts in the aircraft and aerospace industries. The mechanism of anodic material removal was initially found by Michael Faraday which played a significant role in the evolution of electroplating, electropolishing, and other related techniques. Material removal is done through a process called electrolysis, in which atoms from the workpiece surface are taken off one by one. In the ECM process, the electrodes are placed very close to each other, which is different from the electrolysis process. This will make the rate at which metal ions dissolve faster and easier to control. The current flow and the metal removal rate from the anode will be faster if the gap is smaller. The quick formation of metal hydroxide and gas bubbles in this instance both impede the machining gap and serve as a hindrance to the flow of the current as shown in Figure 8.1.

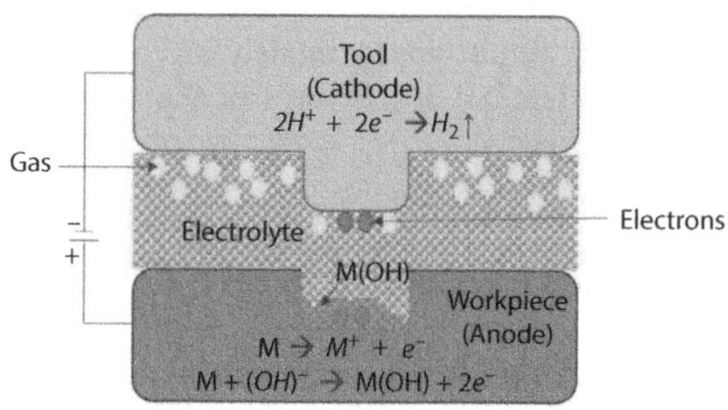

Figure 8.1 Basic working principle of electrochemical machining.

ECM technique is independent of material hardness; however, the dissolution rate is dependent on machining time, the valency of generated ion, machining current and atomic weight of workpiece metal. As a result, the numerous ECM process parameters include applied voltage, electrolyte type, machining current, concentration, inter-electrode, gap, and flow rate, etc. [1]. In recent years, ECM has received a lot of attention in the fabrication of micro parts due to its ability to solve existing problems. When the ECM process is used in microscopic machining, it is referred to as electrochemical micromachining (EMM) [2].

In electrically conductive materials of any physical characteristics, EMM can remove material of highly complicated geometries with high precision and accuracy without altering the size and shape of the tool or creating a heat-affected zone on the final machined micro product.

8.3 Importance of Electrochemical Micromachining

The term micromachining refers to the removal of material in micron levels between 1 μm and 999 μm. Chemical machining and electrochemical machining are not thermal oriented when compared to other nonconventional machining process, but chemical machining offers a difficulty to control the process in the micromachining domain. Electrochemical micromachining (EMM) has an advantage of better precision control, flexible, high machining rate, reliable and environmentally acceptable. EMM is used in machining of chemically resistant material like stainless steels, super alloys, copper alloys and titanium. EMM traces wide applications in the areas of electronic industries, sensors, microelectron mechanical systems (MEMS) and biomedical applications [4, 5]. EMM has a higher rate of controllable machining and as well as have high precision. Electrochemical micro machining provides the safer environment as it produces a sludge of neutral salts during electrochemical dissolution. Here in EMM, an interelectrode gap of order 20 μm approximately are maintained to modulate the process parameters. So, it has most of the applications in micro machining of metallic parts.

8.4 Fundamentals of Electrochemical Micromachining

EMM operates based on Faraday's law of electrolysis. The negative shape of the micro tool electrode forms the final shape of the workpiece approximately. As the interelectrode gap decreases the shape accuracy increases.

8.4.1 Electrochemistry of Electrochemical Micromachining

In the EMM process, the workpiece acts as the anode and the micro tool serves as the cathode, which are separated by a constant-flowing electrolyte. Many electrochemical processes occur at the cathode, the anode, and the electrolyte. Oxidation potential will occur at the cathode and anode. The factors that increase the oxidation potential depends on nature of metal being machined.

a) Nature of electrolyte
b) Electrolyte temperature
c) Current density
d) Type of materials

Reactions
Cathode reactions
As shown in Figure 8.1,

Reaction causing hydrogen gas at the cathode.

$$2H^{+} + 2e^{-} \dashrightarrow H_2\uparrow \text{ (acidic electrolyte)}$$

$$2H_2O + 6e^{-} \rightarrow 2\,H_2\uparrow + O_2\uparrow \text{ (alkaline lectrolyte)}$$

Neutralisation of positive charged metal ions

$$M^{+} + e^{-} \rightarrow \mathrm{M}$$

Anodic reactions
Reaction leading to the evolution of oxygen or hydrogen gases.

$$2H_2O \rightarrow O_2\uparrow + 4H^{+} + 4e^{-} \text{ (acidic electrolyte)}$$

$$4(OH)^{-} \rightarrow 2H_2O + O_2\uparrow + 4e^{-} \text{ (alkaline electrolyte)}$$

Dissolution of metal ions

$$\mathrm{M} \rightarrow M^{+} + e^{-}$$

$$\mathrm{M} + (OH)^{-} \rightarrow \mathrm{M(OH)} + 2e^{-}$$

8.4.2 Mechanism of Material Removal

Faraday introduced two fundamentals' laws, which govern their phenomena of electrolysis. Those are as follows:

a) The rate of deposited or dissolved is directly proportional to the quantity of electricity that is transmitted to the electrolyte.
b) The amount of material that is deposited or dissolved also depends on the substance's electrochemical equivalence (ECE), which is a ratio of atomic weight to valency.

Material removal rate (MRR) is given by in Eq. (8.1),

$$Q_{th} = \frac{Ita}{vF} \tag{8.1}$$

Where, I - Current supply
a - atomic weight of the elements
v - valence of metal elements
t - machining time
F- Faraday's constant.

8.5 Major Factors of EMM

The parameters which influence the EMM [6] are as follows,

i. Nature of power supply
ii. Inter electrode gap
iii. Temperature, concentration, and electrolyte flow

8.5.1 Nature of Power Supply

The EMM involves voltage in the range of 1 to 10 Volts. The ideal current density for the EMM process is 100 A/cm^2. Although electrolyte can partially eliminate this contamination, this causes the interelectrode gap to become contaminated with reaction products. As a result, the contamination builds up on the microtool, changing its form and impairing material removal from the workpiece. Additionally, changes in electrical resistance, temperature rise, and electrolyte composition all have an impact on machining precision [7]. Due to pulse current DC voltage, the interelectrode gap can be effectively cleansed during the current intervals. The duration of the time pulse ought to be sufficient to guarantee the successful completion of the electrolyte's removal from the small end gap, hence the current efficiency is determined by the current density. When the current density is decreased, this current efficiency drops rapidly in pulsed DC voltage, which is much greater than in continuous DC voltage, where the drop in current efficiency is slow. When the current density is decreased, this current efficiency decreases. As a result of this, the precision of the workpiece can be improved when it is subjected to pulsed DC voltage.

8.5.2 Interelectrode Gap (IEG)

In the micromachining process, the interelectrode gap plays a critical role in stable metal removal. However, the interelectrode gap in micromachining is measured in microns. The primary factor, which defines the accuracy and precision of micromachining is the gap width. The interelectrode spacing should be kept between 15 and 20 μm to help with excellent precision and surface smoothness. When IEG and tool location are monitored in pulsed EMM systems, the gap width is significantly reduced during the pulse-off period. The magnitude of the interelectrode gap directly affects the accuracy of machining.

8.5.3 Temperature, Concentration, and Electrolyte Flow

Working material is properly dissolved during the EMM process when the correct temperature concentration is maintained on the electrolyte flow.

a) Temperature and pressure
The temperature of the electrolyte at the IEG's entrance and exit must be monitored. If the temperature rises, it indicates a decrease in specific resistance. Insoluble sludge material can sometimes increase the resistance between the workpiece and the micro tool. This will have a direct impact on the flow rate or pressure. As a result, as the temperature rises, the electrode reaction accelerates, and the voltage requirement decreases.

b) Concentration
The resistance between the interelectrode gaps decreases as the electrolyte concentration increases. However, raising the concentration increases the possibility of salt crystallization, which inhibits the passage of current

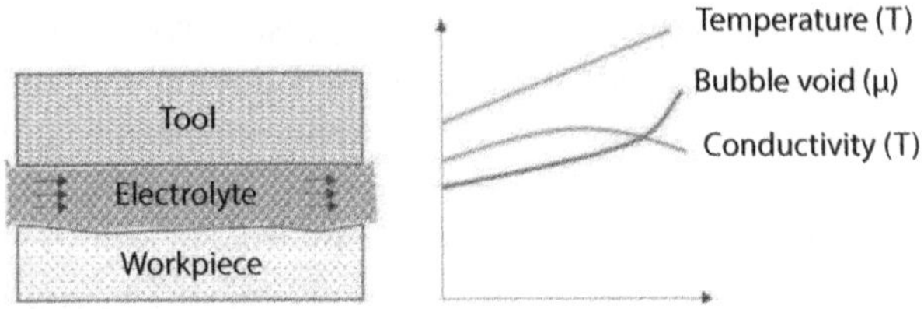

Figure 8.2 Electrolyte characteristics vary by machining length.

between the interelectrode gap. As a result, dilute electrolytes are used. The machining precision can be improved by dilute electrolyte and EGG.

c) Electrolyte flow

The electrolyte flows through the interelectrode gap between tool and workpiece as shown in Figure 8.2. There are different delivery systems for electrolyte such as electrolytic jet, slotted jet, channel flow and multi nozzle systems. Hydrodynamics of the electrolyte flow and cavitation processes are altered using a jet to create the required shape forms. In IEG, laminar electrolyte flow is the better option for improving surface quality.

d) Microtool feed rate

The inter electrode spacing should be kept consistent during the EMM process for greater machining accuracy and precision. In order to prevent a short circuit from happening during machining, the micro tool feed rate should be selected. For this, this was always kept equal to linear MRR. Minimizing sparks or short circuits between the micro tool and workpiece determines the maximum micro tool feed rate for a certain machining condition [8].

8.6 Jet Electrochemical Micromachining

The Jet EMM process is considered for this chapter from the various classification as belonging to the category of maskless EMM since there will not be any photo resist mask on the surface of the material. The classification of electrochemical micromachining process in Figure 8.3. Here, controlled localised metal dissolution is feasible. Electrolyte jets are significant in jet EMM. The steam containing the electrolyte passes through the nozzle, which in turn produces the cathode as shown in Figure 8.4.

The electrolyte jet assembly as shown in Figure 8.5, it consists of nozzle and the diameter of the nozzle is 50 to 200 μm. Electrolyte is pumped out

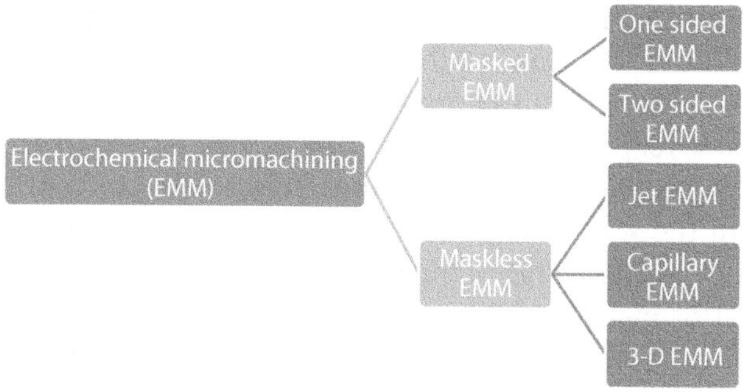

Figure 8.3 Classification of electrochemical micromachining process.

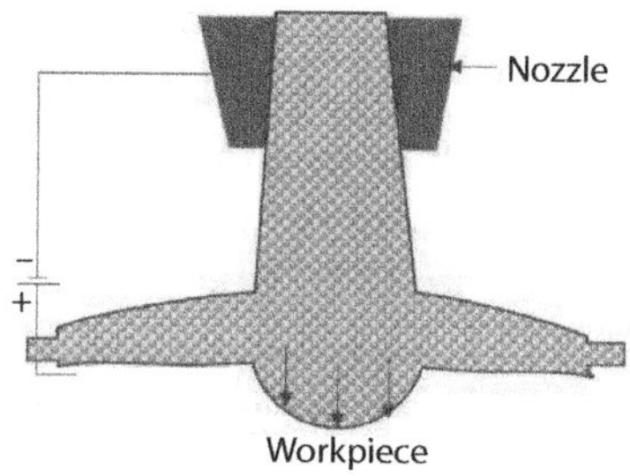

Figure 8.4 Nozzle jet assembly setup.

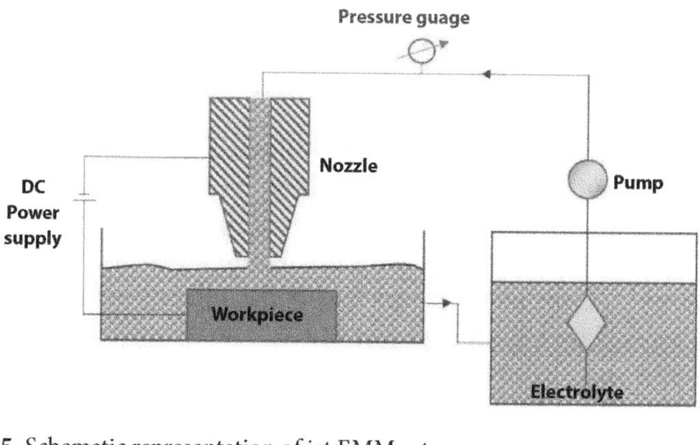

Figure 8.5 Schematic representation of jet EMM setup.

after passing through a flow metre, pressure gauge, filter, pH metre, and flow control valve. As it moves through the nozzle, this flow maintains a high pressure throughout the process. It moves across the surface of the workpiece at a very fast speed. The nozzle can be moved in a z-direction because it is flexible. Therefore, the inter electrode gap continues to be preserved all the way through the machining process. The high voltage power supply with a voltage range of 500 volts is used in the jet EMM. The removal of metal and the precision of the profile are both affected by factors such as nozzle diameter cell voltage, standoff distance, electrolyte type, pressure, and concentration. Jet EMM can cut tougher, thinner materials.

8.7 Laser as Assisting Process

In the case of micromachining, laser has been adaptively partnered with other machining processes such as vibration, electric field, external magnetic field, and fluids to increase quality and characteristics. Laser improves MRR when utilised as a secondary technique in conjunction with ECM. Both techniques are used in laser-assisted ECM to overcome their individual limitations. Whereas ECM has limited precision, thermal stress causes LBM defects. LAECMMM makes use of photon energy, whereas ECM makes use of ion energy. By using an electrolyte jet coincident with the laser beam during laser drilling, the thickness of the recast layer (the layer which is formed after the solidification, when the laser welding is done) is reduced. LAJECM can quickly and precisely produce micromachined holes. For noncontact machining applications like LAJECM, the ECM technique, which employs an electrolyte jet and a laser beam, is suitable because it can melt a small area without generating thermal stress or surface defects.

Pajak [10] utilised an electrolyte and a laser beam that emerged from a nozzle with a 0.1 mm diameter. By combining the energies of the laser beam, electrolyte jet, and flow in this case, electrochemical dissolution is accomplished. Separate studies on the temperature distribution in the workpiece and IEG found that the electrolyte temperature in various metal sections was 1.75 to 3.0 times greater in the laser-heated zone. Wang [11] came up with an ECM method that uses both a laser and a shaped tube. In this technique, a hole in a specifically designed tubular electrode directs the electrolyte jet and laser beam to the machining zone. The laser beam must first pass through a sequence of reflections within this tubular electrode before it can reach the machining zone. The aspect ratio, precision and MRR of the holes are produced by this approach, which is significantly higher than those produced by standard ECM methods.

8.8 Laser-Assisted Jet Electrochemical Micromachining (LA-JECM)

8.8.1 Working Principles of LAJECM

Energy from ions (as in ECM process) and energy from photons (as in laser beam process) are combined concurrently by LAJECM. An electrolyte jet and laser are coupled to help with electrochemical dissolution at a particular material surface. The jet of electrolyte and the laser beam are oriented coaxially, which results in a tool-electrode that does not require physical contact. The primary method of material removal is electrochemical dissolution, which also is assisted by the simultaneous action of a low power laser beam. The laser utilizes its thermal activation to facilitate disintegration rather than remove any material. The kinetics of electrochemical reactions are sped up by the application of thermal energy, which results in a quicker dissolution. Additionally, it helps dissolve the oxide layer that is present on some materials in some electrolytes and prevents effective dissolving. The basics of hybrid LAJECM are shown in Figure 8.6.

As a result of the direct contact that the electrolytic jet has with the surface of the workpiece, the material starts to dissolve from a significant

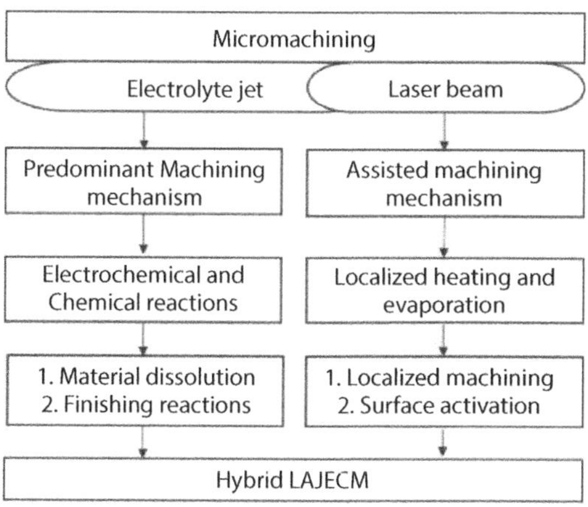

Figure 8.6 Electrolyte jet and laser functions in LAJECM.

area of the surface. On the other hand, the rate of the process of dissolution is sped up at the region where the laser beam and the electrolytic jet are in contact with one another. The main purpose of using lasers in electrochemistry is to help and direct the electrochemical energy, which allows rapid electrochemical dissolution. Electrochemical dissolution is by far the most common method of material removal. This method is aided by thermal energy from a laser that is delivered to the workpiece, and it is the most prevalent method. This electrolytic jet also contributes to the process of clearing away the reaction products from the inter-electrode gap where it is located.

The laser beam may be readily focused on the surface of the workpiece, which is one of the benefits of the LAJECM technique. As a result, dissolution can be expedited in any direction that is desired, in combination with the cleansing electrolyte jet. Because it reduces the amount of stray machining action, the so-called "localization effect" improves accuracy [10].

When it comes to machining holes or slots, LAJECM does a more thorough job of removing material along the axis than it does in the radial direction as shown in Figure 8.7. According to Arrhenius law, the electrolyte temperature rises in the machining region directly under the focused laser spot, increasing electrolyte conductivity and resulting in increased current and current density, which promote the reaction in the localised area [7]. Second, because an increase in temperature reduces the amount of energy required to activate a process, electrochemical reactions can be started more easily. Decreased electrode polarisation potential improves the diffusion process for transporting reaction products and allows for

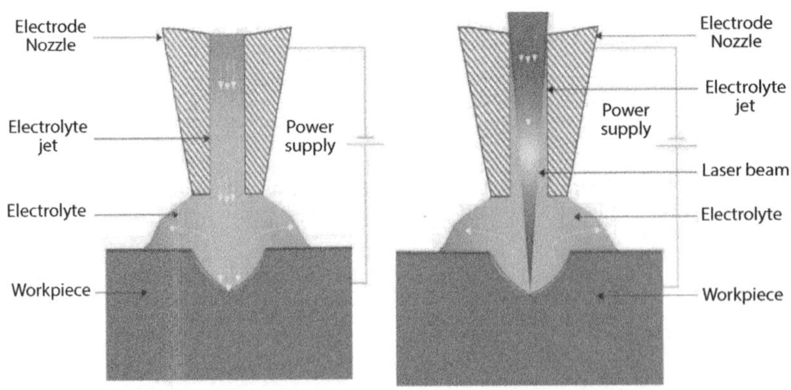

Figure 8.7 (a) Jet-ECM (b) LAJECM.

larger current densities. Reduced electrode polarisation potential improves the diffusion process for transporting reaction products and permits for larger current densities.

However, LAJECM itself has some drawbacks. Fundamentally, it is necessary to keep the laser beam coaxial with the jet of electrolyte and always focused on a single point on the workpiece. The hydrodynamic characteristics of electrolytic jet and bubbles formation at the cathode are difficult factors in LAJECM. As increase the turbulence in the electrolyte due to gas formation at the cathode, which causes the laser-jet position to shift. Flow variations can cause a jet to no longer be coaxial with the laser beam. Due to localization impact not occurring at the required position, the process has to be realigned. However, it is difficult to focus the required spot size when laser beam is subjected to scattering and diffraction.

In order for LAJECM to be effective, the properties of workpiece, such as its electrical conductivity, absorption and thermal conductivity, must be present. When used to materials with strong electrical conductivity, a high absorption factor, and low thermal conductivity, the LAJECM method performs well.

8.8.2 Mechanism of Material Removal

The primary material removal methods of LA-JECM are shown in Figure 8.8. The workpiece's surface is greatly affected by the laser beam's effect on it in terms of both chemical reactions. The energy of the incident laser beam reaches the workpiece as a second source of heat [12].

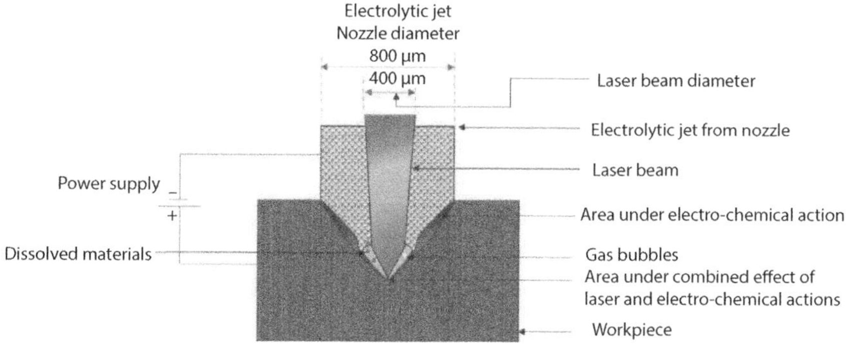

Figure 8.8 LAJECM process.

As a result, the workpiece's surface where the laser beam energy hits it becomes thermally activated. This increases the current density in the affected area and speeds up the dissolving of metallic materials. The localization effect caused by the laser beam has an impact on electrochemical machining.

Figures 8.9 and 8.10, respectively, describe the localisation effect and the processes of material dissolution in the LAJECM. The following reasons are provided below to support that [12]:

1. The laser pulse hits the outside workpiece surface, transmitting heat energy and raising the temperature (T_1). The current density (J_1) increases as the temperature rises. In Arrhenius' law, a greater current density results in a faster rate of electrochemical dissolution as well as material removal.

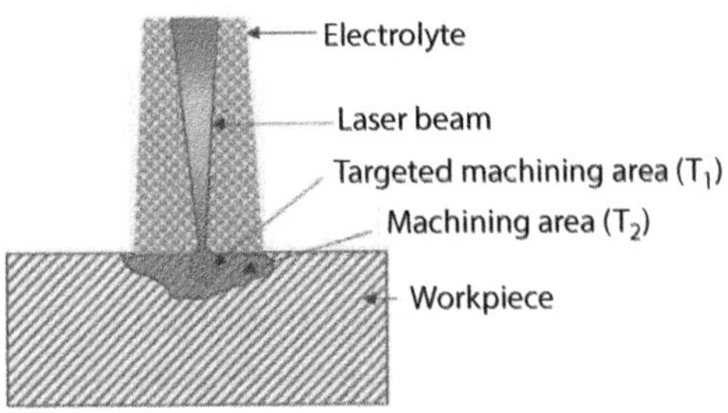

Figure 8.9 Localization effect in LA-JECM.

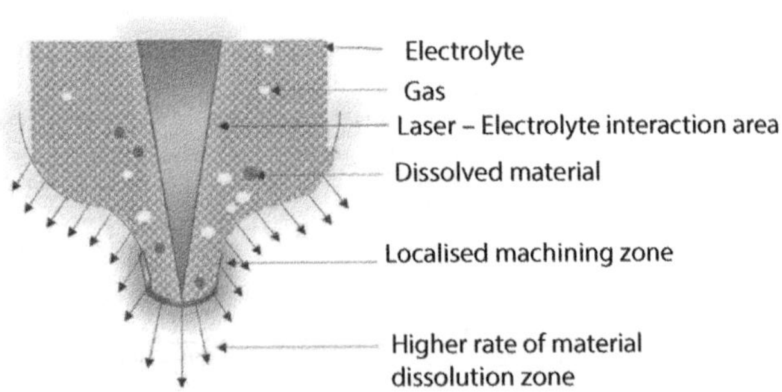

Figure 8.10 Localized machining zone in LAJECM.

2. Laser source is raising the electrolyte temperature (T_1) and improve the conductivity (K_1). High current is produced by increased conductivity, which raises the density of the current. Rise in current density speeds up electrochemical processes and enhances material removal rate.

$$T_1 > T_2$$

$$J_1 > J_2$$

$$K_1 > K_2$$

$$Ea_1 > Ea_2$$

3. The required of activation energy (Ea_1) for electrochemical processes is decrease as temperature rises with the aid of a laser beam, resulting in the quick initiation of electrochemical reactions.
4. Due to diffusion in the machining area, the temperature rises at machining zone.

LA-JECM has greater machining rates than JECM because of the laser beam's effect on localization. In addition to increasing the quality and accuracy of machining, laser beam aided JECM also allows for more efficient machining in some regions. Due to the higher density of current in the machining zone, the LAJECM procedure is more precise than the JECM method. The laser exposed the region with a larger electric field as a result, greatly reducing the stray machining impact. Since more material is removed in the longitudinal direction than in the lateral direction, the taper is consequently reduced.

8.8.3 Materials

The electrical conductivity, thermal conductivity, absorption factor, and oxidation tendency are significant material properties for LAJECM process. The material properties of few materials are given in Table 8.1.

Table 8.1 Significant material properties for LAJECM process [13].

Materials	Absorption factor (At 20°C)	Electrical conductivity [$m\Omega^{-1}$ cm^{-1}]	Thermal conductivity [$W\ m^{-1}\ K^{-1}$] (At 25°C)
Stainless Steel	0.8–0.9	11.4–13	14–17
Aluminium Alloy	0.03–0.1	217.3–222.2	144–150
Hastelloy C276	0.88–0.95	8.26–8.70	10.4–11.2
High Carbon Steel	0.8–0.9	50–58.8	47–53
Titanium Alloy	0.3–0.5	5.88–5.95	7.1–7.3

8.8.4 Theoretical and Experimental Method for Process Energy Distribution

Assuming that electrochemical reactions of dissolution are carried out by the energy obtained through the workpiece. The heat energy from the laser causes the temperature to rise, which in turn speeds up the reaction [14]. However, this also causes an increase in the standard enthalpy, which effects the change in thermodynamic potential [10]. Finally, it decreases the activation energy of the process. In Figure 8.11, the energy distribution in the LAJECM is depicted graphically.

Assuming an electric field that is constant in space and time, the E_1 component represents the ions energy (ECM process).

$$E_1 = V\int_0^{t_f} J(t)A(t)dt \tag{8.2}$$

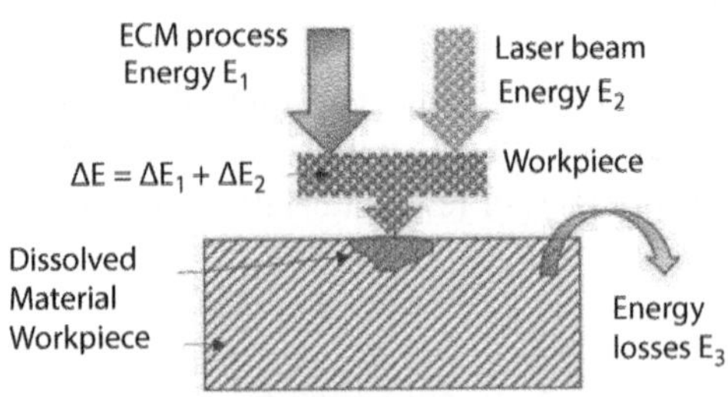

Figure 8.11 Mechanism for energy distribution.

Gaussian distribution of laser beam energy in a single mode makes up the E_2 component.

$$E_2 = \beta \int_0^{t_f} \left\{ \int_0^{\infty} qexp(-kr^2) 2\pi r \, dr \right. \tag{8.3}$$

E_3 component is thermal energy that the electrolyte convectively absorbs from the workpiece.

$$E_3 = \int_0^{t_f} \alpha \Delta T \left(\iint_A dA \right) dt \tag{8.4}$$

Where,

V- Voltage supply
J- Current density
β- workpiece heat absorption factor
A- Area under machining
ΔT- Temperature difference between an electrolyte layer and material surface
q- Laser power distribution (single mode)
t- Time
t_f- Final time
r_s- Laser spot radius
k- Gaussian distribution factor ($k=r_s^{-2}$)
α- convection coefficient.
r-Laser dot radius

The rate of a reaction depends on the energy ΔE (ΔE = ΔE1 + ΔE2 – ΔE3) that is provided to the process.

$$\Delta E = \Delta U = \Delta H$$

$$\Delta H \Rightarrow \Delta G \Rightarrow \Delta E_a$$

where ΔE - Total Energy
ΔU - Change of internal energy in workpiece
ΔH - Change of enthalpy
ΔG - Change of Gibbs' energy
ΔE_a - Activation energy

Convection from the workpiece raises the temperature of the electrolyte. Newton's law simplifies the convection component even when electrolyte cleanses the workpiece through forced turbulent flow. When the appropriate convection coefficient (α) is included, analysis of the electrolyte turbulent flow proves that Newton's Law accurately predicts electrolyte temperature growth. The E3 energy component plays an important role in increasing the electrolyte temperature, which increases its conductivity even if it is lost from the aspect of the workpiece.

The equation of Arrhenius's Law which relates the direct process parameters such as activation energy and current density is given in Eq. (8.5). Again, a higher absolute amount of activation energy and higher temperature lead to a higher current density, which speeds up the dissolving process.

$$J = C_a K_R \exp \frac{E_a}{RT} \tag{8.5}$$

Where,

J- Current density

K_R- Reductor concentration

C_a- Anode process constant.

8.8.5 LAJECM Process Temperature

Locally increasing the machining zone temperature improves LAJECM's efficiency and accuracy. In a different study, FEA modelling is used to investigate the distribution of temperature in the workpiece and Interelectrode gap for localised heating [15]. The laser localized heating zone in different materials, the temperature is 1.75 to 3.25 % higher than the electrolyte temperature as shown in Figure 8.12. Temperature rises in a localised are is influenced by the thermal conductivity coefficient of the material. The formation of bubbles in the electrolyte is caused by temperatures higher than 100°C. A higher MRR is the direct result of a higher temperature in the localised area. Generally, the higher material removal rate and precision may be accomplished the higher temperature at the localised area.

8.8.6 Material Removal Rate and Taper Angle

The Material removal rate is determined by comparing the sample's weight before and during the machining process. The material removal rate (MRR), expressed in milligrams per minute, is calculated using the

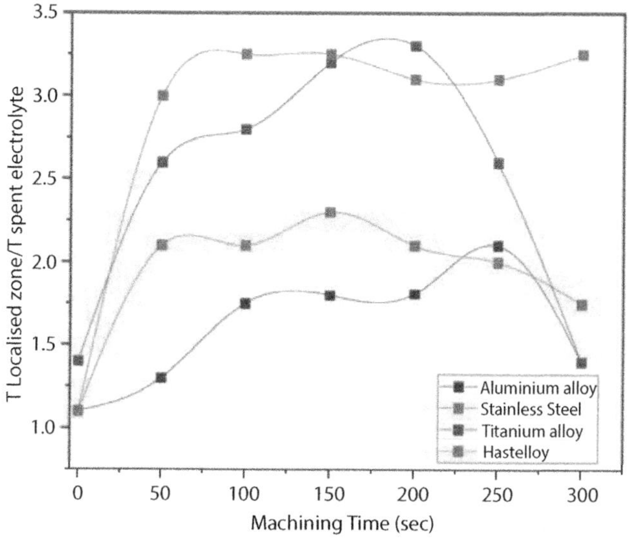

Figure 8.12 Electrolyte temperature and laser-localized area temperature as a function of machining time [15].

relationship shown in Eq. (8.8). From the collected reading, taper angle is determined [12] based on the connection described in Eq. (8.7).

$$\text{Material Removal Rate(MRR)} = \frac{W_b - W_a}{T_m} \tag{8.6}$$

$$\text{Taper Angle(TA)} \tan\theta = \frac{D_T - D_B}{2t} \tag{8.7}$$

Where,

T_m- Machining Time
W_a- Weight of the sample after micromachining
t- sample thickness
W_b- Weight of the sample before micromachining
D_T- diameter of machined hole at top cross section
D_B- diameter of machined hole at bottom at bottom cross section.

8.8.7 LAJECM and JECM Comparison

It can be seen quite clearly from Figure 8.13 that the rate of material removal accelerates with increasing electrolyte concentration. The concentration of

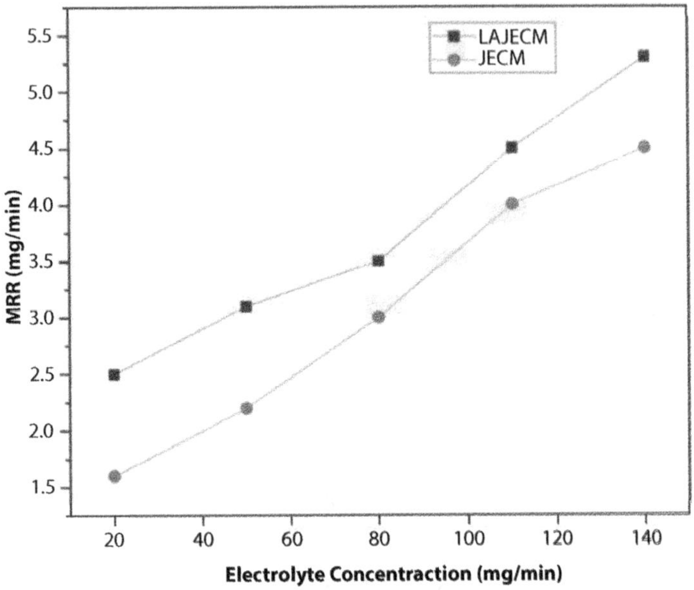

Figure 8.13 Influences of electrolyte concentration of LAJECM on MRR [12].

the electrolyte significantly affects the rate at which material is removed from the workpiece in both the LAJECM and the JECM machining processes. The number of charge carriers grows as electrolyte concentration rises, speeding up the material removal rate. Figure 8.13 makes it abundantly evident that the laser's aid speeds up electrochemical reactions, which speeds up material removal.

It can be seen rather clearly from Figure 8.14 that the taper gets steeper as the duty cycle gets higher. As the duty cycle is increased, the current density in the machining region also increases. This leads to an increase in the rate of dissolution as the length of the dissolving process also increases. The effect of laser assistance is far more significant at low duty cycles than at high duty cycles, as seen in Figure 8.4. Therefore, LA-JECM machining at low duty cycles is advised.

8.8.8 Machining Precision

8.8.8.1 Geometry Precision

The effectiveness of machining localization affects shape precision in LAJECM. The LAJECM taper is narrower than the JECM taper. Due to a

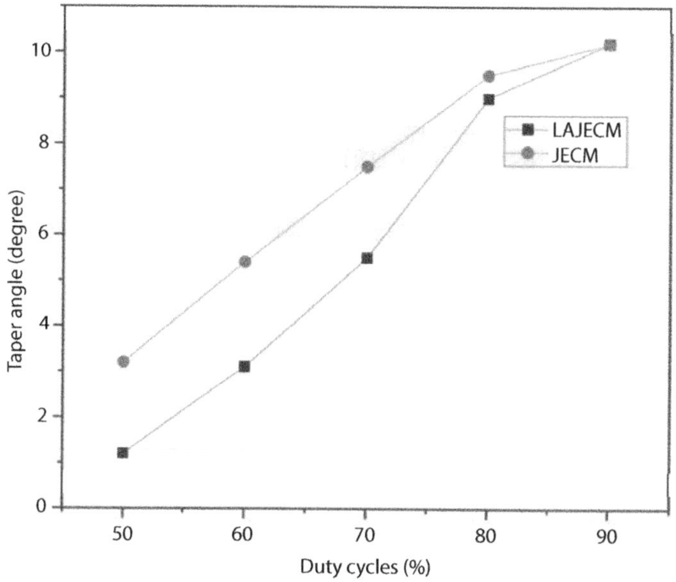

Figure 8.14 Influences of duty cycles of LAJECM on taper angle [12].

greater material removal rate, LAJECM may generate bigger holes in the same machining time [14].

$$\text{Taper} = \frac{D_T - D_B}{D_T} \tag{8.8}$$

$$\text{Taper reduction for LAJECM} = \frac{\text{Taper}_{\text{JECM}} - \text{Tapper}_{\text{LAJECM}}}{\text{Taper}_{\text{JECM}}} \tag{8.9}$$

$$\text{Overcut} = \frac{\dfrac{(D_T + D_B)}{2 - D_J}}{D_J} \tag{8.10}$$

The method of measuring the taper and the overcut is illustrated in Figure 8.15. The taper, taper reduction for LAJECM and overcut are calculated from Eq. (8.8) to (8.9). LAJECM dissolution is faster in the axial direction. However, the laser-heated electrolyte causes increased overcut when it exits the cavity by striking the cavity walls. Considering overcut,

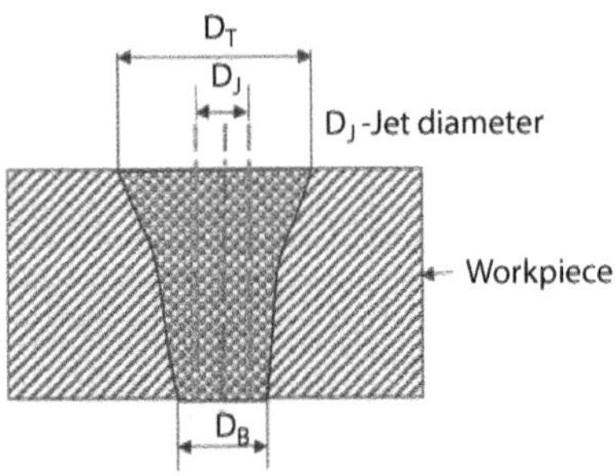

Figure 8.15 Measurements of taper and overcut.

taper and the same machining time for the process, the hole formed in LAJECM is typically larger; the entry and exit diameters are also closer, resulting in a smaller taper.

8.8.8.2 *Profile Surface Roughness*

In Figure 8.16, a variety of SEM images of surface roughness profile and cross section of hole for Titanium alloy and Hastelloy are displayed. When compared in terms of profile roughness, these two materials are very different from one another. The surface of the titanium alloy is rough, whereas the surface of the Hastelloy is electropolished, thus the Hastelloy surface is shining. The brittle oxide that develops on the titanium's surface breaks apart during the machining process, leaving sharp holes open to the flow of the electrolyte. When compared to the surface of titanium alloy, the Hastelloy material has a substantially smoother texture.

The mean values of the surface roughness (Ra) are compared for both JECM and LAJECM as shown in Figure 8.17. According to the results, the typical profile roughness in LAJECM is significantly lower than that of JECM. Localization causes more concentrated, accurate dissolution and enhanced smoothness in LAJECM.

8.9 Applications of LAJECM

Generally, the LAJECM uses to create holes and cavities in the field of automobile and aerospace applications. Due to enhanced laser beam guidance and its higher absorption, the LAJECM utilize in Microgroove shaping [8].

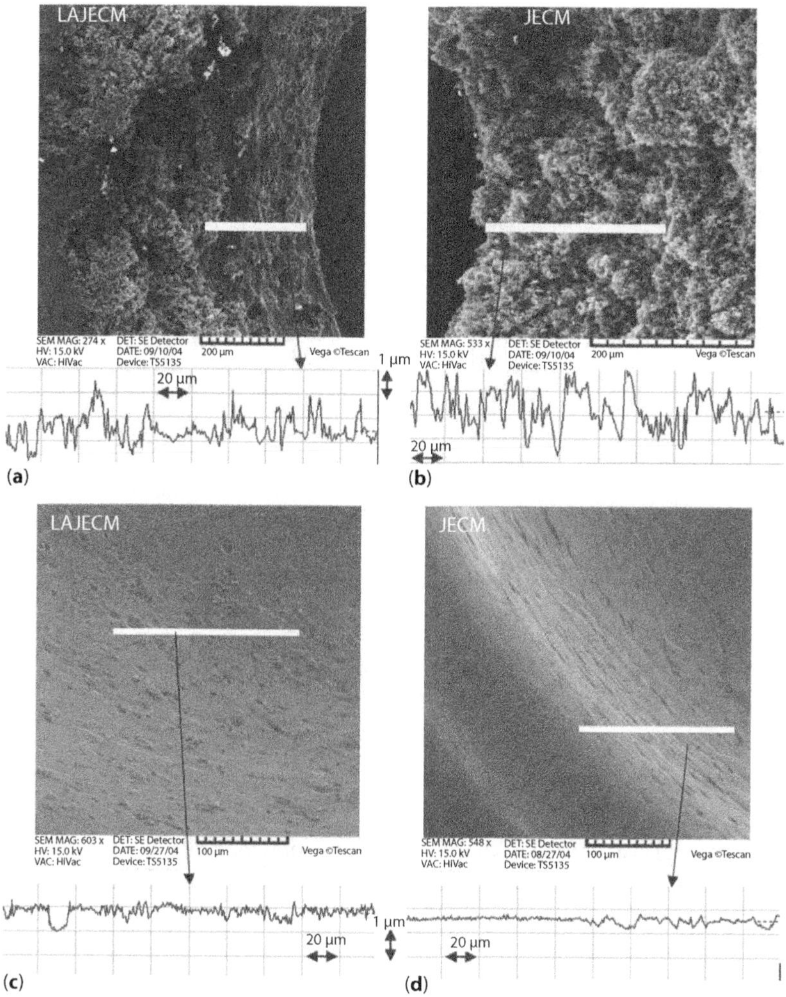

Figure 8.16 SEM analysis for cross-section views of hole and surface roughness profile (a) LAJECM for Ti alloy (b) JECM Ti alloy (c) LAJECM for Hastelloy (d) JECM for Hastelloy (Reproduced with permission) [14].

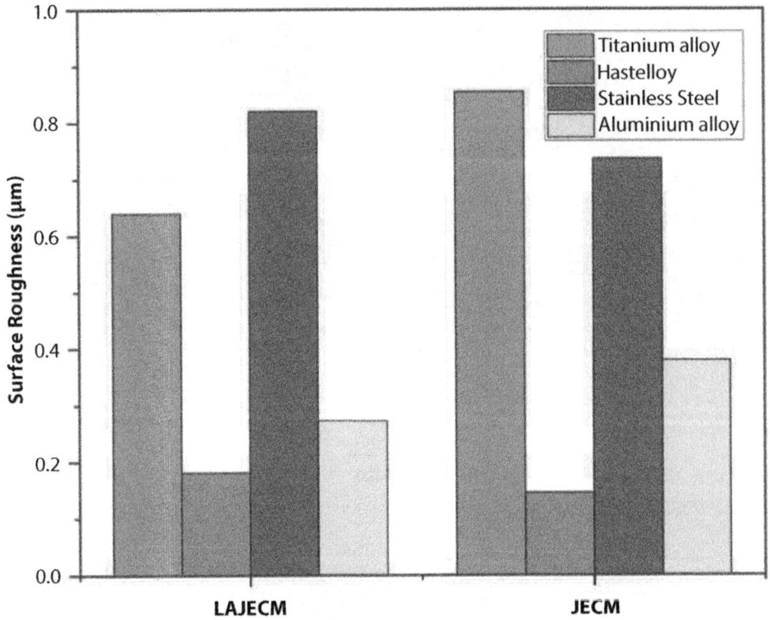

Figure 8.17 Surface roughness for titanium alloy, hastelloy, stainless steel, and aluminum alloys [14].

References

1. Bhattacharyya, B. and Sorkhel, S.K., Investigation for controlled electrochemical machining through response surface methodology-based approach. *J. Mater. Process. Technol.*, 86, 200, 1999.
2. Bhattacharyya, B., Mitra, S., Boro, A.K., Electrochemical machining: New possibilities for micromachining. *Robot. Comput. Integr. Manuf.*, 18, 283, 2002.
3. Datta, M., Microfabrication by electrochemical metal removal. *IBM J. Res. Dev.*, 42, 655, 1998.
4. Rajurkar, K.P., Zhu, D., McGeough, J.A., Kozak, J., De Silva, A., New developments in electro-chemical machining. *CIRP Ann.*, 48, 567, 1999.
5. Landolt, D., Principles of electrochemical machining. *Chemie Ing. Tech.*, 47, 996, 1975.
6. Bhattacharyya, B., Munda, J., Malapati, M., Advancement in electrochemical micro-machining. *Int. J. Mach. Tools Manuf.*, 44, 1577, 2004.
7. Kozak, J., Rajurkar, K.P., Wei, B., Modelling and analysis of pulse electrochemical machining (PECM). *J. Eng. Ind.*, 116, 316, 1994.

8. Bhattacharyya, B. and Munda, J., Experimental investigation on the influence of electrochemical machining parameters on machining rate and accuracy in micromachining domain. *Int. J. Mach. Tools Manuf.*, 43, 1301, 2003.
9. Zhang, H., Xu, J., Wang., J., Investigation of a novel hybrid process of laser drilling assisted with jet electrochemical machining. *Opt. Lasers Eng.*, 47, 1242, 2009.
10. Pajak, P.T., De Silva, A.K.M., McGeough, J.A., Harrison., D.K., Modelling the aspects of precision and efficiency in laser-assisted jet electrochemical machining (LAJECM). *J. Mater. Process. Technol.*, 149, 512, 2004.
11. Wang, Y. and Zhang, W., Theoretical and experimental study on hybrid laser and shaped tube electrochemical machining (laser-STEM) process. *Int. J. Adv. Manuf. Technol.*, 112, 1601, 2021.
12. Malik, A. and Manna, A., Multi-response optimization of laser-assisted jet electrochemical machining parameters based on gray relational analysis. *J. Braz. Soc. Mech. Sci. Eng.*, 40, 147, 2018.
13. Shackelford, F. and Alexander, W. (Eds.), *CRC Materials Science and Engineering Handbook*, CRC Press, Boca Raton, 2000.
14. Pajak, P.T., De Silva, A.K.M., Harrison, D.K., McGeough, J.A., Precision and efficiency of laser-assisted jet electrochemical machining. *Precis. Eng.*, 30, 288, 2006.
15. De Silva, A.K.M., Pajak, P.T., McGeough, J.A., Harrison, D.K., Thermal effects in laser assisted jet electrochemical machining. *CIRP Ann.*, 60, 243, 2011.

9

Ultrasonic Vibration-Assisted Microwire Electrochemical Discharge Machining

Sandip Kunar[1,2*], Kagithapu Rajendra[1,2], Devarapalli Raviteja[1,2], Norfazillah Talib[3], S. Rama Sree[2,4] and M.S. Reddy[1,2]

[1]Department of Mechanical Engineering, Aditya Engineering College, Surampalem, India
[2]Jawaharlal Nehru Technological University Kakinada, Kakinada, East Godavari, India
[3]Department of Manufacturing Engineering, Faculty of Mechanical and Manufacturing Engineering, Universiti Tun Hussein Onn Malaysia, Batu Pahat, Johor, Malaysia
[4]Department of Computer Science and Engineering, Aditya Engineering College, Surampalem, India

Abstract
The necessity for glass microstructure is rising as microelectromechanical systems (MEMSs) develop so quickly. Ultrasonic vibration is used in the microwire electrochemical discharge machining (WEDM) to attain superior quality of machining for glass microstructures. This means that by increasing the appropriate ultrasonic amplitude, the width of the gas film was decreased, lowering the critical voltage, and improving the machining stability. Then, a series of tests are performed using a micro helical electrode with 100 μm diameter to examine the influence of duty factor, ultrasonic amplitude, voltage, and frequency on the micro slit width. The investigational findings demonstrate that by including an appropriate ultrasonic amplitude and machining quality are greatly enhanced. The average slit width was decreased to 128.63 μm when the amplitude was 5.25 μm. Finally, using the ideal machining conditions, high-aspect ratio microcantilever and micro planar coil structures were successfully created on the glass plate. It has been demonstrated that microstructure with high-aspect ratio on brittle materials can be accomplished using ultrasonic vibration-assisted micro wire WEDM.

**Corresponding author*: sandip.sandip.kunar@gmail.com

Sandip Kunar, Golam Kibria, Prasenjit Chatterjee and Asma Perveen (eds.) Hybrid Micromachining and Microfabrication Technologies: Principles, Varieties and Applications, (205–218) © 2023 Scrivener Publishing LLC

Keywords: WEDM, helical electrode, ultrasonic vibration, glass microstructures, high-aspect ratio

9.1 Introduction

Due to their superior qualities, glass microstructures are becoming more and more in demand as microelectromechanical systems (MEMSs) evolve quickly. However, it is extremely challenging to machine complicated microstructure owing to the rigidity, fragility, and nonconducting qualities of glass [1–3]. A type of noncontact machining known as wire electrochemical discharge machining can overcome the rigidity, breakability, and nonconductive properties of glass [4, 5]. Numerous academics have been actively working in electrochemical discharge machining (ECDM) recently and have shown some success. By measuring and documenting the current signal, the mechanism of increasing temperature and material loss were analysed [6]. The effects of duty factor and frequency by ECDM were investigated during glass machining [7]. According to the experimental findings, thermal damage decreases as pulse frequency and duty factor increase. The microstructure of ECDM ceramics was investigated to explore the effects of tool feed rate, pulse closure and opening duration on the output performance [8]. The effect of electrolyte concentration, and field orientation was investigated on the output performance during glass micro structuring using ECDM [9]. It was revealed that the way of bubble movement was prejudiced by the direction of magnetic field. Electrochemical drilling using ultra voltage was utilized to show the impact of rotating electrode on the gap flow field [10]. ECDM is explored utilizing an ultrasonic tool electrode, studied the parameters influencing the thickness of the gas film and formed a mathematical model [11]. A list of the techniques researched and compiled for enhancing the machining rate in ECDM [12]. The high-speed rotated electrodes can significantly increase machining accuracy and efficiency by boosting electrolyte performance [13]. A new technique for rotational electrochemical etching is used to describe the etching mechanism using the helical electrode [14]. The tube electrode tool was used to distribute the reactive products in the narrow gap and the experimental results demonstrated the mechanism of elimination of products from the narrow gap [15].

WEDM, a subfield of ECDM, has received extensive scholarly study and has a significant impact on the micromachining of nonconductive and brittle materials. The global WEDM research landscape improved the WEDM machining settings [16]. Brass wire having 200 μm diameter was exploited to machine the glass fibre composite with size of machined slit ranging from 220 to 223 μm [17]. The glass is cut by diamond wire with the assistance of WEDM [18]. The coated steel wire having 0.15 mm is used to cut quartz. The effect of line speed, and voltage is investigated on slit width and material removal [19]. The machining of silica nanocomposites was investigated and deliberated the influences of wire speed and pulse width on the roughness and machining rate. There has been important advancement in the study of both WEDM and ECDM, but no study was done on the utilization of accelerated vibration in WEDM [20]. To increase the drilling depth and intake quality, ultrasonic vibration is used to directly apprise the electrolyte movement during drilling [21]. To enhance surface smoothness and machining rate during ECDM drilling, ultrasonic vibration support was implemented [22].

The creation of gas film on the helical electrode surface during WEDM method was first simulated, demonstrating that the gas film thickness was reduced by combining an appropriate ultrasonic amplitude, thereby dropping the critical voltage. Next, the machining capability was improved. Then, many studies were conducted utilizing micro helical electrode having 100 μm diameter to determine how the duty factor, feed rate pulse frequency, and ultrasonic amplitude impacted the slit width. According to the investigational findings, applying ultrasonic amplitude significantly improved the machining quality. Lastly, utilizing the standard machining conditions, high-aspect ratio microplanar coil and microcantilever structures were effectively produced on the glass plate.

9.2 Experimental Setup

In this study, micro-WEDM is subjected to ultrasonic vibration, and the experimental platform exhibits great stability, antivibration, and motion control precision. Figure 9.1 depicts the experimental platform's physical layout. The ultrasonic generator, ultrasonic motorised spindle, machine body, power unit, electrode system, frequency converter, testing system, and monitoring system make up the experimental platform. Granite, a material that can withstand corrosion and absorb vibration, serves as the machine's basis. The machine tool has a motion accuracy of 0.1μm per step, which is sufficient for micromachining. The ultrasonic generator produces the vibration, which has the range of frequency between 20,000 and 31,000 Hz and a constantly variable amplitude. The ultrasonic generator allows for measurement and controls the vibration amplitude. When cutting, the power supply serves as a crucial energy control terminal.

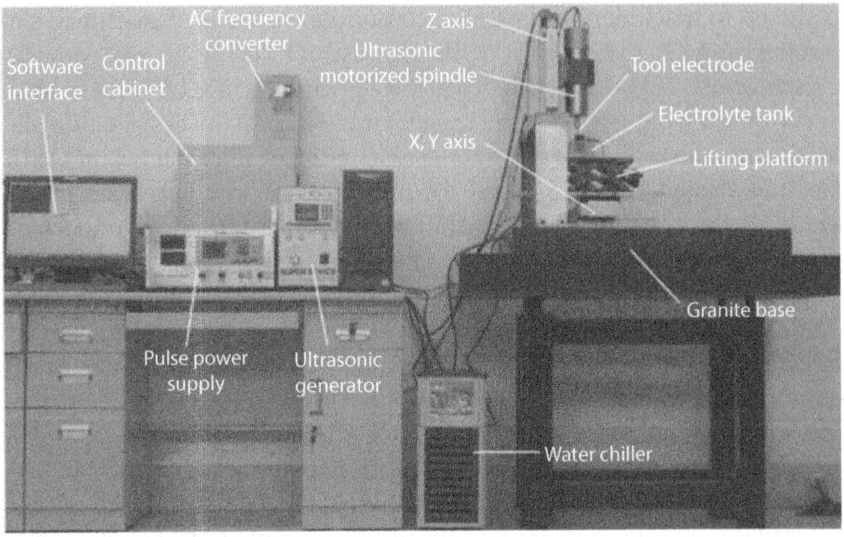

Figure 9.1 Experimental platform [23].

A rectangular pulse power unit was used in this experiment. The duty factor, voltage, and frequency are all regulated to regulate the output energy. The machining speed of helical electrode is regulated by the frequency converter to precisely regulate the spindle speed. The tool, auxiliary fixture manual lifting table, electrolyzer, and auxiliary anode graphite are the primary components of the electrode system. The tool electrode has a diameter of 0.1 mm and is made of helical electrode and tungsten wire. The circuit's Hall current sensor captures the current signal, which can then be seen online using an oscilloscope to view the cutting state of the machine. After machining, the morphology and size were measured and examined using a laser confocal microscope and electron scanning microscope.

9.3 Results and Discussion

With the aid of ultrasonic vibration, the electrode was used to conduct the investigational work on the glass using WEDM. The focus of the following research is the impact of ultrasonic vibration during generation of slit width. Analysis is done on how the cutting slit width is affected by frequency, feed rate, ultrasonic amplitude, voltage, and duty factor. Finally, utilising the optimised machining settings, the machining of glass complicated microcantilever beam structure and micro planar coil structure is finished by the selection of constant machining settings, including spindle speed, electrolyte concentration, and others.

For ECDM, the creation of a gas film is a crucial procedure. Electrolysis creates hydrogen bubbles, which gather, mix, and expand on the cathode tool's surface before forming bubble films. Another theory holds that the electrolyte will also evaporate and generate bubbles due to the high temperature that results from the tool electrode's discharge. This research focuses on the gas film created by electrolyzed hydrogen bubbles. The gas film's quality and thickness significantly influence the micro ECDM's quality and rate of material removal. The reproducibility of ECDM can also be enhanced by a steady gas film.

9.3.1 Influence of Ultrasonic Amplitude on Micro Slit Width

Figure 9.2 illustrates the difference between the machining outcome without and with ultrasonic pulsation aid. It is revealed that the ultrasonic vibration decreases the diameter of the micro slit. The effect of various amplitudes on slit width is investigated considering the advantageous effects of ultrasonic vibration. Only the amplitude is altered in the experiment with other constant machining parameters i.e., spindle speed of 3000 r/min, feed rate of 1 μm/s, frequency of 1000 Hz, voltage of 35 V, and duty ratio of 70% and the amplitude is 7 μm at its highest. Figure 9.3 displays the average slit width for various amplitudes. When the ultrasonic pulsation is not applied until the amplitude raises up to 5.25 μm, the average slit

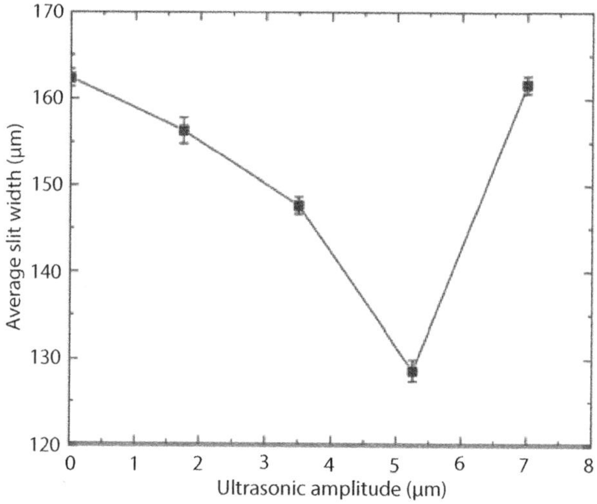

Figure 9.2 Effect of amplitude on micro slit [23].

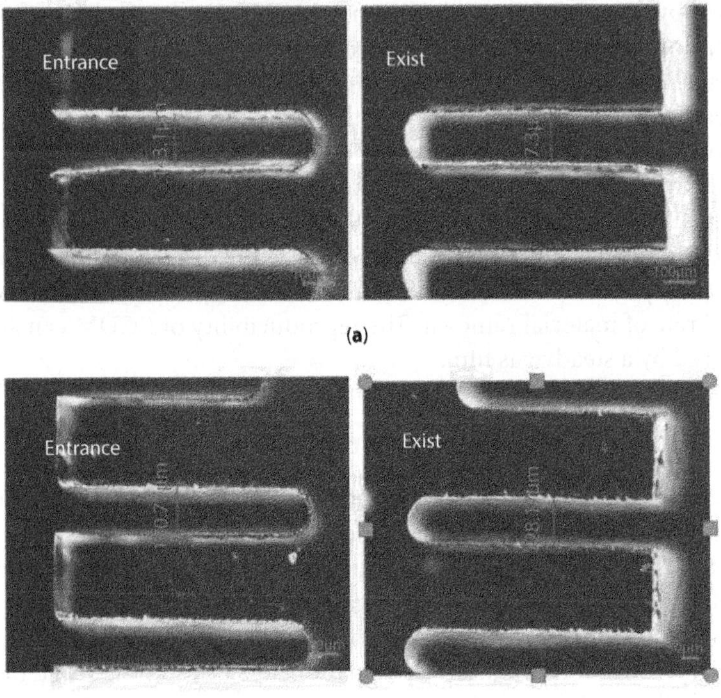

Figure 9.3 Micro slits (a) without ultrasonic vibration and (b) with ultrasonic vibration [23].

width steadily declines from 162.38 to 128.63 μm. The slit width grows to 161.68 μm when the amplitude exceeds 5.25 μm and rises to 7 μm. This demonstrates that vibration decreases the slit width after machining in the slight amplitude range, and it enhances the slit width when the amplitude surpasses the critical value.

The impact of various amplitudes on the slit width after machining reveals that the ultrasonic amplitude does not necessarily correlate with higher machining quality, but rather can enhance it within a given range. When the amplitude is zero, it is revealed that the number of single pulse spark discharges is significantly lower than ultrasonic-assisted sparks. Additionally, the highest current of a spark discharge with ultrasonic vibration is greater than it is without it. The gas sheet can be made uniformly thin by applying ultrasonic pulsation to the tool electrode. The critical voltage of the breakdown film lowers as the thickness of the gas film does, and consequently, so does the highest current of spark discharge. A greater frequency and more stable discharge are made possible by a superior film condition. The narrower slit with greater dimensional accuracy is then machined.

The peak current progressively declines as the amplitude rises from 1.75 μm to 5.25 μm. The layer thickness reduces as the amplitude rises. Both the peak discharge current and the related critical discharge voltage fall. Thus, the cutting slit width is decreased in the modest energy and high-density spark discharge. However, the slit width grows as the amplitude rises from 5.25 μm to 7 μm. The creation of the gas film will degrade with increasing amplitude, as seen by the current waveforms with amplitudes of 5.25 μm and 7 μm, lengthening the time until the gas film first forms.

9.3.2 Influence of Voltage on Micro Slit Width

In WEDM, machining voltage is a crucial factor. To fuel the many bubbles produced by electrolysis, pressure is required on the one hand, and voltage is required to fuel the spark discharge caused by the breaking of the gas film. The single variable technique is utilised to conduct the investigation to examine how pulse voltage affects the slit width after machining. Each workpiece in the experiment has a thickness of 300 μm and is composed of quartz glass. The parameters are used in the experiments i.e., frequency of 1000 Hz, voltage of 33 V to 38 V, duty factor of 70%, spindle speed of 3000 rpm, tool electrode amplitude of 5.25 μm, and feed rate of 1 μm/s.

Figure 9.4 displays the average slit width for various machining voltages. In the experiment, there are two groups: one that receives support

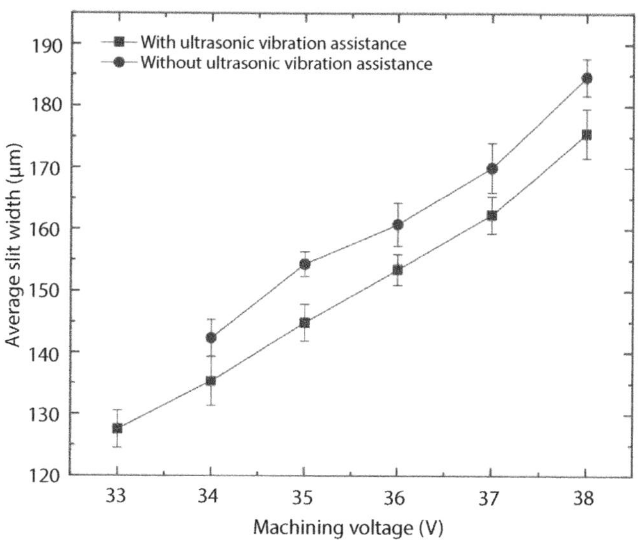

Figure 9.4 Influence of voltage on micro slit [23].

from ultrasonic vibration, and the other that does not. Whether or not there is ultrasonic aid, the average slit width grows as the voltage increases. The outcomes demonstrate that, after employing ultrasonic vibration to the electrode, the average slit width may be decreased; the drop is near to 10 μm. In addition, the lowest voltage that can be machined without ultrasonic vibration is 34 V, whereas the lowest voltage that can be machined with ultrasonic vibration is 1 V lower. The critical value of WEDM can be lowered with the aid of ultrasonic vibration. Even though there is just a 1 V reduction in the lowest machinable voltage, the quality of the machining will be higher if less discharge energy is utilized to eliminate the material.

9.3.3 Effect of Duty Ratio on Micro Slit Width

The duty factor, frequency, and voltage serve as the primary heat sources that determine how much material is removed from WEDM during discharge. Experiments were conducted on glass with a thickness of 300 μm at a fixed 1000 Hz frequency to investigate the impact of the duty ratio on the slit width. Figure 9.5 illustrates the average slit width after machining under various duty ratios.

Due to the ideal conditions that an increase in duty factor will create to produce a gas film and an increase in spark discharges, a single pulse's discharge times will increase dramatically. Since spark discharge produces heat for only a very brief period, cutting glass is mostly done to eliminate

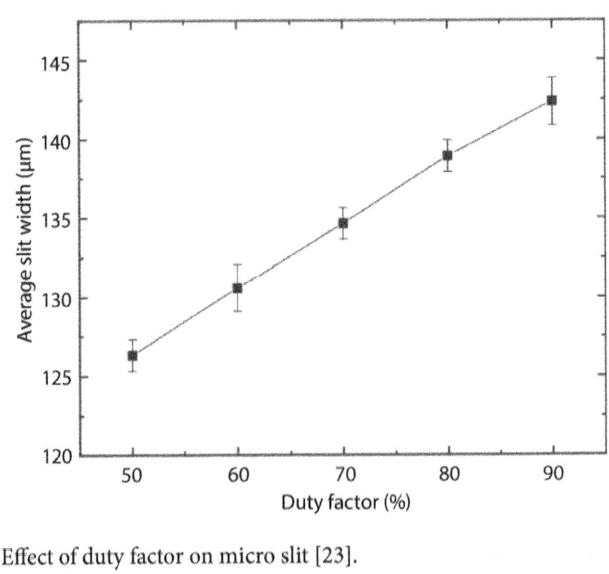

Figure 9.5 Effect of duty factor on micro slit [23].

that heat. The heat transfer to the glass surface is incredibly restricted since it dissipates in the surrounding electrolyte in such a brief period. Therefore, a single pulse must have adequate discharge periods to deliver heat to the workpiece constantly. Consequently, the cutting machining has a minimum duty factor. In this experiment, if the duty ratio is less than 50%, the tool will immediately break. When duty ratio is considered for its impact on slit width, it can be determined that a duty ratio shouldn't be too high to prevent the slit width from growing. If the duty ratio is too low, the tool electrodes may fracture because not enough heat can build up to constantly remove the workpiece material.

9.3.4 Influence of Frequency on Slit Width

It has been revealed via numerous studies that regulating the length and width of a single pulse allows one to achieve the influence of duty factor and frequency on machining condition. The duty factor in the experiment is set at 70% with the intention of examining the impact of pulse frequency. Glass is 300μm thick with parameters i.e., feed rate of 1 μ/s, amplitude of 5.25 μm, spindle speed of 3000 rpm, and voltage of 35 V. Considering that the easier the machining is, the greater the matching pulse period is, the lower the frequency. As a result, the regular cutting cannot be done because the frequency range starts at 500 Hz and enhances by 500 Hz until it extends 2500 Hz. Figure 9.6 depicts the average slit width at various frequencies in accordance with the experimental investigation.

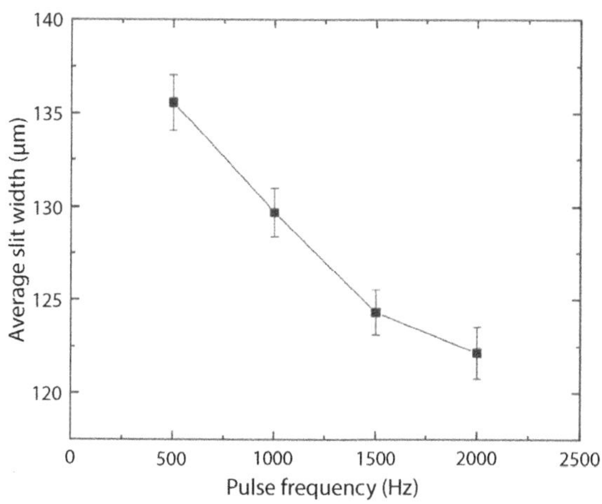

Figure 9.6 Effect of frequency on micro slit [23].

9.3.5 Analysis of Micro Slits

The primary machining parameters are designed to carry out the usual structural cutting considering different parameters affect the slit width. After parameter adjustment, 300 μm thick glass was used for the micro array slit cutting with the following parameters: 34 V voltage, 5.25 μm amplitude, 70% duty factor, 1000 Hz frequency, 1 m/s feed rate, and 3000 rpm spindle speed. The entrance slit diameter of the intricate micro planar coil construction, as depicted in Figure 9.7, is 124.1 μm.

To validate the machining capability using ultra-sonic vibration-assisted micro wire WEDM, the microcantilever structure is generated with greater aspect ratio, as revealed in Figure 9.8. The structure has a span of 1500 μm and an aspect ratio of 42, as displayed in Figure 9.8(a), and the structure has a span of 1194 μm and aspect ratio of 13 as revealed in Figure 9.8(b).

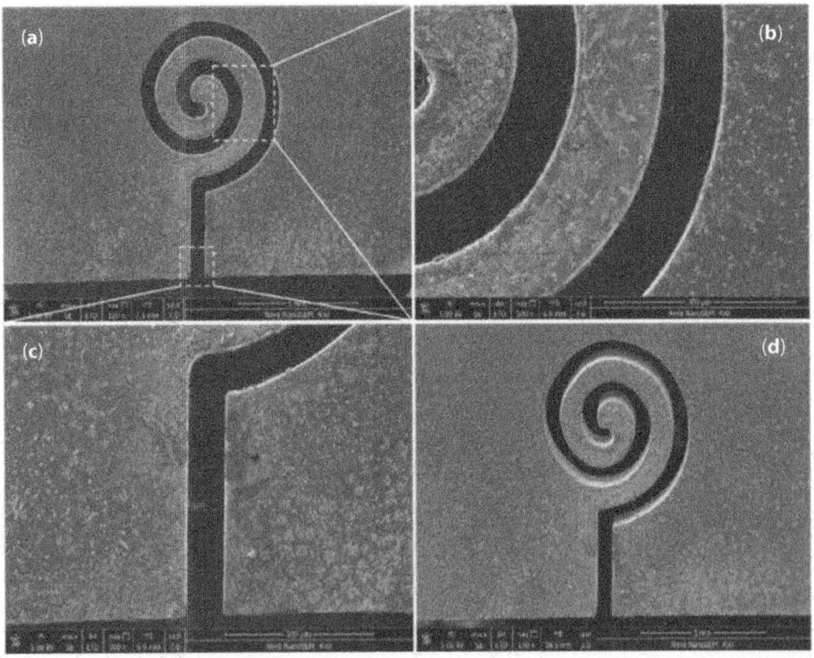

Figure 9.7 Microplanar coil structure and entrance slit width. (a) Front view; (b) enlarged view of spiral part; (c) enlarged view of entrance; (d) rear view [23].

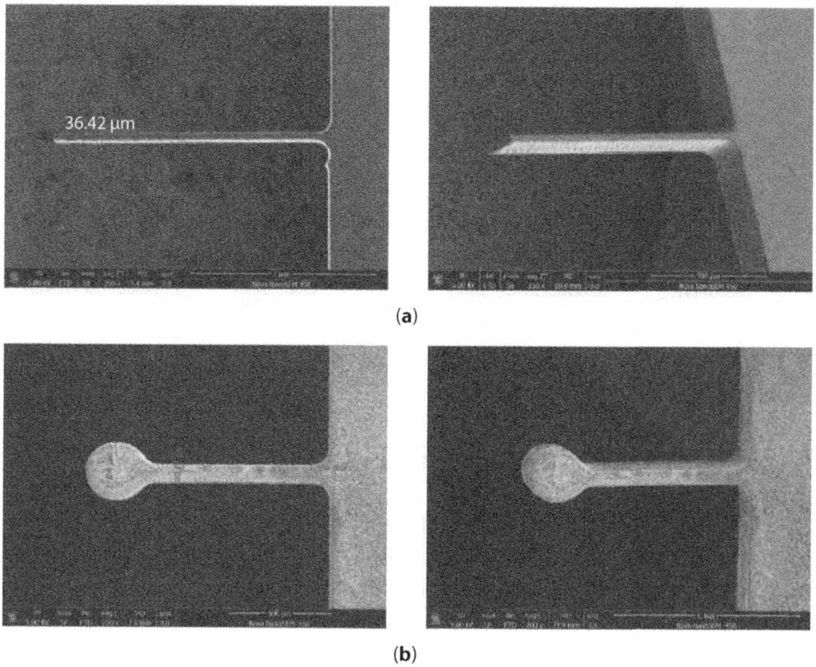

Figure 9.8 Structure of glass microcantilever (a) columnar microcantilever (b) disc-free end microcantilever [23].

9.4 Conclusions

In this study, simulation and experiment are used to examine the microstructure of glass cut by helical electrode WEDM assisted by ultrasonic vibration. The following are the conclusions:

- When employing ultrasonic pulsation to the tool, the thickness of the gas film at electrode surface is enhanced and reduced, which lowers the voltage for electrochemical discharge, improves machining condition, and boosts localization. The gas layer thickness may not be consistent if the amplitude is too great because the bubbles are difficult to cling to.
- The cutting test was conducted using a micro helical electrode that had 100 μm diameter. Deep analysis was done on how the current waveform affected the key input parameters.

Finally, under the impact of the primary machining parameters, the parameter combination is optimized.
- The microcantilever structure with a higher aspect ratio of 42 and 13 and the micro planar coil structure with a 124.1-μm entrance slit width are both cut utilizing the optimum machining settings. The findings demonstrate that hard and brittle materials may essentially be machined with high-aspect ratios using ultrasonic vibration-assisted WEDM cutting.

References

1. Zheng, Z.P., Cheng, W.H., Huang, F.Y., Yan, B.H., 3D microstructuring of Pyrex glass using the electrochemical discharge machining process. *J. Micromech. Microeng.*, 17, 960–966, 2007.
2. Vulto, P., Huesgen, T., Albrecht, B., Urban, G.A., A full-wafer fabrication process for glass microfluidic chips with integrated electroplated electrodes by direct bonding of dry film resist. *J. Micromech. Microeng.*, 19, 077001, 2009.
3. Schwartzentruber, J. and Papini, M., Abrasive waterjet micro-piercing of borosilicate glass. *J. Mater. Process. Technol.*, 219, 143–154, 2015.
4. Wüthrich, R. and Fascio, V., Machining of non-conducting materials using electrochemical discharge phenomenon—An overview. *Int. J. Mach. Tools Manuf.*, 45, 1095–1108, 2005.
5. Tsuchiya, H., Inoue, T., Miyazaki, M., Wire electro-chemical discharge machining of glasses and ceramics. *Bull. Japan. Soc Precis. Eng.*, 19, 73–74, 1985.
6. Kulkarni, A., Sharan, R., Lal, G.K., An experimental study of discharge mechanism in electrochemical discharge machining. *Int. J. Mach. Tools Manuf.*, 42, 1121–1127, 2002.
7. Kim, D.J., Ahn, Y., Lee, S.H., Kim, Y.K., Voltage pulse frequency and duty ratio effects in an electrochemical discharge microdrilling process of pyrex glass. *Int. J. Mach. Tools Manuf.*, 46, 1064–1067, 2006.
8. Nguyen, K.H., Lee, P.A., Kim, B.H., Experimental investigation of ECDM for fabricating microstructures of quartz. *Int. J. Precis. Eng. Manuf.*, 16, 5–12, 2015.
9. Hajian, M., Razfar, M.R., Movahed, S., An experimental study on the effect of magnetic field orientations and electrolyte concentrations on ECDM milling performance of glass. *Precis. Eng.*, 45, 322–331, 2016.
10. Liu, Y., Li, M., Niu, J., Lu, S., Jiang, Y., Fabrication of taper free micro-holes utilizing a combined rotating helical electrode and short voltage pulse by ECM. *Micromachines*, 10, 28, 2019.

11. Elhami, S. and Razfar, M.R., Analytical and experimental study on the integration of ultrasonically vibrated tool into the micro electro-chemical discharge drilling. *Precis. Eng.*, 47, 424–433, 2017.
12. Goud, M., Sharma, A.K., Jawalkar, C., A review on material removal mechanism in electrochemical discharge machining (ECDM) and possibilities to enhance the material removal rate. *Precis. Eng.*, 45, 1–17, 2016.
13. Liu, Y., Xu, X., Guo, C., Niu, J., Analysis on machining performance of nickel-base superalloy by electrochemical micro-milling with high-speed spiral electrode. *Micromachines*, 10, 476, 2019.
14. Xiong, Q., Wang, H., Wang, X., Li, M., Liu, Y., Lv, Z., Experimental investigation on rotating electrochemical etching of a micro spiral cylindrical electrode. *Micromachines*, 10, 704, 2019.
15. Zhang, Y., Wang, C., Wang, Y., Ji, L., Tang, J., Ni, Q., Effects of helical tube electrode structure on mixed machining product transfer in micromachining channel during tube electrode high-speed electrochemical discharge machining. *Micromachines*, 10, 634, 2019.
16. Sunder, S. and Vaishya, R.O., A review on travelling wire electro chemical discharge machining. *Int. J. Adv. Res. Eng. Sci. Technol.*, 3, 2394–2444, 2016.
17. Malik, A. and Manna, A., An experimental investigation on developed WECSM during micro slicing of e-glass fibre epoxy composite. *Int. J. Adv. Manuf. Technol.*, 85, 2097–2106, 2016.
18. Wang, J., Jia, Z., Guo, Y.B., Shape-cutting of quartz glass by spark discharge-assisted diamond wire sawing. *J. Manuf. Process.*, 34, 131–139, 2018.
19. Oza, A.D., Kumar, A., Badheka, V., Arora, A., Traveling wire electrochemical discharge machining (TW-ECDM) of quartz using zinc coated brass wire: Investigations on material removal rate and kerf width characteristics. *Silicon*, 11, 2873–2884, 2019.
20. Yadav, P., Yadava, V., Narayan, A., Experimental investigation for performance study of wire electrochemical spark cutting of silica epoxy nanocomposites. *Silicon*, 12, 1023–1033, 2019.
21. Han, M.S., Min, B.K., Lee, S.J., Geometric improvement of electrochemical discharge micro-drilling using an ultrasonic-vibrated electrolyte. *J. Micromech. Microeng.*, 19, 065004, 2009.
22. Rusli, M. and Furutani, K., Performance of micro-hole drilling by ultrasonic-assisted electro-chemical discharge machining. *Adv. Mater. Res.*, 445, 865–870, 2012.
23. Chen, Y., Feng, X., Xin, G., Experimental study on ultrasonic vibration assisted WEDM of glass microstructures with a high aspect ratio. *Micromachines*, 12, 125, 2021.

10

Study of Soda-Lime Glass Machinability by Gunmetal Tool in Electrochemical Discharge Machining and Process Parameters Optimization Using Grey Relational Analysis

Pravin Pawar[1]*, Amaresh Kumar[2] and Raj Ballav[2]

[1]*Department of Mechanical Engineering, National Institute of Technology Goa, Farmagudi, Ponda, Goa India*

[2]*Department of Production & Industrial Engineering, National Institute of Technology Jamshedpur, Jharkhand, India*

Abstract

The electrochemical discharge machining (ECDM) is a nontraditional integrated method can be useful for cutting all types of materials. In this chapter, the influence of input factors, i.e., concentration of electrolyte, voltage, and rotation speed on average material removal rate (MRR), average hole diameter, as well as average machined depth on drilling on soda-lime glass is analyzed. The gunmetal material tool electrode was used as cathode in electrochemical discharge machining for experimentation, and it is optimized by Taguchi methodology in addition with Grey relational analysis. Based on the current investigation, the optimum input factors were determined. The results illustrate the maximum average material removal rate, raised average hole diameter as well as maximum average machined depth which were found to be 75 V applied voltage, 20% concentration of electrolyte, and 50 rpm rotational speed.

***Keywords*:** ECDM, soda-lime glass, gunmetal, grey relational analysis, Taguchi, optimization

**Corresponding author*: pravinpawar@nitgoa.ac.in

Sandip Kunar, Golam Kibria, Prasenjit Chatterjee and Asma Perveen (eds.) Hybrid Micromachining and Microfabrication Technologies: Principles, Varieties and Applications, (219–242) © 2023 Scrivener Publishing LLC

10.1 Introduction

The soda-lime glass is widely utilized for circuit substrates, mirrors, filters, photo masks, data storage disks, chemical apparatus, microelectromechanical systems, touch screens, microfluidic devices, microcapillary electrophoresis, and camera lens [1]. Due to its wide applications, it is required to process by using traditional or nontraditional removal processes. The material removal process on soda-lime glass is a tough task by conventional material removal processes because of its properties. It has noteworthy properties like brittleness, nonporosity, optical transparency, durability, high chemical homogeneity, biocompatibility, various reflective indices, high hardness, corrosion resistance, isotropy, attractive appearance, temperature stability, and high electrical resistivity [1, 2]. The ECDM (EDM+ECM) is an amalgam nontraditional material removal technology useful for cutting of all types of material [3] Kurafuji *et al.* first used the ECDM process to cut glass material in the year 1968 [4]. This machining process is used for micromachining of microelectromechanical systems, texturing in microfluidic devices, making of micro-filters and digital micro-mirror devices [5]. The drawbacks of this process are accuracy, quality, and machining efficiency [6]. In this process high applied voltage machining conditions and higher concentration of electrolyte can give higher material removal rate [7]. Arab *et al.* found that lesser surface roughness cathode electrode showed a smaller gas film therefore it creates hole with small overcut [8]. Singh *et al.* observed that overcut and taper of hole increases when machining voltage is increased [9]. Laio *et al.* detected inclusion of sodium dodecyl sulfate in the chemical solution which conceivable improve the intensity of current due to this reason the superior quality of holes made on quartz material [10]. Similarly, Dhanvijay *et al.* found that flowing electrolyte has given continuous stream of electrolyte which causes the rise in hydrogen vapor bubbles and hence rise in MRR [11]. Sarkar *et al.* observed concentration of electrolyte that was the utmost significant parameter for MRR [12]. Pawar *et al.* previously experimented on workpiece, i.e., soda-lime glass in the ECDM process by using brass and copper tool materials and optimized parameters, which were investigated with the help of response surface methodology [13, 14] A gravity feed ECDM fabricated machine setup has been utilized to machined soda-lime glass through gunmetal cathode electrode [15]. The average material removal rate, average machining depth, as well as average hole diameter were examined through input parameters like tool rotation, concentration of electrolyte, and voltage. The experiments were carried out by using the Taguchi L_{27} method, as well as optimized input factors by grey relational analysis.

10.2 Experimental Conditions

The fabricated ECDM setup (Figure 10.1) is utilized to make holes on workpiece (soda-lime glass) through gunmetal electrode. The X, Y, and Z axes slide were controlled through handle and workpiece material is transferred upward direction due to counterweight. The gunmetal electrode was coupled to the Z-axis single slide table. This tool electrode was attached to the motor shaft which is located on the Z axis single slide table, and its rotational motion is operated by the Arduino Uno board which was interfaced to the notebook computer. The DC voltage is provided among two electrodes, i.e., anode and cathode [15]. The conical-shaped gunmetal electrode (Cathode) with a diameter of 3-mm diameter and anode electrode (stainless steel 416 material) is having size of 7.5 mm radius and 100 mm length electrodes, which were utilized for this work. The NaOH was utilized as a working chemical (electrolyte), the size of the workpiece material (soda-lime glass) is 150×150×3 mm^3 and the machining time is 25 minutes for one experiment. The Taguchi L_{27} method is employed for experimentation, and the reason behind this is three input factors and its three different levels having two ways of correlation therefore it has eighteen all degrees of freedom. Hence, Taguchi L_{27} technique was utilized for the current study which gives 26 degrees of freedom [16]. The removal of material observed

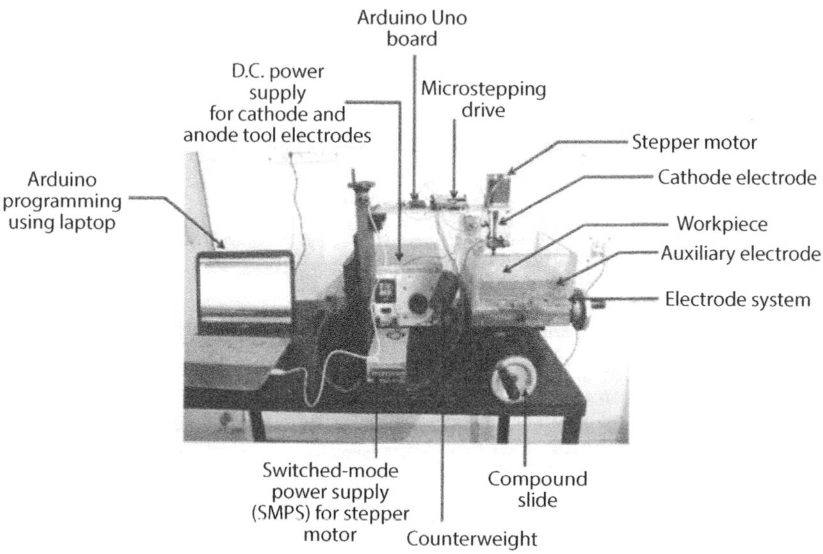

Figure 10.1 Electrochemical discharge (ECDM) machine setup.

Table 10.1 Input parameters and different levels.

	Levels	**1**	**2**	**3**
Parameters	Voltage (V)	55	65	75
	Rotation (RPM)	0	20	50
	Electrolyte Concentration (%)	10	15	20

because of high heat produced among the workpiece and the cathode electrode. The process parameters with their levels shown in Table 10.1.

The signal denotes the impact of each factor on the output responses and noise denotes the degree of the influence of uncertainty on average output results. For the current study, the diameter of hole criteria is nominal the best. The targeted nominal value is considered for hole diameter is 3.25 mm. Likewise, average material removal rate and average machined depth were considered as large is better criteria and following equations were used for calculations [17].

For higher the better:

$$S/N = -10\log_{10}\frac{1}{n}\sum_{i=1}^{n}\frac{1}{y_i^2} \tag{10.1}$$

For nominal the best:

$$S/N = -\log_{10}\frac{1}{n}\sum_{i=1}^{n}\sigma^2 \tag{10.2}$$

where, y_i is the measured value and n denotes the number of the measurements

Figure 10.2 and Figure 10.3 show experimental results of 1 and replicated results 2 on soda-lime glass through gunmetal electrode (cathode). Table 10.2 represents the microscopic images of machined workpiece material (soda-lime glass) by gunmetal electrode. The Average material removal rate, Average machined depth as well as Average diameter of hole were investigated by varying input parameters. The experimental observations are depicted in Table 10.3.

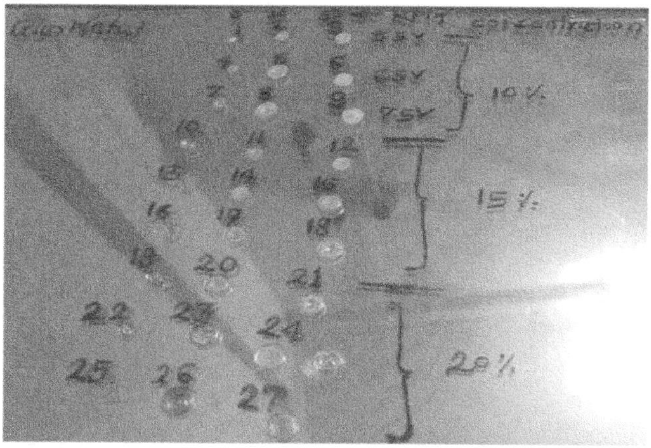

Figure 10.2 Experiments results 1 of soda-lime glass through gunmetal electrode.

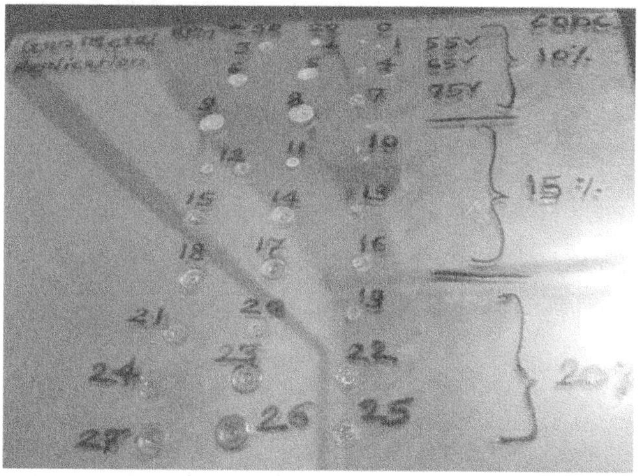

Figure 10.3 Experiments results 2 of soda-lime glass through gunmetal electrode.

10.3 Analysis of Average MRR of Workpiece (Soda-Lime Glass) Through Gunmetal Electrode

The ANOVA statistical method is utilized for analyzing the importance of each input factor. The response table shows that S/N ratio criteria is higher the better for Average MRR is applied to investigate rank of each process parameter.

Table 10.2 Microscopic images of each experimental condition results.

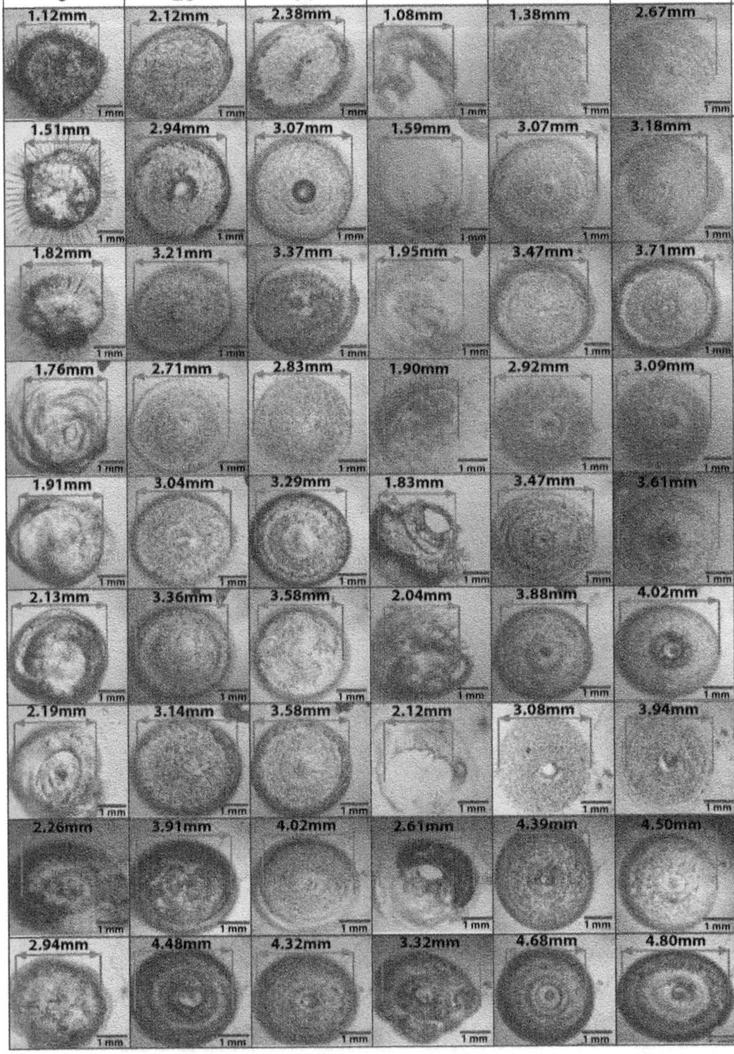

Experimental outcomes 1 of Soda-lime glass through Gunmetal cathode tool			Experimental outcomes 2 of Soda-lime glass through Gunmetal cathode tool				
Rotation (rpm)			Rotation (rpm)			Volt. (V)	Con. (%)
0	**20**	**50**	**0**	**20**	**50**		
1.12mm	2.12mm	2.38mm	1.08mm	1.38mm	2.67mm	55	10
1.51mm	2.94mm	3.07mm	1.59mm	3.07mm	3.18mm	65	
1.82mm	3.21mm	3.37mm	1.95mm	3.47mm	3.71mm	75	
1.76mm	2.71mm	2.83mm	1.90mm	2.92mm	3.09mm	55	15
1.91mm	3.04mm	3.29mm	1.83mm	3.47mm	3.61mm	65	
2.13mm	3.36mm	3.58mm	2.04mm	3.88mm	4.02mm	75	
2.19mm	3.14mm	3.58mm	2.12mm	3.08mm	3.94mm	55	20
2.26mm	3.91mm	4.02mm	2.61mm	4.39mm	4.50mm	65	
2.94mm	4.48mm	4.32mm	3.32mm	4.68mm	4.80mm	75	

10.3.1 ANOVA for Average MRR

Table 10.4 illustrates the ANOVA table of Average MRR of gunmetal tool electrode in which desirability criteria of larger is better. The maximum F-value 58.19 is observed for concentration of electrolyte therefore this is an utmost important factor as related to voltage and tool rotation speed.

Table 10.3 Observations of soda-lime glass material through gunmetal electrode.

Run	Elec. conc. (%)	Voltage (V)	Rotation (rpm)	MRR 1 (mg/min)	MRR 2 (mg/min)	Average MRR (mg/min)	Hole dia. 1 (mm)	Hole dia.2 (mm)	Average hole dia. (mm)	Machined depth 1 (mm)	Machined depth 2 (mm)	Average machined depth (mm)
1	10	55	0	0.1	0	0.05	1.12	1.08	1.1	0.07	0.06	0.065
2	10	55	20	0.15	0.1	0.125	2.12	1.38	1.75	0.09	0.08	0.085
3	10	55	50	0.17	0.15	0.16	2.38	2.67	2.525	0.10	0.11	0.105
4	10	65	0	0.14	0.14	0.14	1.51	1.59	1.55	0.12	0.14	0.13
5	10	65	20	0.25	0.26	0.255	2.94	3.07	3.005	0.20	0.22	0.21
6	10	65	50	0.27	0.27	0.27	3.07	3.18	3.125	0.29	0.28	0.285
7	10	75	0	0.17	0.18	0.175	1.82	1.95	1.885	0.21	0.20	0.205
8	10	75	20	0.31	0.33	0.32	3.21	3.47	3.34	0.32	0.39	0.355
9	10	75	50	0.32	0.35	0.335	3.37	3.71	3.54	0.35	0.37	0.36
10	15	55	0	0.16	0.13	0.145	1.76	1.90	1.83	0.15	0.11	0.13
11	15	55	20	0.20	0.20	0.2	2.71	2.92	2.815	0.26	0.23	0.245
12	15	55	50	0.21	0.22	0.215	2.83	3.09	2.96	0.24	0.21	0.225
13	15	65	0	0.20	0.20	0.2	1.91	1.83	1.87	0.21	0.19	0.2
14	15	65	20	0.28	0.30	0.29	3.04	3.47	3.255	0.35	0.37	0.36

(*Continued*)

Table 10.3 Observations of soda-lime glass material through gunmetal electrode. (*Continued*)

Run	Elec. conc. (%)	Voltage (V)	Rotation (rpm)	MRR 1 (mg/min)	MRR 2 (mg/min)	Average MRR (mg/min)	Hole dia. 1 (mm)	Hole dia.2 (mm)	Average hole dia. (mm)	Machined depth 1 (mm)	Machined depth 2 (mm)	Average machined depth (mm)
15	15	65	50	0.30	0.33	0.315	3.29	3.61	3.45	0.41	0.42	0.415
16	15	75	0	0.22	0.21	0.215	2.13	2.04	2.085	0.25	0.20	0.225
17	15	75	20	0.32	0.39	0.355	3.36	3.88	3.62	0.39	0.50	0.445
18	15	75	50	0.35	0.42	0.385	3.58	4.02	3.8	0.46	0.52	0.49
19	20	55	0	0.23	0.21	0.22	2.19	2.12	2.155	0.27	0.20	0.235
20	20	55	20	0.33	0.28	0.305	3.14	3.08	3.11	0.39	0.31	0.35
21	20	55	50	0.34	0.33	0.335	3.58	3.94	3.76	0.42	0.38	0.4
22	20	65	0	0.26	0.25	0.255	2.26	2.61	2.435	0.37	0.31	0.34
23	20	65	20	0.51	0.49	0.5	3.91	4.39	4.15	0.98	0.83	0.905
24	20	65	50	0.50	0.48	0.49	4.02	4.50	4.26	0.92	0.75	0.835
25	20	75	0	0.34	0.30	0.32	2.94	3.32	3.13	0.61	0.42	0.515
26	20	75	20	0.61	0.58	0.595	4.48	4.68	4.58	1.21	1.04	1.125
27	20	75	50	0.55	0.62	0.585	4.32	4.80	4.56	1.08	1.17	1.125

Table 10.4 ANOVA for Average MRR of Soda-Lime Glass Workpiece by Gunmetal Electrode.

Source	DF	Adj SS	Adj MS	F-value	P-value	% Contribution
Voltage (V)	2	0.13287	0.066433	41.41	0.000	27.82
Electrolyte concentration (%)	2	0.18674	0.093369	58.19	0.000	39.10
Rotation speed (rpm)	2	0.12587	0.062936	39.23	0.000	26.36
Error	20	0.03209	0.001604			6.72
Total	26	0.47757				100

Table 10.5 Response table for Average MRR.

Level	Voltage (V)	Electrolyte concentration (%)	Tool speed (rpm)
Level 1	−15.230	−14.984	−15.251
Level 2	−11.019	−12.167	−10.489
Level 3	-9.360	-8.457	-9.869
Delta	5.870	6.527	5.382
Rank	2	1	3

Due to increasing concentration which cause rapid electron ions movement between cathode and anode. Hence, it creates more hydrogen bubbles developed nearby the cathode electrode and therefore more sparks generated and higher local temperature and hence MRR increases [18]. The p values of voltage, tool rotation speed and conc. of electrolyte are lower than 0.05 which means all parameters are significant. The present results show that good agreement with the previously available results, which were reported for silicon wafer and glass materials [19, 20].

Table 10.5 shows, larger the better (criteria) of Average MRR represents the first rank or concentration of electrolyte then afterward the voltage and tool speed.

10.3.2 Influence of Input Factors on Average MRR

Figure 10.4 states the graphical representation of the Average MRR rises markedly with the raise in machining voltage, conc. of electrolyte and electrode speed. The higher Average MRR is achieved at a condition of voltage is 75 V, Conc. of electrolyte is 20%, and tool speed is 50 rpm. The same type of results for MRR were obtained previously for silicon wafer and quartz glass materials [19, 21].

10.4 Analysis of Average Depth of Machined Hole on Soda-Lime Glass Through Gunmetal Electrode

The statistical analysis is done by using an ANOVA table for Average machined depth and larger is better criteria is applied for Average machined depth and results which are shown in the response table.

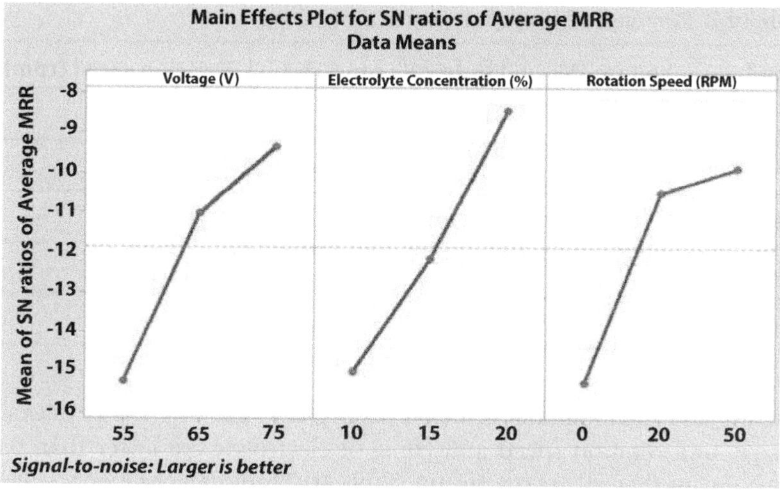

Figure 10.4 Average MRR of soda-lime glass workpiece by gunmetal electrode.

10.4.1 ANOVA for Average Machined Depth

Table 10.6 illustrates the analyzed results of Average machined depth in which desirability criteria is larger the better. It denotes the highest F value 26.30 was found for the concentration of electrolyte; therefore, it is the utmost significant factor related to voltage and tool speed. The machining depth was enhanced with rise in conc. of electrolyte from 10% to 20%. The increasing trend of the electrolyte rises the conductivity of electrolytes which causes the chemical etching process. Therefore, it helps to produce more bubbles formation and thus, more sparks and high temperatures

Table 10.6 ANOVA table for Average machined depth.

Source	DF	Adj SS	Adj MS	F-value	P-value	Percentage contribution
Voltage (V)	2	0.5101	0.25505	13.57	0.000	23.11
Electrolyte concentration (%)	2	0.9887	0.49434	26.30	0.000	44.79
Rotation speed (rpm)	2	0.3328	0.16639	8.85	0.002	15.08
Error	20	0.3760	0.01880			17.03
Total	26	2.2075				100

Table 10.7 Response table of Average machined depth.

Level	Voltage (V)	Electrolyte conc. (%)	Rotation speed (rpm)
Level 1	−15.253	−15.394	−14.178
Level 2	-9.386	−11.064	-9.005
Level 3	-6.837	-5.020	-8.294
Delta	5.870	10.374	5.883
Rank	2	1	3

generate and thus material removal takes place [22–24]. The p-values of voltage, tool rotation speed and conc. of electrolyte are lower than 0.05 which means they all are significant. Table 10.7 indicates the Conc. of electrolyte is the first rank after that machining voltage and electrode speed.

10.4.2 Influence of Input Factors on Average Machined Depth

Figure 10.5 represents the Average machined depth which is enhanced considerably with the enhance in voltage, Conc. of electrolyte and speed of electrode. The higher Average machined depths were achieved at the machining voltage 75 V, conc. electrolyte 20%, and electrode speed 50 rpm. The same output responses were observed in previous literature for glass materials [19, 21].

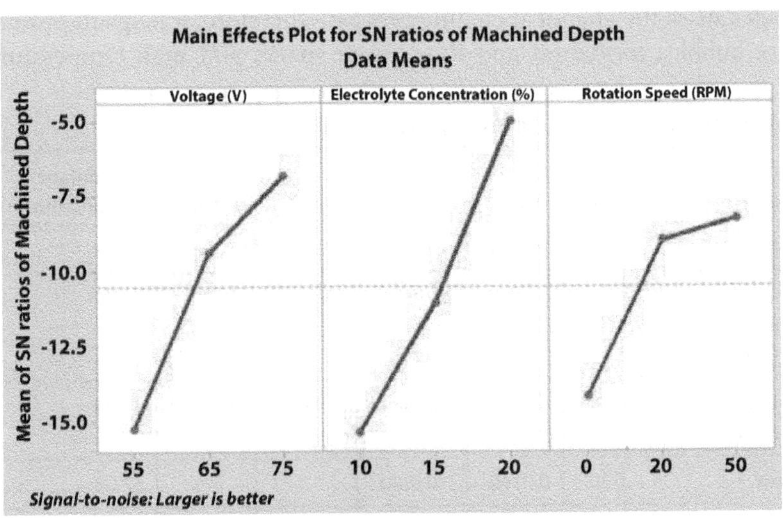

Figure 10.5 Main effect plot for average machined depth.

10.5 Analysis of Average Diameter of Hole of Soda-Lime Glass Through Gunmetal Electrode

The statistical analysis is explained with the help of ANOVA Table 10.8 for Average hole diameter and nominal is better criteria are set for Average hole diameter.

10.5.1 ANOVA for Average Hole Diameter

Table 10.8 represents the higher F value 141.05 is found for tool speed (rpm) which denotes it is the most significant parameter related to voltage and concentration of electrolyte. The p-values of voltage, tool rotation speed and conc. of electrolyte are smaller than 0.05 which means all parameters are significant. The experimental results of the current research work match earlier published article results on diameter of hole for glass material [20, 23, 25, 26]. Table 10.9 shows tool speed is the first rank then after conc. of electrolyte and voltage.

10.5.2 Influence of Input Factors on Average Hole Diameter

Figure 10.6 represents the Average hole dia. of is enhanced considerably with the raise in tool speed, electrolyte Conc., and voltage. The same output responses were observed in previous literature for glass materials [20, 24].

Table 10.8 ANOVA table for Average hole diameter.

Source	DF	Adj SS	Adj MS	F-value	P-value	Percentage contribution
Voltage (V)	2	4.0977	2.04887	46.71	0.000	17.52
Electrolyte concentration (%)	2	6.0410	3.02051	68.86	0.000	25.83
Rotation speed (rpm)	2	12.3734	6.18670	141.05	0.000	52.90
Error	20	0.8773	0.04386			3.75
Total	26	23.3894				100

Table 10.9 Response table for Average hole diameter (Nominal is better).

Level	Voltage (V)	Electrolyte conc. (%)	Rotation speed (rpm)
Level 1	2.445	2.424	2.004
Level 2	3.011	2.854	3.292
Level 3	3.393	3.571	3.553
Delta	0.948	1.147	1.549
Rank	3	2	1

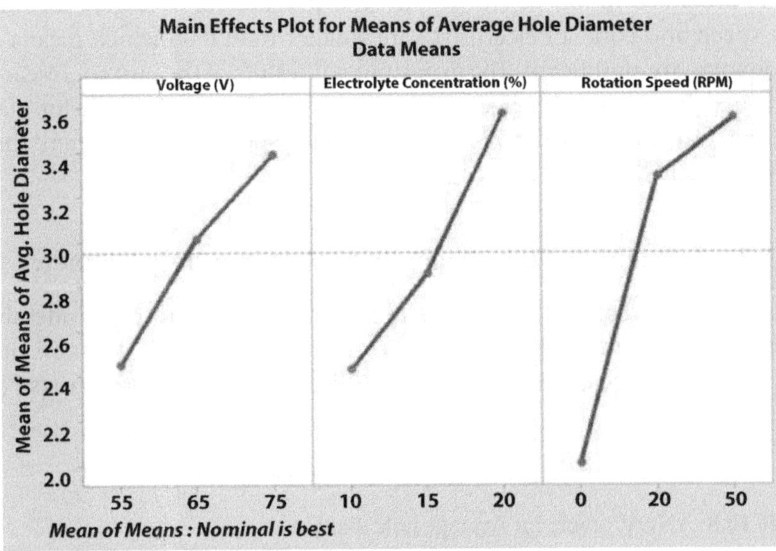

Figure 10.6 Main effect plot for Average hole diameter.

10.6 Grey Relational Analysis Optimization of Soda-Lime Glass Results by Gunmetal Electrode

Prof. Deng firstly presented the grey relational analysis theory in the year 1982. It is used in an incomplete, uncertain information for multi-objective optimization of parameters [27, 28]. The "grey" term is nothing but the combination of black and white. Hence, black designates required information is neither exactly obtainable on the other hand white expresses anticipated data is exactly obtainable. Grey theory indicates the correlation among white with black [29, 30]. In this research work, the desirability criterion 'higher the better' is set of average MRR and average machined depth because higher values are required. For average hole diameter "nominal the best" criterion is chosen.

10.6.1 Methodology of Grey Relational Analysis

Step i: Determine the calculated data and convert into the new data. After that, remove the different measurements of the new data and convert into data that can be associated with all data.

Step ii: Compute the data order for reference. After that reference order was selected based on the actual application.

Step iii: Calculate the comprehensive variance and the minimum and maximum values of the related parameter.

Step iv: Find out the grey correlation degree.

Step v: Find out grey degrees in the order from smaller to larger or larger to smaller [31].

10.6.2 Data Pre-Processing

The data are transferred from initial order to a proportionate order. Therefore, the result values are normalized in the extent of zero to one. The following equations are utilized to find out value of grey relational analysis [32].

10.6.3 Grey Relational Generating

The experimental data order and its analyzed data order in the form of as x_0 (k) and x_i (k), i = 1, 2, 3...., m; k=1, 2, 3..., n consequently. Whereas, m is the number of total experiments, n is the no. of all investigational data. The $x_i^*(k)$ is the grey relation generating number. If the aim of experimental results value order is larger is better therefore the original experimental results order is normalized with reference to equation 3 is shown below [27, 29, 32]. Equation 3 is utilized for to find out normalized results for average material removal rate and average machined depth. Equation 10.5 is utilized to find out normalized results of diameter of hole.

$$x_i^*(k) = \frac{x_i^0(k) - \min x_i^0(k)}{\max x_i^0(k) - \min x_i^0(k)} \tag{10.3}$$

Likewise, if it is lower the better criteria, the original order is normalized as equation 10.4.

$$x_i^*(k) = \frac{\max x_i^0(k) - x_i^0(k)}{\max x_i^0(k) - \min x_i^0(k)} \tag{10.4}$$

However, if target value the original order is normalized is calculated using equation 10.5. OB is known as target value.

$$x_i^*(k) = 1 - \frac{\left|x_i^0(k) - OB\right|}{\max\left\{x_i^0(k) - OB; OB - \min x_i^0(k)\right\}} \tag{10.5}$$

10.6.4 Deviation Sequence

Here, Δ_{0i} (k) is a deviation order among $x_0^*(k)$ reference sequence and $x_i^*(k)$ corresponding sequence. The deviation order is evaluated by using equation 10.6.

$$\Delta_{0i}(k) = \left|x_0^*(k) - x_i^*(k)\right| \tag{10.6}$$

The greatest deviation and the lowest deviation are assessed as in equation 10.7 and equation 10.8.

$$\Delta_{\max} = \max_{\forall j \in i} \max_{\forall k} \left|x_0^*(k) - x_j^*(k)\right| \tag{10.7}$$

$$\Delta_{\max} = \max_{\forall j \in i} \max_{\forall k} \left|x_0^*(k) - x_j^*(k)\right| \tag{10.8}$$

$$\sum_{k=1}^{n} \beta_k = 1$$

Here, the Grey relational grade $\gamma\left(x_0^*, x_i^*\right)$ represents the level of relationship among the reference sequence and corresponding sequence. The greatest grey relational grade represents improve product quality and hence grey relational grade gives information about the influencing factor [33, 34].

10.6.5 Grey Relational Coefficient

It gives information about how close $x_0^*(k)$ is to $x_i^*(k)$. The greater the grey relational coefficient the nearer $x_0^*(k)$ and $x_i^*(k)$. It is assessed by using eq. 10.9 and identification coefficient ζ considering a value 0.5 is put in equation 10.9.

$$\gamma\left[x_0^*(k), x_i^*(k)\right] = \frac{\Delta \min + \xi \Delta max}{\Delta x_i(k) + \xi \Delta max} \tag{10.9}$$

whereas, ξ is the distinguishing coefficient in [0, 1].

$$0 < \gamma\left[x_0^*(k), x_i^*(k)\right] \leq 1 \tag{10.10}$$

10.6.6 Grey Relational Grade

It is the summation of grey relational coefficients and these values are determined using equation 10.11.

$$\gamma\left(x_0^*, x_i^*\right) = \sum_{k=1}^{n} \beta_k \gamma\left[x_0^*(k), x_i^*(k)\right] \tag{10.11}$$

The rank of each experiment was assessed with the help of grey relational grade equation number 11. The grey relational analysis values of voltage, electrolyte conc. and speed of electrode at level 1 to level 3 were determined and illustrates in Table 10.10. The mean of selected grey relational grade for electrolyte concentration level 1 was assessed by using following equation 10.12 similarly find out all values of voltage and tool speed of level 1 to level 3, which are depicted in Table 10.11. The greatest values of average grey relational grade indicate the most influencing factor [33, 34].

$$Electrolyte..Conc..(L.1) = \frac{0.3333 + 0.3618 + 0.3952 + 0.3623 + 0.4455 + 0.4625 + 0.3838 + 0.4964 + 0.5156}{9} \tag{10.12}$$

The Average grey relational grades values are representing in Table 10.11, which shows optimized parameters for all output responses of gunmetal

Table 10.10 Grey relational analysis table for soda-lime glass material through gunmetal electrode.

No.	Normalised value			Deviation sequence			Grey relation coefficient			Grade	Rank
	Average MRR (mg/min)	Average machined depth (mm)	Average hole diameter (mm)	Average MRR (mg/min)	Average machined depth (mm)	Average hole diameter (mm)	Average MRR (mg/min)	Average machined depth (mm)	Average hole diameter (mm)		
1	0.0000	0.0000	0.0000	1.0000	1.0000	1.0000	0.3333	0.3333	0.3333	0.3333	27
2	0.1376	0.0189	0.1868	0.8624	0.9811	0.8132	0.3670	0.3376	0.3807	0.3618	26
3	0.2018	0.0377	0.4095	0.7982	0.9623	0.5905	0.3852	0.3419	0.4585	0.3952	21
4	0.1651	0.0613	0.1293	0.8349	0.9387	0.8707	0.3746	0.3475	0.3648	0.3623	25
5	0.3761	0.1368	0.5474	0.6239	0.8632	0.4526	0.4449	0.3668	0.5249	0.4455	15
6	0.4037	0.2075	0.5819	0.5963	0.7925	0.4181	0.4561	0.3869	0.5446	0.4625	14
7	0.2294	0.1321	0.2256	0.7706	0.8679	0.7744	0.3935	0.3655	0.3923	0.3838	23
8	0.4954	0.2736	0.6437	0.5046	0.7264	0.3563	0.4977	0.4077	0.5839	0.4964	11
9	0.5229	0.2783	0.7011	0.4771	0.7217	0.2989	0.5117	0.4093	0.6259	0.5156	8
10	0.1743	0.0613	0.2098	0.8257	0.9387	0.7902	0.3772	0.3475	0.3875	0.3707	24
11	0.2752	0.1698	0.4928	0.7248	0.8302	0.5072	0.4082	0.3759	0.4964	0.4269	18
12	0.3028	0.1509	0.5345	0.6972	0.8491	0.4655	0.4176	0.3706	0.5179	0.4354	16
13	0.2752	0.1274	0.2213	0.7248	0.8726	0.7787	0.4082	0.3643	0.3910	0.3878	22

(Continued)

Table 10.10 Grey relational analysis table for soda-lime glass material through gunmetal electrode. (*Continued*)

No.	Normalised value			Deviation sequence			Grey relation coefficient			Grade	Rank
	Average MRR (mg/min)	Average machined depth (mm)	Average hole diameter (mm)	Average MRR (mg/min)	Average machined depth (mm)	Average hole diameter (mm)	Average MRR (mg/min)	Average machined depth (mm)	Average hole diameter (mm)		
14	0.4404	0.2783	0.6193	0.5596	0.7217	0.3807	0.4719	0.4093	0.5677	0.4829	12
15	0.4862	0.3302	0.6753	0.5138	0.6698	0.3247	0.4932	0.4274	0.6063	0.5090	9
16	0.3028	0.1509	0.2830	0.6972	0.8491	0.7170	0.4176	0.3706	0.4109	0.3997	20
17	0.5596	0.3585	0.7241	0.4404	0.6415	0.2759	0.5317	0.4380	0.6444	0.5381	6
18	0.6147	0.4009	0.7759	0.3853	0.5991	0.2241	0.5648	0.4549	0.6905	0.5701	5
19	0.3119	0.1604	0.3032	0.6881	0.8396	0.6968	0.4208	0.3732	0.4178	0.4040	19
20	0.4679	0.2689	0.5776	0.5321	0.7311	0.4224	0.4844	0.4061	0.5421	0.4775	13
21	0.5229	0.3160	0.7644	0.4771	0.6840	0.2356	0.5117	0.4223	0.6797	0.5379	7
22	0.3761	0.2594	0.3836	0.6239	0.7406	0.6164	0.4449	0.4030	0.4479	0.4319	17
23	0.8257	0.7925	0.8764	0.1743	0.2075	0.1236	0.7415	0.7067	0.8018	0.7500	3
24	0.8073	0.7264	0.9080	0.1927	0.2736	0.0920	0.7219	0.6463	0.8447	0.7376	4
25	0.4954	0.4245	0.5833	0.5046	0.5755	0.4167	0.4977	0.4649	0.5455	0.5027	10
26	1.0000	1.0000	1.0000	0.0000	0.0000	0.0000	1.0000	1.0000	1.0000	1.0000	1
27	0.9817	1.0000	0.9943	0.0183	0.0000	0.0057	0.9646	1.0000	0.9886	0.9844	2

Table 10.11 Response table for the average grey relational grade.

	Level 1	Level 2	Level 3	Max-min	Rank
Voltage (V)	0.4159	0.5077	**0.5990**	0.1831	2
Electrolyte concentration (%)	0.4174	0.4578	**0.6473**	0.2299	1
Rotation speed (rpm)	0.3974	0.5532	**0.5720**	0.1746	3

tool material are level 3 of voltage, the concentration of electrolyte and tool speed for output responses of Average material removal rate, Average machining depth and Average hole diameter. The most dominating factor for all responses using gunmetal tool material is electrolyte concentration followed by machining voltage and tool speed.

10.7 Conclusion

The developed ECDM machine was utilized to drill holes on soda-lime glass through gunmetal tool material. From current research work, the grey relational multiobjective optimization method outcomes confirmed that the electrolyte concentration was the utmost powerful parameter for the average material removal rate, average machined depth and average diameter of hole. The optimum process factors for the average material removal rate, average depth of hole and average hole diameter are tool rotation speed 50 rpm, voltage 75 V, the electrolyte conc. is 20%.

Acknowledgments

The authors are grateful to Production and Industrial Engineering Department, National Institute of Technology Jamshedpur India for providing continuous support during this research work.

References

1. Pawar, P., Ballav, R., Kumar, A., Review on material removal technology of soda-lime glass material. *Indian. J. Sci. Technol.*, 10, 1, 2017.

2. Jahan, M.P., Perveen, A., Rumsey, A.M., A review on the conventional, non-conventional, and hybrid micromachining of glass. *Mach. Sci. Technol.*, 23, 264, 2019.
3. Goud, M., Sharma, A.K., Jawalkar, C., A review on material removal mechanism in electrochemical discharge machining (ECDM) and possibilities to enhance the material removal rate. *Precis. Eng.*, 45, 1, 2016.
4. Kurafuji, H. and Suda, K., Electrical discharge drilling of glass. *CIRP Ann.*, 16, 415, 1968.
5. Pawar, P., Ballav, R., Kumar, A., FEM analysis of different materials based on explicit dynamics ANSYS in electrochemical discharge machine, in: Dixit, U., Kant, R. (eds) *Simulations for Design and Manufacturing, Lecture Notes on Multidisciplinary Industrial Engineering*, Springer, Singapore, pp. 231–258, 2018.
6. Wuthrich, R. and Fascio, V., Machining of non-conducting materials using electrochemical discharge phenomenon—an overview. *Int. J. Mach. Tool. Manuf.*, 45, 1095, 2005.
7. Singh, T. and Dvivedi, A., Developments in electrochemical discharge machining: A review on electrochemical discharge machining, process variants and their hybrid methods. *Int. J. Mach. Tools Manuf.*, 105, 1, 2016.
8. Arab, J., Kannojia, H.K., Dixit, P., Effect of tool electrode roughness on the geometric characteristics of through-holes formed by ECDM. *Precis. Eng.*, 60, 437, 2019.
9. Singh, M., Singh, S., Kumar, S., Experimental investigation for generation of micro-holes on silicon wafer using electrochemical discharge machining process. *Silicon*, 12, 1683, 2020.
10. Laio, Y.S., Wu, L.C., Peng, W.Y., A study to improve drilling quality of electrochemical discharge machining (ECDM) process. *Procedia CIRP*, 6, 609, 2013.
11. Dhanvijay, M.R., Kulkarni, V.A., Doke, A., Experimental investigation and analysis of electrochemical discharge machining (ECDM) on fiberglass reinforced plastic (FRP). *J. Inst. Eng. (India) C*, 100, 763, 2019.
12. Sarkar, B.R., Doloi, B., Bhattacharyya, B., Investigation on electrochemical discharge micro-machining of silicon carbide. *IJMFMP*, 4, 29, 2017.
13. Pawar, P., Kumar, A., Ballav, R., Parametric analysis of electrochemical discharge drilling on soda-lime glass material using Taguchi L27 orthogonal array method. *Strojnícky Casopis J. Mech. Eng.*, 69, 115, 2019.
14. Pawar, P., Kumar, A., Ballav, R., Experimental study on the MRR, machined depth and hole diameter for soda-lime glass by electrochemical discharge machining process with copper tool. *Int. J. Mod. Manuf. Technol.*, 12, 131, 2020.
15. Pawar, P., Kumar, A., Ballav, R., Development and manufacturing of arduino based electrochemical discharge machine. *J. Mach. Eng.*, 18, 45, 2018.

16. Pinar, A.M., Uluer, O., Kırmaci, V., Optimization of counter flow Ranque–Hilsch vortex tube performance using Taguchi method. *Int. J. Refrig.*, 32, 1487, 2009.
17. Roy, A., Nath, N., Nedelcu, D., Experimental investigation on variation of output responses of as cast TiNiCu shape memory alloys using wire EDM. *Int. J. Mod. Manuf. Technol.*, 9, 90, 2017.
18. Chak, S.K. and Rao, P.V., Trepanning of Al_2O_3 by electro-chemical discharge machining (ECDM) process using abrasive electrode with pulsed DC supply. *Int. J. Mach. Tools Manuf.*, 47, 2061, 2007.
19. Paul, L. and Hiremath, S., Evaluation of process parameters of ECDM using grey relational analysis. *Procedia Mater. Sci.*, 5, 2273, 2014.
20. Jain, V.K., Choudhury, S.K., Ramesh, K.M., On the machining of alumina and glass. *Int. J. Mach. Tools Manuf.*, 42, 1269, 2002.
21. Goud, M. and Sharma, A.K., On performance studies during micromachining of quartz glass using electrochemical discharge machining. *J. Mech. Sci. Technol.*, 31, 1365, 2017.
22. Wuthrich, R., Spaelter, U., Wu, Y., Bleuler, H., A systematic characterization method for gravity-feed micro-hole drilling in glass with spark assisted chemical engraving (SACE). *J. Micromech. Microeng.*, 16, 1891, 2006.
23. Wei, C., Ni, J., Hu, D., Electrochemical discharge machining using micro-drilling tools. *Trans. NAMRI/SME*, 38, 105, 2010.
24. Razfar, M.R., Ni, J., Behroozfar, A., Lan, S., An investigation on electrochemical discharge micro-drilling of glass, in: *ASME 2013 International Manufacturing Science and Engineering Conference collocated with the 41st North American Manufacturing Research Conference*, pp. V002T03A013–V002T03A013, 2013.
25. Yang, C.T., Ho, S.S., Yan, B.H., Micro hole machining of borosilicate glass through electrochemical discharge machining (ECDM). *Key Eng. Mater.*, 196, 149, 2001.
26. Gao, C., Liu, Z., Li, A., Study of micro drilling on pyrex glass using spark assisted chemical engraving. *Micro Nanosyst.*, 6, 26, 2014.
27. Kuo, Y., Yang, T., Huang, G.W., The use of grey relational analysis in solving multiple attribute decision making problems. *Comput. Ind. Eng.*, 55, 80, 2008.
28. Porwal, R.K., Yadava, V., Ramkumar, J., Multiobjective optimization of hole drilling electrical 248 discharge micromachining process using grey relational analysis coupled with principal component analysis. *J. Inst. Eng. (India) C*, 94, 317, 2013.
29. Ertugrul, I., Oztas, T., Ozcil, A., Oztas, G.Z., Grey relational analysis approach in academic performance comparison of university: A case study of Turkish universities. *Eur. Sci. J.*, 12, 128, 2016.
30. Dhal, P.R., Datta, S., Mahapatra, S.S., Flexible manufacturing system selection based on grey relation under uncertainty. *Int. J. Serv. Oper. Manage.*, 8, 516, 2011.

31. Wang, X.L., Application of grey relation analysis theory to choose high reliability of the network node. *J. Phys. Con. Ser.* IOP Publishing, 1237, 3, 032056, 2019.
32. Hasani, H., Tabatabaei, S.A., Amiri, G., Grey relational analysis to determine the optimum process parameters for open-end spinning yarns. *J. Eng. Fibers Fabr.*, 7, 81, 2012.
33. Haq, A.N., Marimuthu, P., Jeyapaul, R., Multi response optimization of machining parameters of drilling Al/SiC metal matrix composite using grey relational analysis in the Taguchi method. *Int. J. Adv. Manuf. Technol.*, 37, 250, 2008.
34. Jung, J.H. and Kwon, W.T., Optimization of EDM process for multiple performance characteristics using Taguchi method and grey relational analysis. *J. Mech. Sci. Technol.*, 24, 1083, 2010.

11

Micro Turbine Generator Combined with Silicon Structure and Ceramic Magnetic Circuit

Minami Kaneko* **and Fumio Uchikoba**

College of Science and Technology, Nihon University, Chiba, Japan

Abstract

With the advent of the Internet of Things (IoT) society, demand for miniature power generator to drive communication devices and sensors of IoT module is increasing. Energy-harvesting generators are being actively researched; however, each of these technologies has its own problems to solve with a single fabrication technology. In the case of thermal cycles that use waste heat, organic Rankine cycle power generation systems have been developed that can use lower-grade heat by changing the conventional working fluid to a low boiling point material. However, these systems are based on machining process. In this research, a miniature electromagnetic induction type generator by combining the microelectromechanical systems (MEMS) process, a microfabrication technology, and the multilayer ceramic technology, a miniature electronic components fabrication technology was developed. The result was an output of 3 mVA in a size of less than 10 mm square. In addition, the rotational behavior and phase change of the turbine were observed as the result of using a low boiling point material as the working fluid for the purpose of waste heat harvesting. The results suggest the usefulness of combining different fabrication techniques and materials in realizing miniature power generation systems.

***Keywords*:** Microelectromechanical systems (MEMS), multilayer ceramic technology, Internet of Things (IoT), energy harvesting, electromagnetic induction type generator, turbine, organic rankine cycle, silicon

**Corresponding author*: takatou.minami@nihon-u.ac.jp

Sandip Kunar, Golam Kibria, Prasenjit Chatterjee and Asma Perveen (eds.) Hybrid Micromachining and Microfabrication Technologies: Principles, Varieties and Applications, (243–270) © 2023 Scrivener Publishing LLC

11.1 Introduction

Miniaturization of communication equipment has dramatically progressed with the progress of miniaturization of electronic components. Contemporaneously, the development and popularization of wearable devices as communication terminals have progressed. The wearable devices not only function as general communication terminals but also enable the measurement of body temperature and heart rate for the management of the owner's health condition. Furthermore, the miniaturization of electronic components not only enriches the lives of individuals but also increases the affluence and safety of society. For example, various things are exchanging information, so it is called the Internet of Things (IoT) society has arrived [1]. Machine to Machine (M2M), a system in which machines exchange data directly, has been used. A confined communication machine is frequently used in this system. And it uses a built-in system for automatic data processing and automatic control. However, in the IoT system, the one that did not have a communication function in the past is now connected to the Internet. This enables remote operation, monitoring, data collection, and analysis. Therefore, the fields of utilization are not limited to the industrial field and medical fields. There are also personal uses such as household electrical appliances and the wearable devices mentioned above. It is required to mount many machines communication devices and sensors to realize the IoT society. Moreover, driving the small communication module requires a miniature power source. Because case size is influenced by weight and volume, wearable devices require a high-power density. Currently, button batteries and lithium-ion secondary batteries are commonly used. The button battery scrap is released as a waste product. The lithium-ion secondary battery shows high power density, but miniaturization is difficult and has limitation of theoretical power density. Furthermore, when the battery becomes empty, these batteries require the replaced. Therefore, IoT communication modules require a miniature generator that do not need to be replaced or recharging.

Various research uses the microelectromechanical systems (MEMS) process, which is a silicon microfabrication technology to achieve a miniature power source. The MEMS process is based on the Integrated Circuit (IC) manufacturing process. Also, the miniature components that have a fine and high aspect ratio structure is formed by a photolithography process and a dry etching process [2, 3]. The Massachusetts Institute of Technology group reported on one of the MEMS generator researches, ultra micro gas turbine (UMGT) [4–6]. Many MEMS generators have been researched since then [7, 8]. The UMGT used an electrostatic type, which is the most of power generation method for MEMS power generators. Because the MEMS process has a planar structure, therefore it is suitable to the electrostatic type. Much research on electrostatic MEMS generators is published, and all of them have miniature structures. On another front, the electrostatic generators have the problems of charge saturation and high internal impedance. Therefore, the output current is low. Moreover, the area affected electrostatic capacitance.

For the IoT communication module, at least a milliwatt level output power is required. To respond to the high output power and miniature structure, electromagnetic induction MEMS generators are also being researched [9–12]. The electromagnetic induction generator has the advantage of high output power due to their low output impedance. Conventionally, the electromagnetic induction generator is often employed in main power generation systems. It has a turbine structure as a moving part, a magnet and a magnetic circuit for power generation. Furthermore, a magnetic material core and a winding wire coil are required to construct the magnetic circuit. However, the winding coil is unsuitable for mounting in communication modules. Therefore, development a monolithic miniature magnetic circuit is important. It is difficult to form a three-dimensional structure like the winding coil and using the magnetic material for the core material in the MEMS process. Therefore, the magnetic circuit had been employed planer patterns such as a meandering, a spiral. As an example, the complex three-phase copper conductor coil pattern on a flat surface has reached from the milliwatt to watt level [13, 14]. While planar coils are suitable for MEMS processes, the coil pattern extends in the planar direction. Therefore, the radial area increases as the number of coils increases. Without the magnetic material, a long-length coil is required to catch the magnetic flux. Therefore, the formed coil pattern has small output power by high internal resistance.

A continuous power supply that does not require replacement or charging work is important for the IoT communication module. Therefore, it is required to consider the input energy. Energy harvesting technology is currently receiving the most attention on this issue. This is a power-generating method that uses waste energy in the environment. Various methods such as solar power, heat, and vibration are being researched [15, 16]. However, solar power generation requires a large area, and it is easy to store heat. The vibration power generator has a small-scale structure, but it shows nanowatt level output power [17]. The reuse of waste heat energy is attracting attention because factory waste heat can be used. In one of the waste heat energy harvesting systems, a thermoelectric element is researched [18, 19]. It uses the Seebeck effect and produces no noise or vibration. However, it requires a temperature difference, and it has the problem of keeping the temperature difference. Another approach to utilizing waste heat is a thermodynamic cycle generation system. The Rankine cycle undergoes a phase change mechanically, so it is easy to maintain the thermal cycle. The large one is the method used in thermal power plants, etc., and water is used as the working fluid. Most of the waste heat energy that has not been used yet is below 300°C [20], and research on an organic Rankine cycle (ORC) system that uses a low boiling point material as a working fluid to utilize this has been reported [20–22]. However, the commercial size is meter scale, which is too large to be applied to the IoT device. Research on miniaturization is underway, but it is manufactured by a machining process, and little research has been reported on microfabrication processes.

Some energy issues in IoT communication modules have been listed. The following is a summary of them.

- It is important to develop a sustainable miniature power generation system to replace batteries.
- Although many small generators have been developed using silicon microfabrication technology that uses the MEMS process, milliwatt level power generation is a major barrier.
- It is desirable to use low-grade (300°C or less) thermal energy as input energy, but the structure size is a problem at the ORC power generation using a thermal cycle.

To solve the energy problems of the IoT communication module, it can be said that the key is to combine the technologies and materials that have been researched separately so far.

This chapter suggests a combining of the silicon microfabrication technology and the multilayer ceramic technology. By the combination, a miniature generator with the electromagnetic induction type is realized. Magnetic circuits are designed and fabricated for single-phase and three-phase structures, respectively. Moreover, assuming that it will be utilized as a part of the ORC power generation system, which is the thermal cycle system. The rotation experiments using a working fluid of low boiling point material are demonstrated, and a phase change is observed.

11.2 Concept

The ORC generating system is shown in Figure 11.1. The proposed miniature electromagnetic induction type generator will be applied to a turbine

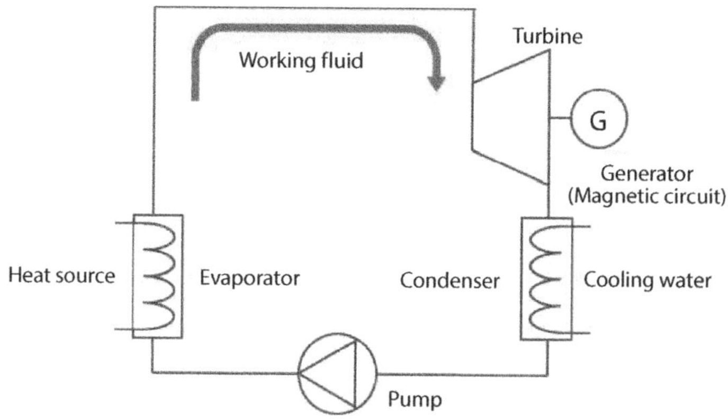

Figure 11.1 Schematic illustration of ORC power generation system.

part and power generation part in this figure. The low boiling point material is the working fluid of the turbine that rotates a rotor blade.

The developed turbine part is made from a single crystal silicon wafer. The photolithography and dry etching processes are used to fabricate the miniature and fine pattern [23]. Furthermore, a carbide shaft through the rotor and structural parts is supported by a miniature mechanical ball bearing. The ball bearing shows a stable rotational motion. The bearing arrangements are changed for each combination of the turbine and magnetic circuit. Moreover, a ring-shape magnet is attached to the shaft, and the magnet is rotated by the rotational motion of the rotor. Through this motion, the magnetic field changes.

The magnetic circuit is made from a magnetic ceramic of Ni-Cu-Zn ferrite and conductive paste of silver. The monolithic three-dimensional coil pattern is fabricated by the multilayer ceramic technology that is usually used for the production process of ceramic electronic elements. Therefore, the mountable magnetic circuit will be realized on the small communication module. The shape of the magnetic core is designed for single-phase and three-phase, respectively.

11.3 Fabrication Technology

11.3.1 Microfabrication Technology of Silicon Material

The MEMS process makes the miniature turbine components. A starting material is a single crystal silicon wafer. Thicknesses are selected for each function. In the first step, for patterns with a deep etching depth, aluminum

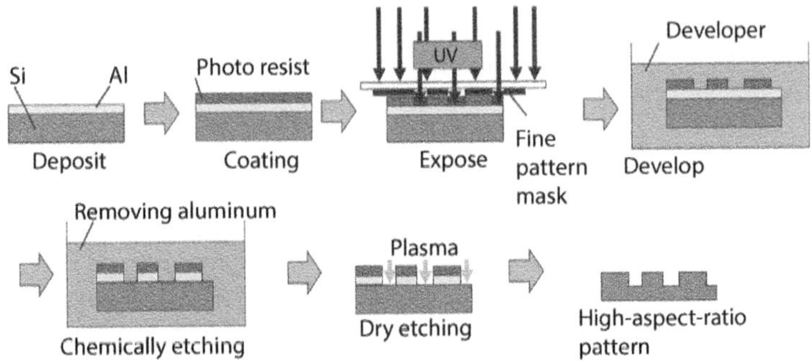

Figure 11.2 Schematic illustration of the fabrication process for the silicon components.

is deposited on the silicon wafer. A photosensitive resist is coated onto the silicon wafer using a spin coater. Coated wafers are dried in a baking furnace to stabilize the photoresist. The coated silicon wafer is exposed through a designed mask pattern. Patterned wafers are developed by soaking the developer. After development, the residual developer solution on the surface is rinsed and the wafer is baked again. The aluminum deposited design is through an aluminum etching process to obtain a wafer with a mask of an arbitrary pattern on its surface. And then, the wafer with a microstructure is dry etching process using a combination of high-aspect-ratio inductively coupled plasma etching and the Bosch process [23]. To make more complex structures, the same process is applied to the reverse side. The schematic illustration of the fabrication process is shown in Figure 11.2.

11.3.2 Multilayer Ceramic Technology

Multilayer ceramic technology is a technology that realizes three-dimensional wiring and high-density wiring by laminating ceramic sheets before firing. Various ceramic and wiring materials are used. In this research, the magnetic circuit is made from a magnetic ceramic of low-temperature co-fired Ni-Cu-Zn ferrite with a permeability of 900. By introducing the magnetic ceramic material, the magnetic core is formed simultaneously. In addition, firing temperature of Ni-Cu-Zn ferrite is around 900°C. Therefore, silver, a low-resistance conductive material with a low sintering temperature, can be used in the coil pattern.

In the first step, the ferrite slurry that is liquid condition ceramic is formed into a ceramic sheet using the doctor blade. The ferrite slurry is made from the ferrite ceramic powder, dispersing agent, plasticizer, toluene, xylene,

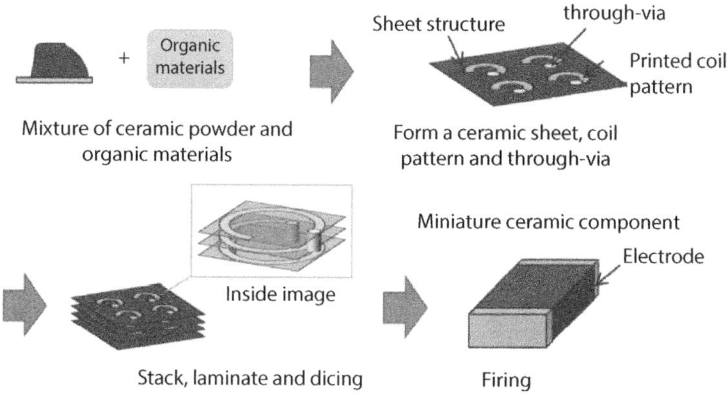

Figure 11.3 Fabrication process of multilayer ceramic technology.

isopropyl alcohol, and binder. These materials are mixed with the ball-mill method. After the dried ceramic sheets are formed through holes, the coil patterns are screen printed. Silver paste is printed for patterning the electrode and conductive coil. The patterned ceramic sheets are achieved by this process. Patterned sheets are laminated to construct the three-dimensional coil and magnetic core inside the magnetic circuit. Through a dicing process, the laminated ceramic shapes into the designed structure. A schematic illustration of the fabrication process is shown in Figure 11.3.

For the single-phase magnetic circuit, the diced coil structure is fired. Moreover, ferrite sheets without printed coil patterns are laminated and fired to form the magnetic material that serves as the magnetic flux path. The coil and the ferrite structures are combined for shaped complex circuit structures like the separated-type magnetic circuit.

Each coil is connected in a Y-connection pattern on the connecting layer to achieve the three-phase multilayer magnetic circuit. The connected layer is stacked. The connecting wire is printed ferrite sheets and blank pattern sheets. Figure 11.4 shows the combined process.

11.4 Designs and Experiments

11.4.1 Designs of Turbine and Magnetic Circuit for Single-Phase Type

Figure 11.5 shows the single-phase type miniature power generation part that combines the silicon turbine and ceramic magnetic circuit. The

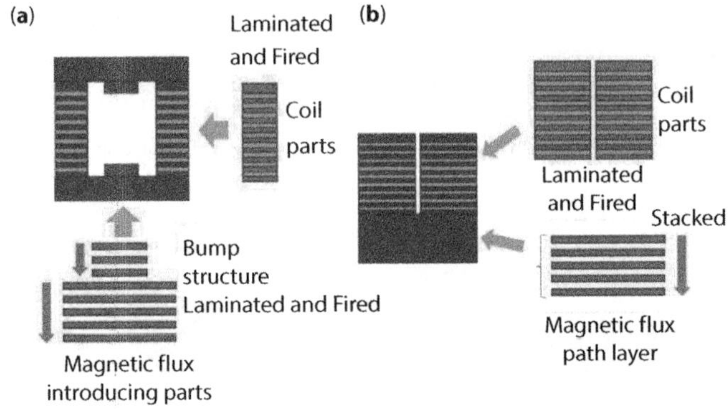

Figure 11.4 Fabrication process for complex structures (a) single-phase coil (b) three-phase coil.

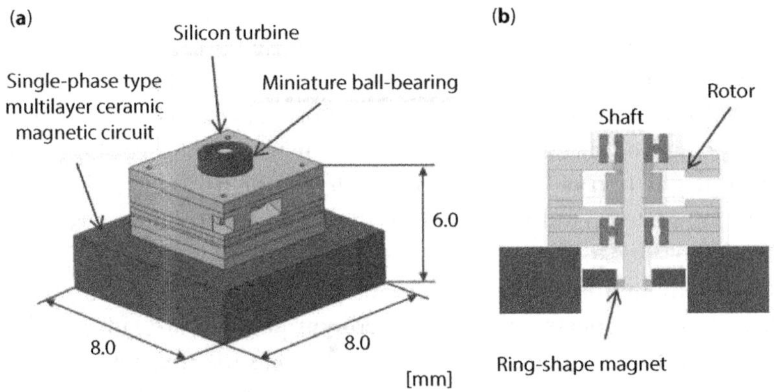

Figure 11.5 Design of single-phase type miniature power generation part that combines the silicon turbine and ceramic magnetic circuit (a) overall view (b) cross-section view.

ring-shape magnet is neodymium with a magnet 2-pole radial direction. Its dimensions are 3.0 mm (outer diameter), 1.0 mm (inner diameter), and 0.5 mm (height), respectively. Therefore, the single-phase type has a shape in which a magnet is placed outside the turbine structure and covered with a magnetic circuit. The overall size is less than 10 mm square.

Figure 11.6 shows the turbine design for the single-phase. The turbine is constructed of seven layers. The turbine is assembled by a laminating process, and an assembled error is suppressed by a pin alignment pattern. Inside the turbine, a ball bearing made from martensitic stainless steel are placed up and down the rotor blade. Dimensions of the ball bearing are as follows, inner diameter 0.6 mm, outer diameter 2.0 mm, and height

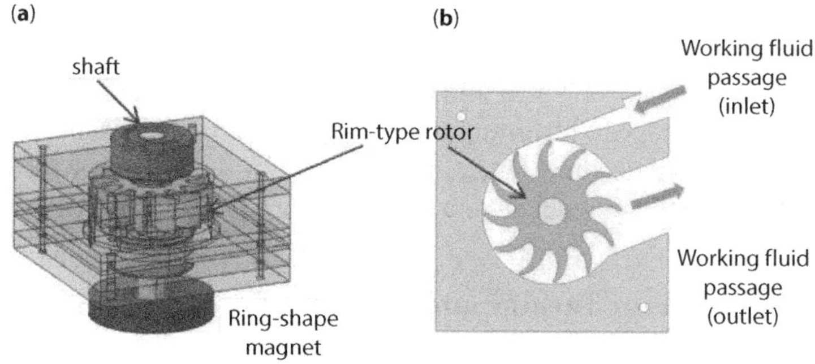

Figure 11.6 Designed single-phase turbine (a) arrangement of parts (b) rotor blade and a flow path pattern.

0.8 mm. Ball bearings are fixed in the top and middle layers. In Figure 11.6 (b), the rotor blade and a flow path are shown. The rotor shape is rim structure, diameter and thickness are 3.0 mm and 850 μm. Rotational motion is generated by the working fluid. A brass tube for the inflow of the working fluid is fixed at the inlet portion of the turbine. The fluid passes through the channels formed on the silicon layer, rotates the rotor blades, and is discharged through an outlet port formed on the side.

The radial magnetizing type 2-pole magnet is arranged at a distance downward from the turbine case. The rim-type rotor is putted in the center of the turbine, the channels form around it. The external dimensions of the turbine are 5.0 mm square and height of 4.6 mm.

Figure 11.7 shows the single-phase type of multilayer ceramic magnetic circuit. The designed circuit has a coil pattern and a magnetic core. The

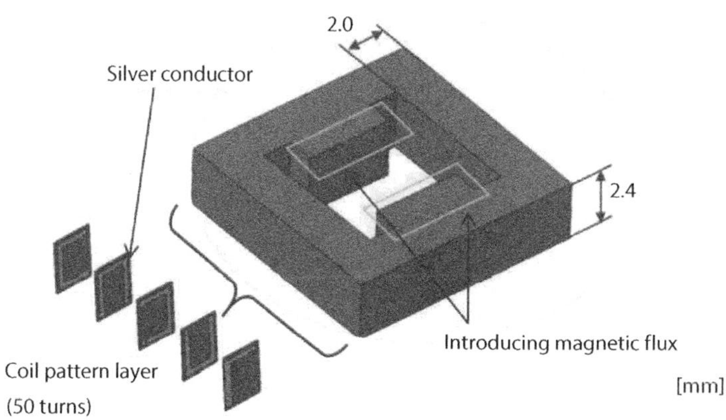

Figure 11.7 Multilayer ceramic magnetic circuit of single-phase.

turn number is 50 turns on one coil, and two coil structures are arranged on both sides. Therefore, the total turn number is 100 turns. The dimensions of coil parts are 2.0 mm and 2.4 mm. Magnetic ceramic structures with a complex shape to introduce magnetic flux connected each coil component. The magnetic circuit is an enclosed type and has a bump structure to be gapless for the magnet arranged in the center.

11.4.2 Designs of Turbine and Magnetic Circuit for Three-Phase Type

To facilitate mounting of the ceramic magnetic circuit on the module substrate in the three-phase type, the coil patterns in the magnetic circuit are arranged upward. And then, an axial direction magnetized ring-shape magnet is employed. The three-phase generator has the turbine structure at the top and the magnetic circuit at the bottom. Figure 11.8 shows the combined three-phase type turbine and magnetic circuit. The dimensions of the combined turbine and magnetic circuit are 5.0 mm (L), 5.0 mm (W), and 7.0 mm (H).

Figure 11.9 shows the designed three-phase turbine. The basic design is the same as the single-phase turbine. The turbine structure has 5.0 mm square, and the height including the magnet is 4.7 mm. Two ball bearings are located above and below the rotor. The shaft has a 2-pole axial direction magnetized ring-shape magnet placed at the tip. In addition, the magnet has a magnetic yoke on the top. And it captures magnetic flux diverging upward. The silicone case around the magnet adjusts the gap between the magnet and the magnetic circuit.

Figure 11.10 shows a designed three-phase type magnetic circuit. Three-phase magnetic circuits are expected to have higher output with fewer turns

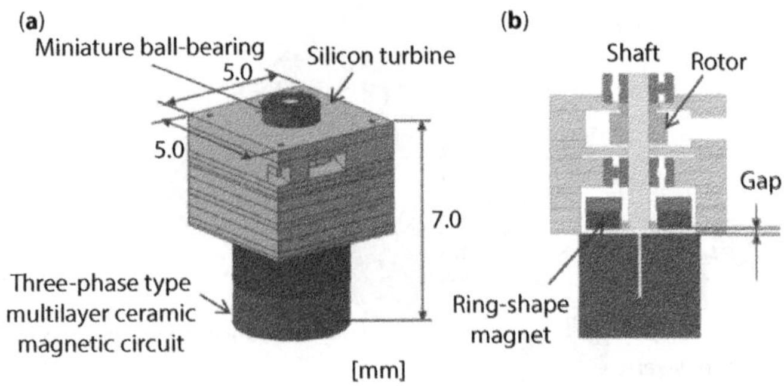

Figure 11.8 Design of combined three-phase type turbine and magnetic circuit (a) overall view (b) cross-section view.

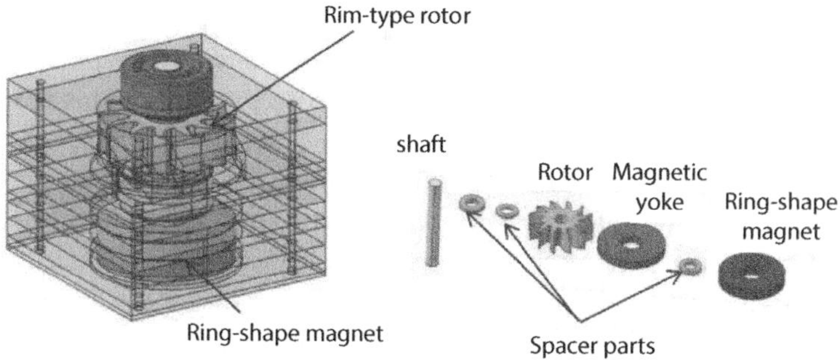

Figure 11.9 Designed three-phase type turbine and components.

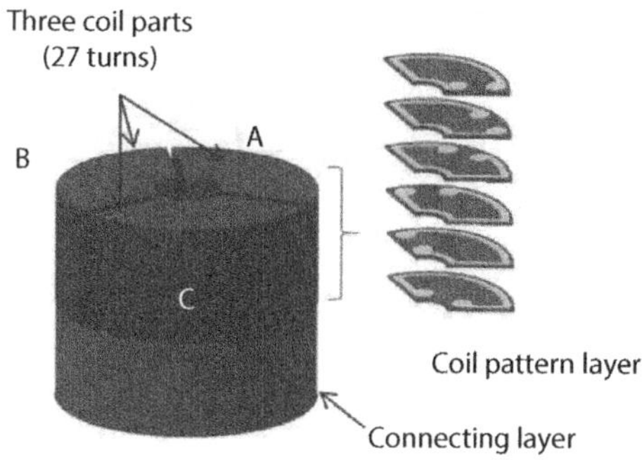

Figure 11.10 Designed three-phase type magnetic circuit.

than single-phase circuits. The developed magnetic circuit has three coil parts with 27 turns each and a connection layer. The three coils are provided with an air gap when they are arranged because it suppresses the magnetic flux leakage. The three-phase circuit is connected with Y-connection. A closed magnetic path is formed by the ceramic magnetic circuit and the magnetic yoke with the rotating magnet combined. Designed dimensions of the magnetic circuit are 5.0 mm (D) and 2.5 mm (H). The three-phase magnetic circuit is adjusted by combining three sintered coil structures so that the DC resistance of each coil is comparable. Gaps are formed between adjacent coils. This design balances the collection of the magnetic flux from the magnet and the suppression of the effects of magnetic flux.

11.4.3 Rotational Experiment and Rotor Blade Design

The fabricated turbine structures demonstrate rotational motion by compressed nitrogen gas. A schematic illustration of the rotational experiment is shown in Figure 11.11. The compressed nitrogen gas filled in a cylinder enters the turbine structure through a flowmeter. Then, the gas rotated the rotor passes out through the outlet path. The rotational motion is observed by the rotating magnet and a combined hall-sensor.

In addition, two types of rotor blade shapes are compared in the rotational experiment because the rotor blade shape affects the operating duration of the turbine generator. Each rotor is proposed for the shape of the blade tip that receives the most load from the working fluid. Figure 11.12 shows the designs of the proposed rotor. One is a type with a sharp tip on the rotor blade, width is 20 μm. Another one has a blunt angle and a width of 180 μm. The diameter and height of each rotor are 3 mm and 0.85 mm, respectively. The flow path is designed for each rotor.

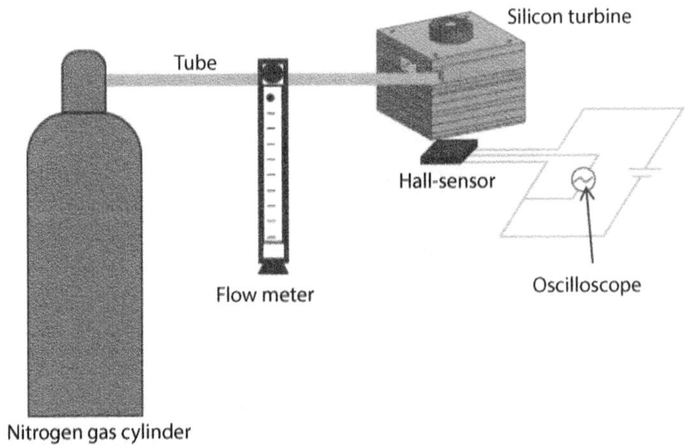

Figure 11.11 Schematic diagram of the experimentation of the rotation.

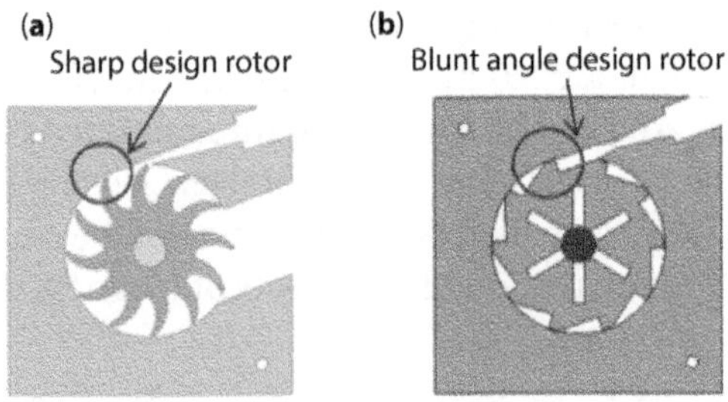

Figure 11.12 Designs of the proposed rotor (a) sharp design (b) blunt angle design.

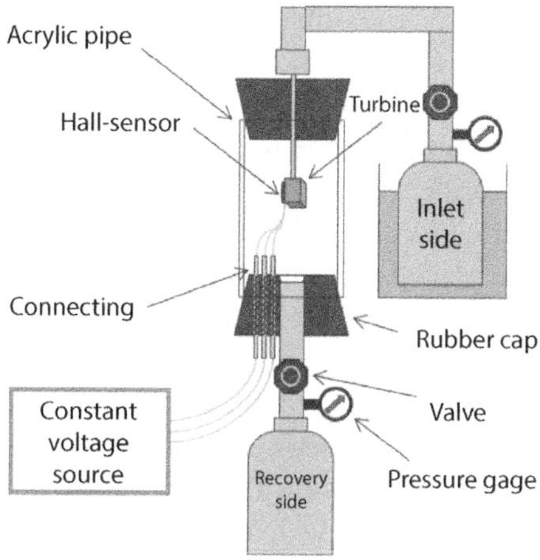

Figure 11.13 State of the rotation experiment using a low boiling point medium.

11.4.4 Low Boiling Point Fluid and Experiment

The proposed silicon turbine and ceramic magnetic circuit are being researched as components of the miniature ORC power generation system. Therefore, HFE7000 (NovecTM7000, 3M) with a boiling point of 34°C as the low boiling point fluid is employed for the rotating experiment. HFE7000 is a fluorine-based solvent that is also used as a heating medium. In addition, HFE7000 is an environmentally friendly material with a low global warming coefficient. Figure 11.13 shows the state of the rotation experiment when a low boiling point material is flowed. The low boiling point material filled in the cylinder is heated from outside the cylinder. The low boiling point fluid expanded by the phase change inside the cylinder flows into the turbine through the piping connected to the turbine and rotates the rotor. The fluid that contributed to the rotation is recovered after being discharged from the turbine, and the inflow amount and the recovered amount are measured. Further, since the thermal cycle is a cycle accompanied by a phase change, a phase change is observed for the low boiling point fluid passing through the turbine.

11.5 Results and Discussion

11.5.1 Fabricated Evaluation

Fabricated silicon components were assembled. Silicon components and the assembled turbine structures of single-phase and three-phase are shown in Figures 11.14 and 11.15. The dimension errors of the components were less than 10 μm. By achieved parts, the high-accuracy components were formed. The length and width of the assembled single-phase type of turbine were 6.2 mm, 5.6 mm, and the height was 4.8 mm. The ring-shaped magnet was arranged under the turbine structure in the single-phase type of turbine, the magnetic flux generated by the magnet is efficiently induced in the magnetic circuit placed around them. Dimensions of the three-phase type turbine structure were 6.2 mm (L), 5.6 mm (W), and 4.5 mm (H). The three-phase turbine had a silicon layer covering the magnet, which kept a 130 μm distance of the magnetic circuit from the magnet.

The single-phase ceramic magnetic circuit is shown in Figure 11.16. The complex monolithic structure that a three-dimensional wiring coil containing a magnetic core was achieved. The sizes of the circuit were 7.4 mm (L), 8.5 mm (W), and 2.4 mm (H). Internal DC resistance at 100 turns was 1.6 Ω. Figure 11.17 shows the three-phase ceramic magnetic circuit. The achieved circuit had an air gap between the coil patterns. And each coil pattern was connected to the magnetic material that was placed at the bottom. Dimensions were 5.5 mm of diameter and 2.6 mm of height. Each fabricated coil had the internal resistance between 0.85 Ω and 0.86 Ω. Inductance values in each phase A, B, C at 1 MHz were 84 μH, 70 μH,

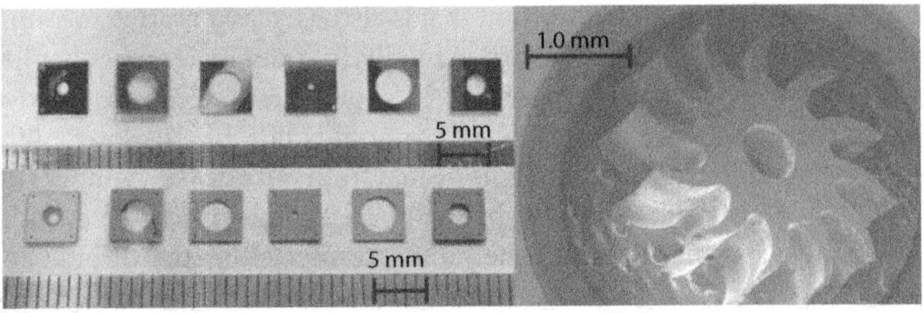

Figure 11.14 Silicon components of MEMS turbine for the single-phase and three-phase.

Figure 11.15 Assembled turbine structures of the single-phase and three-phase.

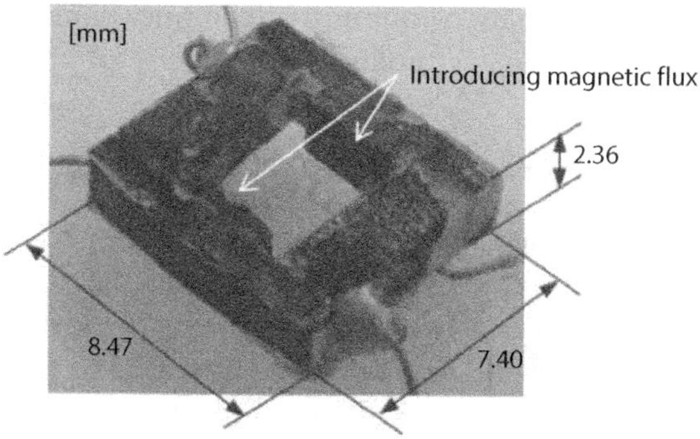

Figure 11.16 Ceramic single-phase magnetic circuit.

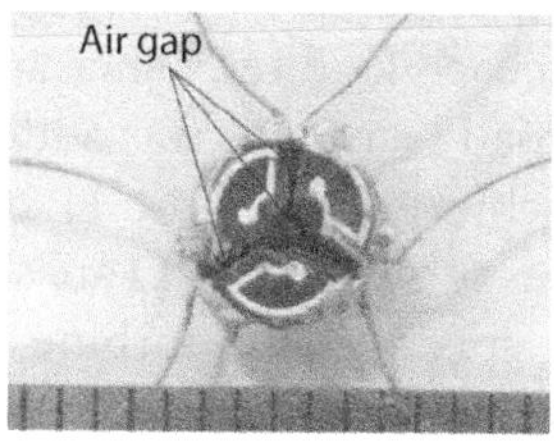

Figure 11.17 Ceramic three-phase magnetic circuit.

and 53 μH. External electrodes were formed on achieved coils, and wires were connected for output power measurement. The miniature electromagnetic induction generator was achieved by combining turbine structures and ceramic magnetic circuits. Figure 11.18 shows the combined generators of single-phase and three-phase. These generators realized the

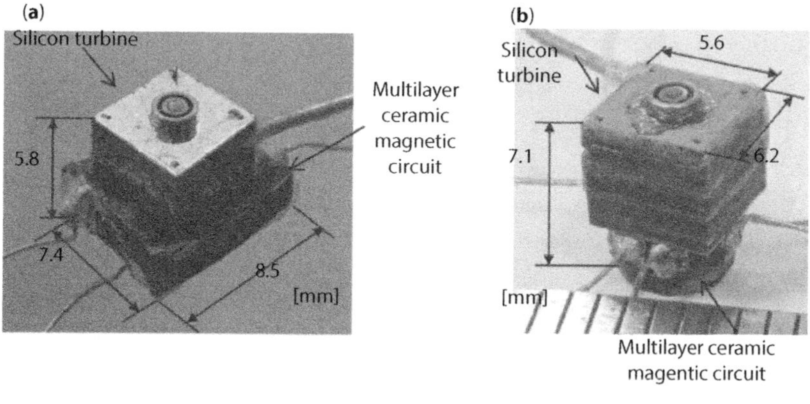

Figure 11.18 Combined turbine generators (a) single-phase generator (b) three-phase generator.

millimetre-scale structure by combining silicon microfabrication technology and multilayer ceramic technology.

11.5.2 Rotational Result

For evaluating the turbine generator, compressed nitrogen gas was flowed to generate rotational motion. A metal tube for gas injection was attached to the turbine structure and the injected gas was regulated with a flow meter. The output voltage and rotation speed were read by the output waveform of the oscilloscope. The maximum speed of the turbine rotor of the single-phase generator was 290,135 rpm, and the experimental conditions at this time were an inlet flow rate of 2.4 l/min and a pressure of 0.3 MPa. The output voltage and output power when the maximum rotational speed conditions were applied at each load resistance are shown in Figure 11.19. As a result, when 8 Ω as load resistance was added, the maximum output power of 2.41 mVA was achieved. In Figure 11.20, each output waveform in the load resistance values of 8 Ω and 1 kΩ are shown. In the three-phase generator, the maximum speed of the turbine rotor was 228,484 rpm, and the experimental conditions at this time were an inlet flow rate of 2.9 l/min and a pressure of 0.3 MPa. Figure 11.21 is the output results that apply various load resistance values and condition of maximum rotation speed. The output phase power then reached its maximum value at load resistance 3 Ω, which was 3.01 mVA. Figure 11.22 shows the respective output waveforms when load resistors 3 Ω and 1 kΩ are connected. By these

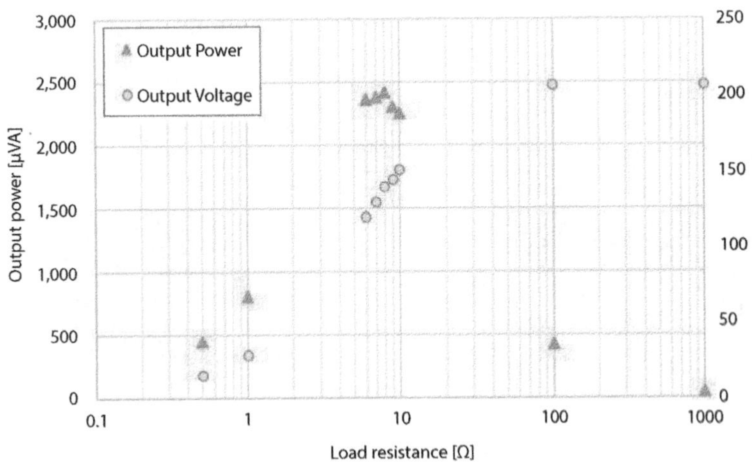

Figure 11.19 Output voltage and output power applied at each load resistance.

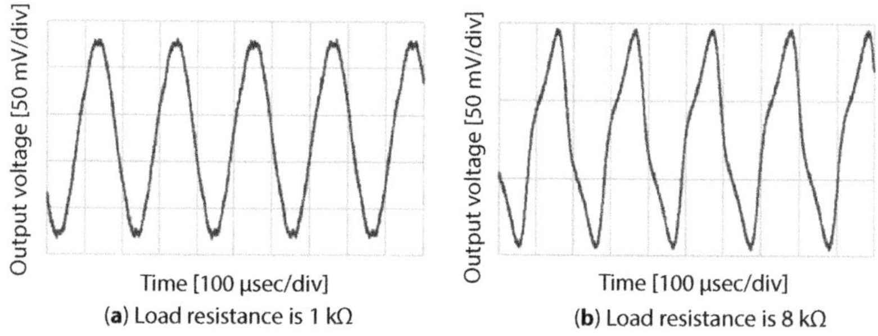

Figure 11.20 Output waveforms in generator of single-phase.

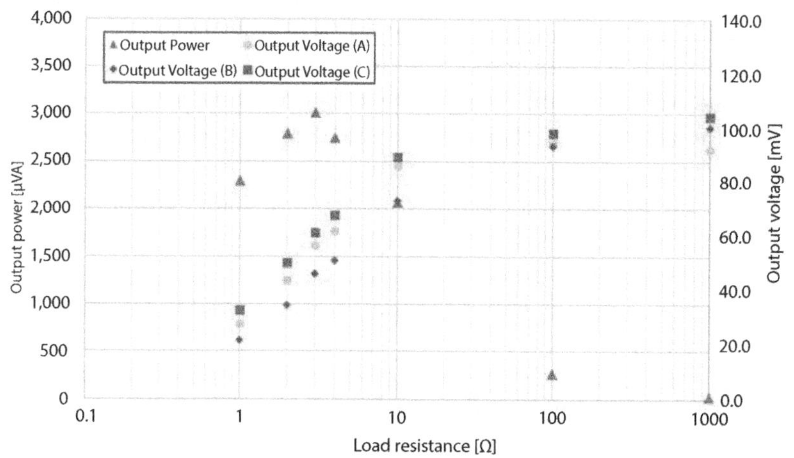

Figure 11.21 Output results that apply each load resistance and maximum rotational speed.

power generation results, the output power of milliwatts was achieved in millimeter-scale structure by combining the silicon microfabrication technology and a multilayer ceramic technology.

The result of the comparison of the theoretical output voltage values and the achieved output voltage values was following. The output value was calculated from equation (11.1). The magnetic flux of the designed multilayer ceramic circuits was analysed. The interlinkage-magnetic-flux from the FEM analysis is denoted by BS in equation (11.1).

$$e = NBS\omega \sin\omega t \tag{11.1}$$

(electromotive force: e, turn number: N, angular velocity: ω)

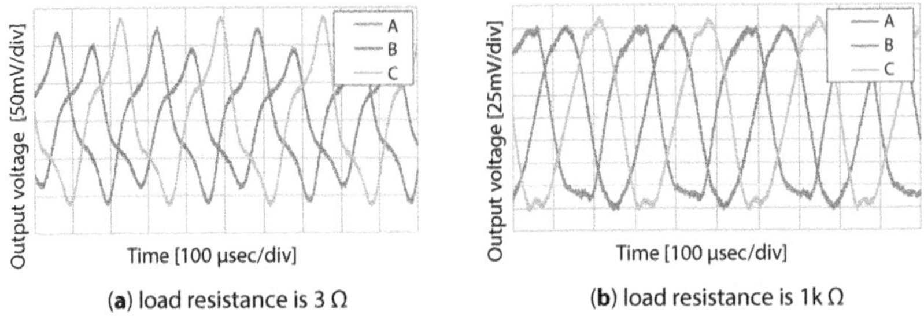

Figure 11.22 Output waveform of three-phase type generator.

From equation (11.1), the single-phase generator had a voltage value of 479 mV, and the measured value at 1 k load resistance was 43% of the calculated value in comparison. One of the causes of the loss is the leakage flux between the flux-introducing structure and the magnet. Figure 11.23 is the analysis results of the designed magnetic structures. Figure 11.23 (a) shows the results of a circuit with a square shape bump and (b) shows the results of a circuit with a round shape bump. In (a), the leakage flux was observed from the corner portion at a distance to the ring-shaped magnet. However, in (b), the distance to the magnet was equally narrow due to the round shape, indicating that the magnetic flux loss was reduced.

In the three-phase type generator, calculation indicated the output of one phase was 149 mV. Phase C showed maximum output voltage of 104 mV when the load resistance was 1 kΩ in the measurement result. Therefore, the fabricated generator achieved about 70% of the calculated value. The output values of the experiments and the calculated values were compared. Equation (11.2) shows the output power at one phase. The interlinkage-magnetic-flux indicated by FEM analysis was assigned to BS. Figure 11.24 shows the analysed three-phase circuit model, and Table 11.1

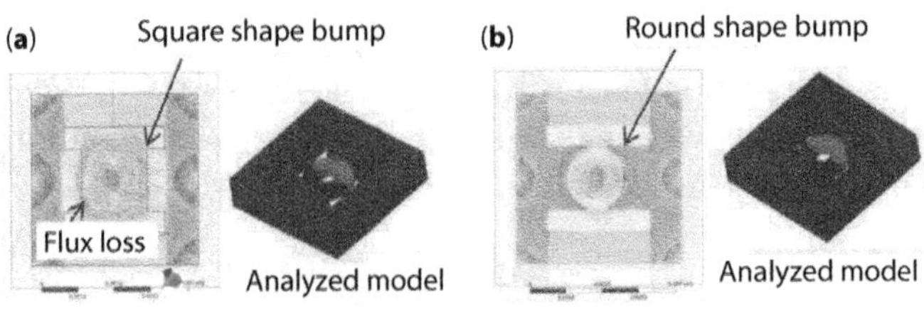

Figure 11.23 Analysis result of designed magnetic structures (a) square shape bump (b) round shape bump.

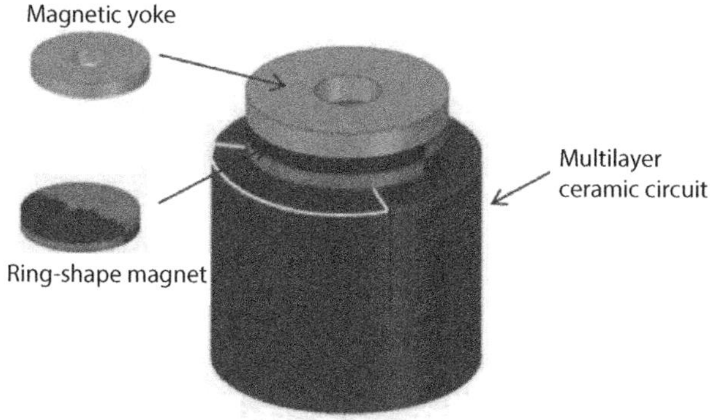

Figure 11.24 Analysis model of three-phase circuit and magnet.

Table 11.1 Each parameter of the analysis.

Magnetic flux density	Permeability		Distance (Magnet-magnetic cicuit)
	Ni-Cu-Zn ferrite	Magnetic yoke	
150 mT	900	4000	130 μm

shows analysing parameters. The analytical result of 0.231 μWb was used in the calculation.

$$P = \frac{R(NBS\omega)^2}{(r_0 + R)^2 + (\omega L_0)^2} \tag{11.2}$$

(load resistance: R, turn number: N, angular velocity: ω, internal resistance: r_0, inductance, L_0)

The maximum power at three-phase was compared, the calculation result was 7.18 mVA and the experiment result was 3.01 mVA. The obtained output power in the experiment was 42% of the calculated value.

The difference in the ratio of voltage and power to the calculated value is the balance of the coil arrangement. Coils with similar resistance values were selected in this experiment and connected to fabricate the three-phase magnetic circuit. The power was calculated by adding the three single-phase powers together. However, experimental results showed different output voltage values for each coil. The maximum output voltage was 104 mV at phase C, but the minimum output voltage at phase A showed under 90 mV. In addition, the differences in inductance values at 1 MHz

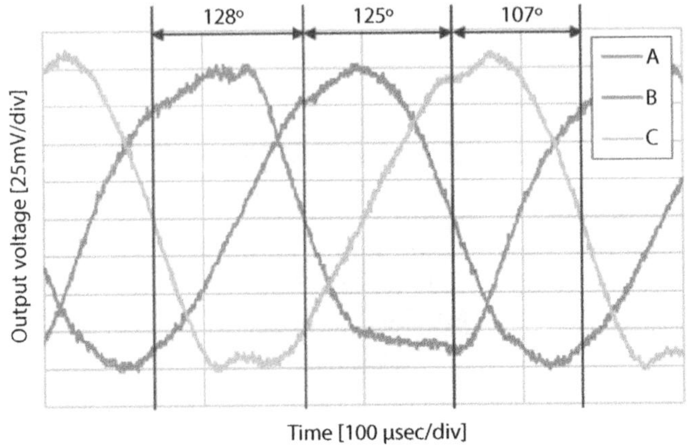

Figure 11.25 Phase difference error of three-phase output waveform.

were shown. Therefore, differences in output values due to differences in these occurred.

In Figure 11.25, the phase difference error is observed. 120° is ideal phase difference for three-phase. In Figure 11.25, a maximum error of the phase difference was 13°. For realizing high output power, assembly process requires the alignment pattern.

11.5.3 Comparison of Rotor Shape and Rotational Motion

Figure 11.26 shows the fabricated rotors. Each of the fabricated rotors was mounted in the three-phase turbine structure, and the rotation rates were compared. The compressed nitrogen gas was employed as working fluid, and it was not combined with the magnetic circuit. Figure 11.27 shows the

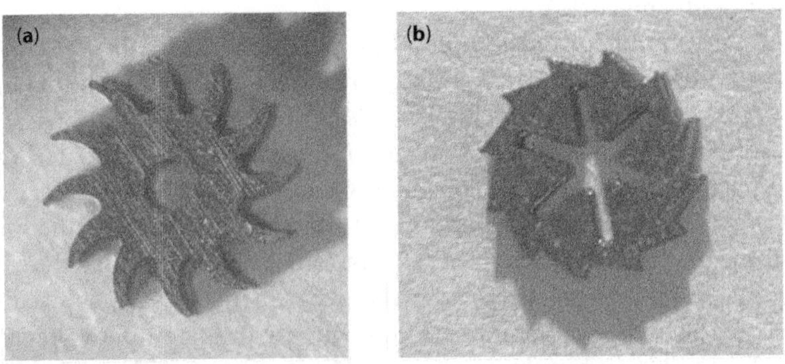

Figure 11.26 Fabricated two type shape rotor (a) sharp design (b) blunt angle design.

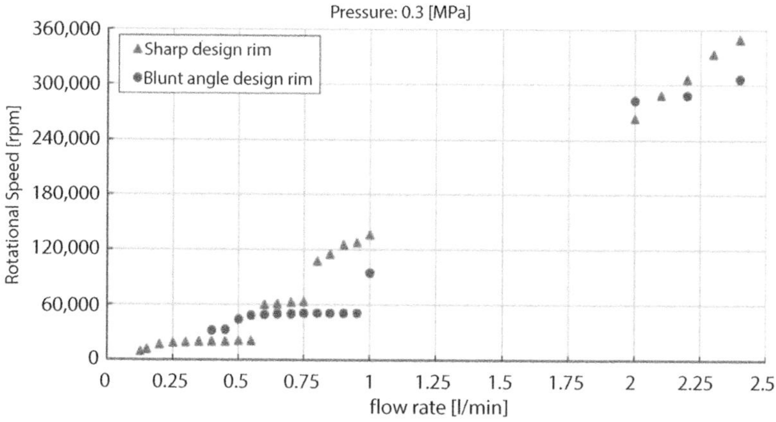

Figure 11.27 Result of rotational speeds at each rotor design.

result of rotational speeds at each rotor design. The maximum rotational rates of each designed rotor were 348,840 rpm (sharp design) and 306,120 rpm (blunt angle design), respectively. The experiment conditions were pressure of 0.3 MPa and flow rate of 2.4 l/min. As a result, the difference in the rotational speed of the two rotors was acceptable.

Moreover, the sharp design rotor damage was observed when the turbine was continuously rotated at the maximum speed by compressed nitrogen gas during the turbine rotation experiment. Figure 11.28 shows the results of the observation of the damaged rotor. However, no damage was observed in the rotor with a blunt angle design. The length of operating

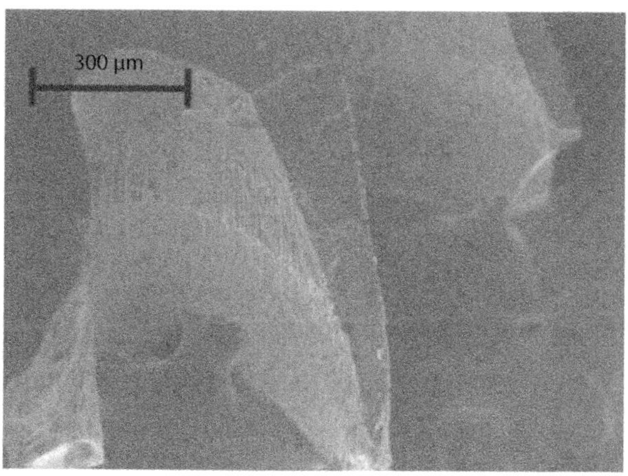

Figure 11.28 Results of the observation of the damaged rotor.

time is also important for the miniature ORC power generation system. Therefore, the blunt angle designed rotor was used for the rotating experiment with the low boiling point fluid.

11.5.4 Phase Change

A rotational experiment was conducted on a fabricated silicon turbine with low boiling point material for the development of the miniature ORC power generation system. The fluid that contributed to the rotational motion was recovered and the results are shown in Tables 11.2 and 11.3. The tables show the measured weights of the inlet and recovery cylinders, from which the recovery rates were calculated.

Figure 11.29 shows the output waveform observed on the oscilloscope during the experiment. From Figure 11.29, it was observed that the silicon turbine rotated at 41,096 rpm when heated at 55°C from the outside to a low boiling fluid. In addition, from Tables 11.2 and 11.3, about 96% of the low boiling fluid that flowed into the turbine was recovered. The low boiling fluid discharged from the turbine during the experiment was in a gas-liquid two-phase state. The uncollected low boiling material was found to remain at the bottom of the acrylic pipe in a liquid phase after the experiment. The weight of the low boiling material remaining in the pipe and the weight of the recovered medium in the cylinder on the recovery side were combined and found to be the same as the weight of the inflow.

The fact that the working fluid discharged from the turbine exhibited a gas-liquid two-phase behavior indicated that a phase change was occurring in the turbine. As a result, we confirmed the phase change using the

Table 11.2 The measurement result of the inlet side.

Inlet side	Weight [g]	Gage pressure [MPa]	Temperature [°C]
Before	1012	0.09	55.4
After	964	0.05	51.0

Table 11.3 The measurement result of the recovery side.

Recovery side	Weight [g]	Gage pressure [MPa]	Temperature [°C]
Before	1129	–0.1	10.4
After	1175	–0.02	8.2

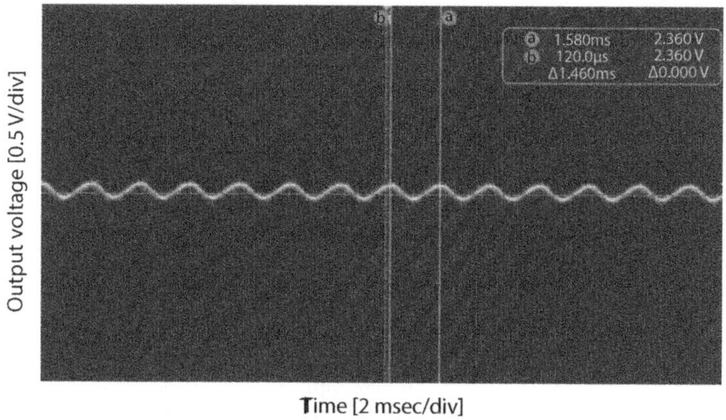

Figure 11.29 Output waveform with low boiling point material.

model and experimental arrangement shown in Figure 11.30. The rotor was removed from the previous turbine, and an acrylic window was formed for the model. A lid was also placed on the bottom. As in the rotational experiment, the low boiling material enclosed in the cylinder was heated from the outside. Heating was also performed using a ribbon heater to maintain the gas phase until just before it flowed into the turbine. The flow path from the cylinder to the turbine was installed so that only the gas phase flowed into the turbine during the inflow. The working fluid was discharged into the atmosphere, and the temperature change was observed with a thermographic camera along with the observation of the phase change.

At an inflow pressure of 0.04 MPa, an applied temperature of 47°C, and room temperature of 22°C, the phase change, and temperature change were observed. Figure 11.31 shows the observed results. "α" in the image is the point where the fluid through the inflow path is injected into the rotor blades.

Figure 11.31 shows that temperature changes were observed before inflow, immediately after inflow, and at point "α" during inflow. This suggests that adiabatic expansion of the low boiling fluid during injection into the rotor lowers the temperature. When comparing the temperatures immediately after the inflow and 40 seconds after the inflow, it was observed that the temperature of the entire turbine compartment was lower immediately after the inflow, whereas only the lower part of the turbine was cooler 40 seconds after the inflow. The gas-liquid two-phase fluid in the turbine could adversely affect the rotation of the rotor in the miniature ORC power generation system. However, because the working fluid is in the liquid phase when it is discharged from the turbine allows

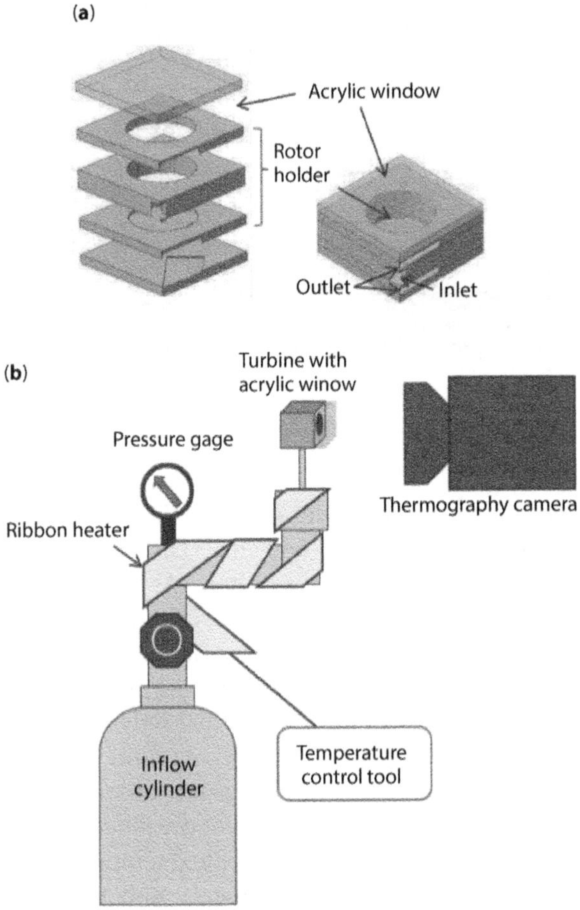

Figure 11.30 Schematic illustration of phase change observation (a) model (b) experimental arrangement.

the condenser to be omitted in the ORC, leading to higher efficiency. It is desirable for the development of a miniature ORC power generation system for the fluid to be in the gas phase when the turbine is rotated and in the liquid phase when the turbine is discharged. To achieve this, the current design, in which the flow velocity is increased by gradually narrowing the inlet path just before it acts on the rotor, should be reviewed, and the turbine should be placed near the heat source to maintain the gas phase and prevent rapid temperature changes inside the turbine. In addition, the design of the flow path to liquefy the working fluid only at the outlet will eliminate the condenser and allow the turbine to rotate at higher speeds, thereby increasing the efficiency of the Rankine cycle.

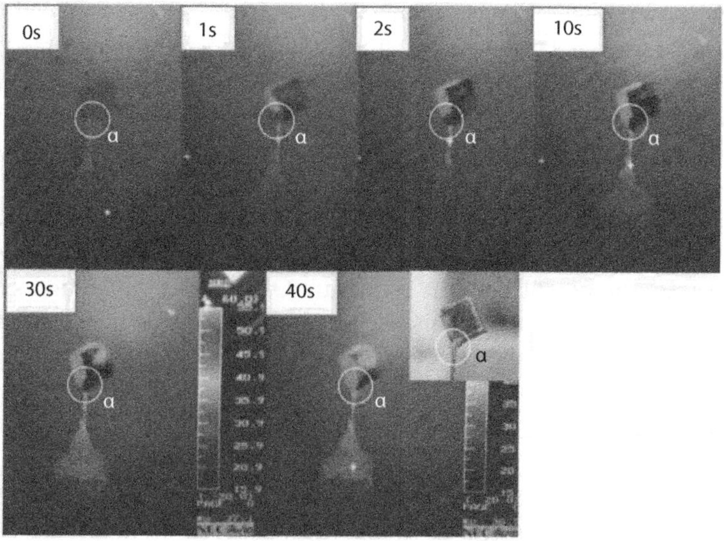

Figure 11.31 Temperature changes results.

11.6 Conclusions

The millimeter-scale compact generator that employed the electromagnetic induction type was realized by combining silicon microfabrication technology and multilayer ceramic technology in this research. The power generation method employed the revolving-field type, in which the magnet is rotated by the rotation of the turbine. The turbine mechanism, which was a component of the generator, used silicon microfabrication technology that enables high-precision, high-aspect-ratio patterns to be formed. Therefore, a miniature turbine with a rotor and flow channels of about 5 mm square was obtained. The optimal shape of the rotor was also studied. For the magnetic circuit, multilayer ceramic technology, which is a mountable passive components fabrication technology, was employed. Two types of ceramic circuits, the single-phase, and the three-phase were obtained. The fabricated magnetic circuits realized the three-dimensional wiring pattern and the complex structure to induce magnetic flux and exhibited low resistance. The single-phase magnetic circuit had a total of 100 coils, with 50 coils on each side, and the resistance was 1.6 Ω. The overall size was less than 10 mm square. The three-phase magnetic circuit was connected by combining three coils of 27 turns in each phase. Resistance values ranged from 0.85 to 0.86 Ω, and the diameter was less than 6 mm. The silicon turbine and ceramic magnetic circuit were combined with compressed nitrogen gas as the working fluid for power generation experiments. Therefore, 2.41 mVA at load resistance of 8 Ω was obtained for the single-phase generator. The three-phase generator produced 3.01 mVA when the load resistances of 3 Ω were connected to each phase. Furthermore, since the system is intended to be used as part of the ORC power generation system, which is a thermal cycle system, the rotational motion using a low boiling working fluid was demonstrated. When the heat of about 50°C was applied externally, the rotational speed of 4,1096 rpm and 96% recovery of the working fluid was realized. These results demonstrate the potential of the miniature ORC power generation system for IoT modules by combining the technologies and materials.

Acknowledgment

This work was supported by the Research Center for Micro-functional Devices, Nihon University, Grant-in-Aid for Promotion of Innovative Research, Institute of Science and Engineering, Nihon University, and Grant-in-Aid for Young Scientists, Nihon University. This work was also supported by Grant-in-Aid for Scientific Research 21K14214.

References

1. Fuqaha, A.A., Guizani, M., Mohammadi, M., Aledhari, M., Ayyash, M., Internet of things: A survey on enabling technologies, protocols, and applications. *IEEE Commun. Surv. Tutor.*, 17, 2347, 2015.
2. Suzuki, K. and Tanigawa, H., Single crystal silicon rotational micromotors. *Proc. IEEE Micro Electro Mechanical Systems*, Nara, pp. 15–20, January 02-30, 1991.
3. Jeon, Y.B., Sood, R., Jeong, J.–H., Kim, S.-G., MEMS power generator with transverse mode thin film PZT. *Sens. Actuators A Phys.*, 122, 16, 2005.
4. Epstein, A.H. and Senturia, S.D., Macro power from micro machinery. *Science*, 276, 1211, 1997.
5. Epstein, A.H., Millimeter-scale, MEMS gas turbine engines. *Proceedings of ASME Turbo Expo 2003 Power for Land, Sea and Air*, Atlanta, pp. GT–2003-38866, pp. 669-696, June 16-19, 2003.
6. Epstein, A.H., Millimeter-scale, micro-electro-mechanical systems gas turbine engines. *J. Eng. Gas Turbines Power*, 126, 205, 2004.
7. Janicek, V. and Husak, M., Designing the 3D electrostatic microgenerator. *J. Electrostat.*, 71, 214, 2013.
8. Wei, M., Rufer, L., Zohar, Y., Wong, M., Design and implementation of an integrated floating-gate electro static power micro-generator. *The 13th*

International Conference on Solid-State Sensors, Actuators and Microsystems, 2005. Digest of Technical Papers. TRANSDUCERS '05, vol. 1, p. 299, 2005.

9. Holmes, A.S., Hong, G., Pullen, K.R., Axial-flux permanent magnet machines for micropower generation. *J. Microelectromech. Syst.*, 14, 54, 2005.
10. Herrault, F., Ji, C.H., Allen, M.G., Ultraminiaturized high-speed permanent-magnet generators for milliwatt-level power generation. *J. Microelectromech. Syst.*, 17, 1376, 2008.
11. Das, S., Arnold, D.P., Zana, I., Park, J.W., Allen, M.G., Multi-watt electric power from a microfabricated permanent-magnet generator. *2005 IEEE 18th International Conference on Micro Electro Mechanical Systems*, FL, pp. 287–290, July 05, 2005.
12. Raisigel, H., Cugat, O., Delamare, J., Permanent magnet planar micro-generators. *Sens. Actuators A Phys.*, 130, 438, 2006.
13. Holmes, A.S., Hong, G., Pullen, K.R., Axial-flux permanent magnet machines for micropower generation. *J. Microelectromech. Syst.*, 14, 54, 2005.
14. Herrault, F., Ji, C.H., Allen, M.G., Ultraminiaturized high-speed permanent-magnet generators for milliwatt-level power generation. *J. Microelectromech. Syst.*, 17, 1376, 2008.
15. Beeby, S.P., Tudor, M.J., White, N.M., Energy harvesting vibration sources for microsystems applications. *Meas. Sci. Technol.*, 17, R175, 2006.
16. Khan, F., Sassani, F., Stoeber, B., Nonlinear behavior of membrane type electromagnetic energy harvester under harmonic and random vibrations. *Microsyst. Technol.*, 20, 1323, 2014.
17. Gokhale, N., Parmar, M., Rajanna, K., Nayak, M.M., Piezoelectric zinc oxide thin film for MEMS application: A comparative study. *3rd International Conference on Sensing Technology*, Tainan, pp. 543–546, Nov. 30-Dec. 3, 2008.
18. Deng, F., Qiu, H., Chen, J., Wang, L., Wang, B., Wearable thermoelectric power generators combined with flexible supercapacitor for low-power human diagnosis devices. *IEEE Trans. Ind. Electron.*, 64, 1477, 2017.
19. Ikoma, K., Munekiyo, M., Furuya, K., Kobayashi, M., Izumi, T., Shinohara, K., Thermoelectric module and generator for gasoline engine vehicles. *Proc. Seventeenth International Conference on Thermoelectrics*, Nagoya, pp. 464–467, May 28-28, 1998.
20. Kosuda, O., Hikichi, T., Kido, O., Nishiyama, N., Development of air-cooled compact organic rankine cycle power generation technology utilizing waste heat. *Energy Procedia*, 129, 559, 2017.
21. Rai, Y., Shibata, N., Wada, D., Miyoshi, K., Akiyoshi, R., Recovery of engine waste heat using the 100-kW class ORC heat innovator®. *IHI Eng. Rev.*, 49, 30–33, 2016.
22. Tocci, L., Pal, T., Pesmazoglou, I., Franchetti, B., Small scale organic rankine cycle (ORC): A techno-economic review. *Energies*, 10, 413, 2017.
23. Bhardwaj, J.K. and Ashraf, H., Advanced silicon etching using high-density plasmas, in: *Proceedings of SPIE Micromachining and Microfabrication Process Technology, Micromachining and Microfabrication Process Technology*, Austin, pp. 224–233, September 19, 1995.

12

A Review on Hybrid Micromachining Process and Technologies

Akhilesh Kumar Singh[1]*, Sandip Kunar[2], M. Zubairuddin[2], Pramod Kumar[3], Marxim Rahula Bharathi B.[1], P.V. Elumalai[2], M. Murugan[3] and Yarrapragada K.S.S. Rao[1]

[1]*Department of Mechanical Engineering, Aditya College of Engineering, Surampalem, Andhra Pradesh, India*
[2]*Department of Mechanical Engineering, Aditya Engineering College, Surampalem, Andhra Pradesh, India*
[3]*Department of Mechanical Engineering, Aditya College of Engineering and Technology, Surampalem, Andhra Pradesh, India*

Abstract
Materials always in demand in everyday life of human beings and industries need as per applications. Every material has its properties. As per the demand for the desired object, raw materials changed through different manufacturing processes (shown in Figure 12.1). The manufacturing process also plays an important role, and today's improvement and development prospects in the manufacturing process help in energy-efficient manufacturing processes. An advanced method was developed for hybrid-micromachining operations per desired object shapes. And still, a researcher is researching to improve the manufacturing and machining process to improve its limitations during work. To maintain a sustainable hybrid-micromachining and hybrid-microfabrication process for industries. This study provides an overview analysis in detail of the various hybrid-micromachining process bibliometric survey, classification and machining mechanisms and their effective utilization of process parameters in the hybrid-micromachining (nanolevel, macrolevel, and microlevel) domain with particular emphasis.

***Keywords*:** Hybrid-micromachining, bibliometric survey, process development, material, applications, types, analysis

**Corresponding author*: akhileshkr.singh@hotmail.com

Sandip Kunar, Golam Kibria, Prasenjit Chatterjee and Asma Perveen (eds.) Hybrid Micromachining and Microfabrication Technologies: Principles, Varieties and Applications, (271–282) © 2023 Scrivener Publishing LLC

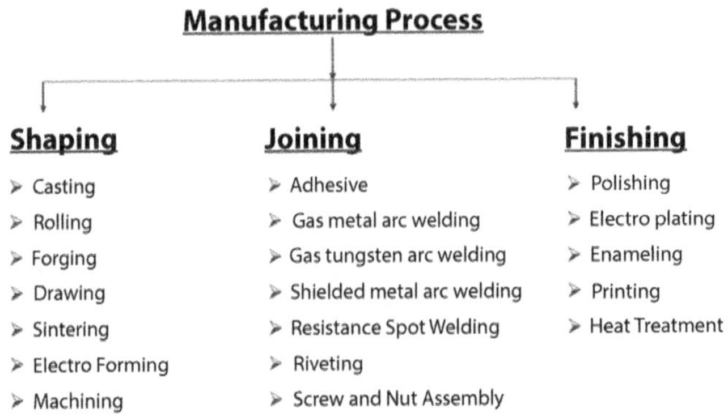

Figure 12.1 Different classes of manufacturing processes.

12.1 Introduction

Manufacturing can be done through 5M processes from the raw materials to the final product [1], as shown in Figure 12.1. Most of the components' desired shapes can be achieved after the primary processes (i.e., casting, forging, smithy, drawing, etc.) [2]. As per the material properties, few materials need one or more hybrid-micromachining and hybrid-microfabrication processes to get the final shapes [3]. Manufacturing maybe consists of conventional manufacturing or non-conventional manufacturing processes [4]. The machining process is vital in getting the final product's desired shape. As per the working principle with the equipment setup, the machining process has a different material removal rate mechanism with energy sources used for material removal, as shown in Figure 12.2 [5]. Machining process as per specific industrial application massive number of various micromachining process invented by the researcher with the help of industrial as per demand [6–13]. Hybrid-micromachining process help to improve the mechanism of material removal rate, improve its performance, and provide relative advantages over traditional and as well as micromachining process and equipment setup.

12.2 Characteristics of Hybrid-Micromachining

Micromachining also known as mechanical micromachining, is a subtractive fabrication or manufacturing technology in the micrometer dimensions range which help to maintains the metallurgical properties. The concept of the material removal process at a micro level with micro tools

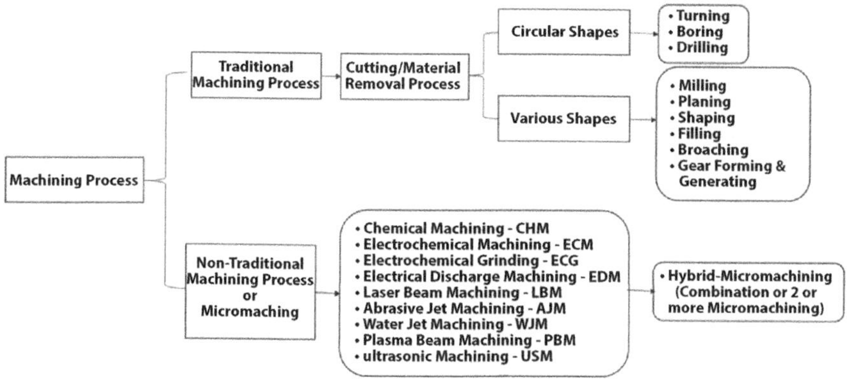

Figure 12.2 Different classes of machining processes.

(i.e., micrometer equal to one-millionth of a 1×10^{-6} m) by hybrid-micromachining is a relatively new/modern or advanced manufacturing process by researchers, engineers, and industries in its work. In recent years, hybrid-micromachining has mainly been used where difficult-to-machine materials, such as materials that are harder, brittle, more rigid or heat-sensitive materials, to meet the required field demands of extreme applications in the machining process [14–17] (microholes, probes and needle shapes can be created) in the aerospace, nuclear weapons, rocket launchers, defense, automotive, etc.

12.3 Bibliometric Survey of Micromachining to Hybrid-Micromachining

Micromachining to hybrid-micromachining technologies allow engineers to work at nanolevel, macrolevel, and microlevel (as shown in Figure 12.6) [14], intricate parts with micro-EDM, micro-USM, micro-EBM, micro-ECM, micro-LASER–assisted machining, and ASJ. The bibliometric survey recognizes the effective utilization of both micromachining and hybrid-micromachining over another machining process to work with complex and brittle materials. Researchers, scientists, and engineers with industries R&D cells developed and modified traditional machining to nonconventional machining. This micromachining to hybrid-micromachining, which work at nanolevel, macrolevel, and microlevel. According to the present bibliometric survey analysis (shown in Figure 12.3(a) and (b)) has been done from the Scopus publication database result regarding micromachining is 30,937 research published

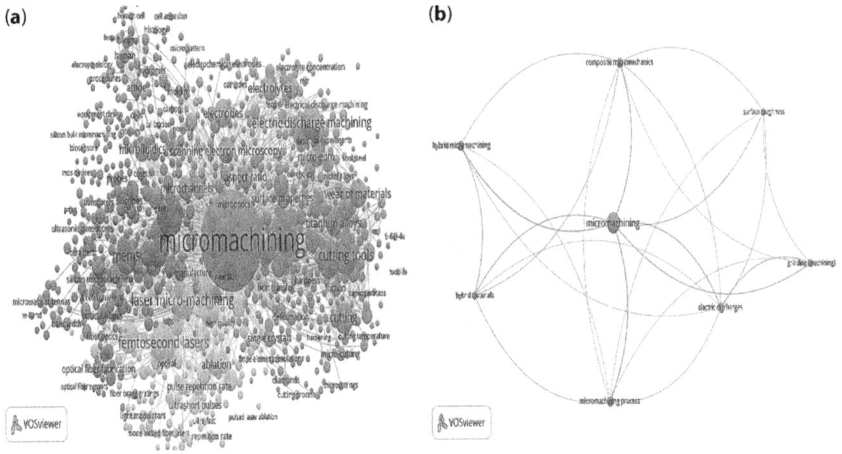

Figure 12.3 Bibliometric analysis of (a) micromachining processes and (b) hybrid-micromachining processes.

from 1955 to 2022; for combine hybrid and micromachining is 10,001 from 1969 to 2022 and for hybrid-micromachining is 37 from 2000 to 2022 research papers published [18]. It has been analyzed that the maximum research work regarding hybrid-micromachining was done in the countries—India, China, Singapore, the US, Belgium, the UK, and other countries, as shown in Figure 12.4. Figure 12.5 shows the documents published on the subject area of hybrid-micromachining uses, which means hybrid-micromachining is very effective in all domains.

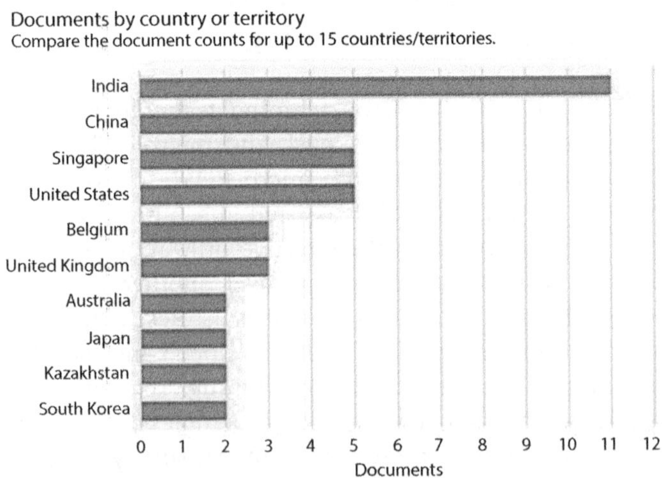

Figure 12.4 Documents by country or territory of hybrid-micromachining processes.

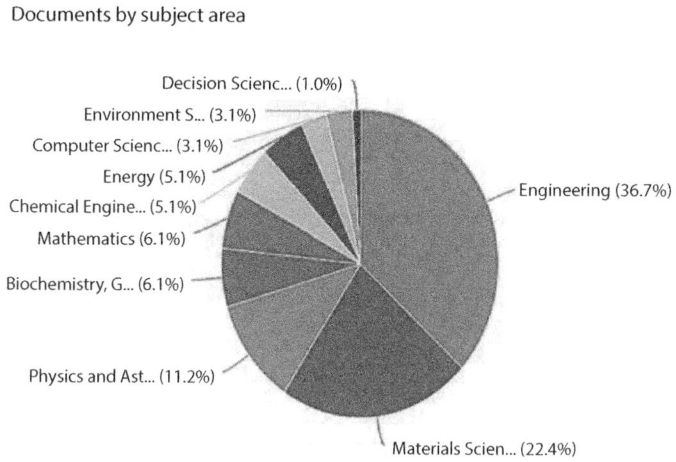

Figure 12.5 Documents by subject area use of hybrid-micromachining processes.

12.4 Material Removal in Microsizes

It has been observed, from bibliometric analysis, that advanced and hybrid-micromachining is used in all fields of science and engineering. It has excellent advantages on the micromachining area in all materials as shown in Figure 12.3, and Figure 12.6 shows the range of operation and surface quality of different machining range [14]. The recent development and achievements in hybrid-micromachining focus on ultraprecision machining and the necessity of a deeper physical understanding of hybrid-machining with range.

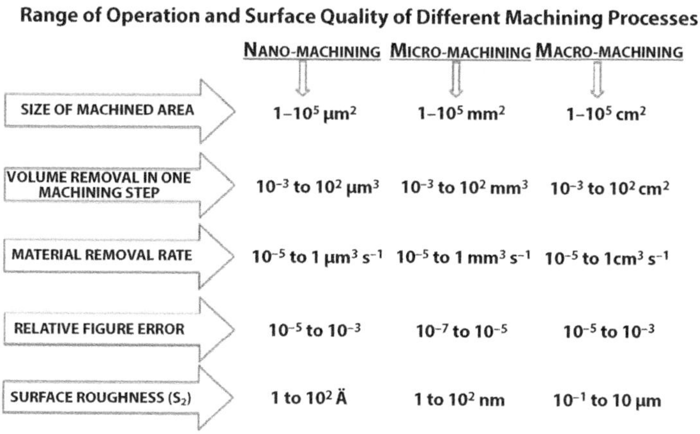

Figure 12.6 Different machining processes have different amount of material removed rate and surface quality.

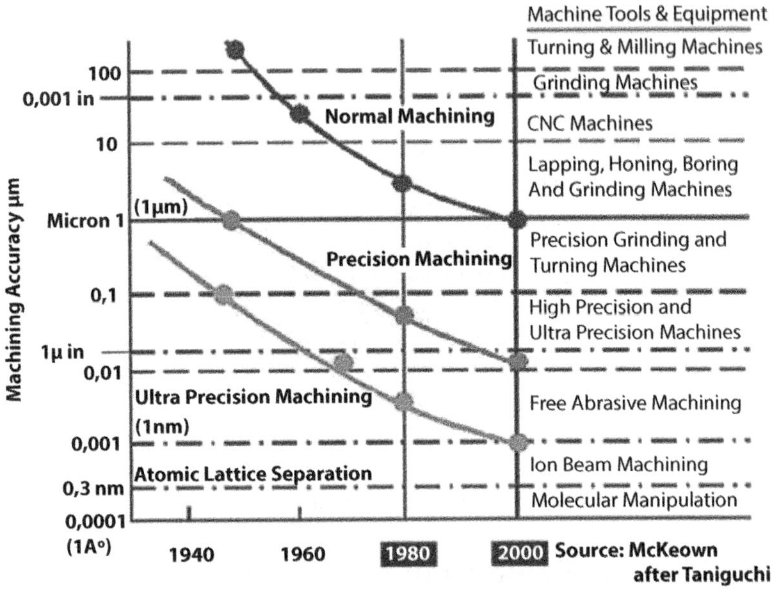

Figure 12.7 Micromachining (Taniguchi equivalent for cutting processes) accuracy and capability over time.

12.5 Nontraditional Hybrid-Micromachining Technologies

To hybrid-micromachining actual 3D-object complex shape and difficult components, machining is possible in non-traditional machining techniques are also being developed at the tiny level. Figure 12.7 shows the machining accuracy [19–21] that will be helpful for hybrid-micromachining with laser de-burring and micromachining, micro milling, micro EDM, micro-ECM, and ECDM, microgrinding and microultrasonic machining are particularly intriguing techniques [19–29, 36–45].

12.6 Classification of Techniques Used for Micromachining to Hybrid-Micromachining

Machining process can be classified by traditional machining and non-traditional or non-conventional machining and hybrid used for all engineering materials [30]. The word "micromachining" refers to small microscopic pieces of material removal or subtractive. And the word "hybrid-micromachining" refers to two and more combination of micromachining. It is

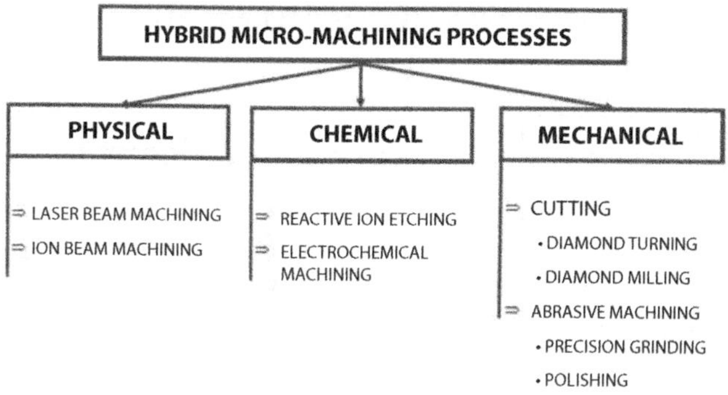

Figure 12.8 Hybrid-micromachining processes.

particularly suited for the micromanufacturing levels at microstructures and microparts, and hybrid-micromachining roughness tolerances can be high geometrical accuracy at manufacturing. Depending on the accuracy of hybrid-micromachining achieved by different machining processes, referred to as precision machining [31].

Hybrid-micromachining processes can be classified according to the combination of micromachining and removal process, i.e., physical, chemical, and mechanical nature (as shown in Figure 12.8). Hybrid-micromachining techniques are primarily used to manufacture small objects and can be classified according to bulk micromachining (i.e., used to create microelectromechanical systems [MEMS]) and surface micromachining (i.e., used to builds within a surface-layer that has been deposited on top of a silicon wafer). Both are used to manufacture a wide array of exceedingly small devices [32–45].

12.6.1 Classification According to Material Removal Hybrid-Micromachining Phenomena

1. Mechanical Force.
2. Ablation.
3. Dissolution.
4. Plastic Deformation.

12.6.2 Classification According to Categories Based on Material Removal Accuracy

1. Conventional Machining.
2. Precision Machining.
3. Ultra-Precision Machining.

12.6.3 Classification According to Hybrid-Micromachining Purposes

1. *AMMPs (Advanced Micro-Machining Processes)*—where different processes are involved for shaping the sizing a part with mechanical, thermal, and electro-chemical AMMPs.
2. *ANFPs (Advanced Nano-Finishing Processes)*—where different processes are involved for surface finishing a part with (i) AFF (Abrasive Flow Finishing), CMP (Chemo-Mechanical Polishing) and elastic-emission machining (EMM) in this all no external control forces required; (ii) M-AF & FP (magnetic—abrasive finishing and float polishing), MR-F & - AFF (magneto rheological—finishing and abrasive flow finishing) in this all external control forces required.

12.6.4 Classification of Hybrid Micromanufacturing Processes

1. Combination of two or more machining processes.
2. Further usage of energy sources assists in material removal.
3. Maybe use of universal tools that can machine two or more different surfaces.

12.7 Materials Are Used and Application of Hybrid-Micromachining

Advanced materials are used to remove the material in the hybrid-micromachining process at high precision, finishing with special tools and fluid materials which have such properties which improve the quality and controlled wear-tear life of devices. Such materials have high strength, corrosion resistance, miniature features, heat resistance, high-hardness, etc., with micro & nano level [14] surface finish from 1μm to 500μm approx. Hybrid-micromachining is used for various materials, including polymers, metals, alloys, and other hard materials. Many of these materials have the drawback of using only single-use and do not allow for easy duplication. Hybrid-micromachining applications include micromilling, microgrinding, microdrilling, chemical-etching, micropunching, manufacturing of injection nozzles, microsurgical-tools, electric circuits, etc.

12.8 Conclusions

Out of several hybrid-micromachining processes, only a few are discussed here because many modern techniques still exist to do the hybrid-micromachining process to the respective material. However, critical shape-precision and dimensional accuracy are required for specific material-to-material with appropriate tools and machining parameters. The proposed hybrid-micromachining processes would be applicable to any particular material based on their limitations, cutting forces, cutting speed, feed rate, surface quality, size effect, economics, temperatures, and environmental impacts to maintain sustainable micromachining processes in the industry.

References

1. Dieter, G.E., Engineering design: A materials and processing approach. 798 pp. McGraw, NY, USA, 2000.
2. Ehmann, K.F., Bourell, D., Culpepper, M.L. *et al.*, *Micromanufacturing*, pp. 1–362, Springer, Dordrecht, 2007.
3. Leuven, K.U., Surface integrity in hybrid machining processes. *Procedia Eng.*, 19, 241, 2012.
4. Manjaiah, M., Narendranath, S., Basavarajappa, S., Review on non-conventional machining of shape memory alloys. *Trans. Nonferrous Met. Soc. China*, 24, 1, 12–21, 2014.
5. Nukman, Y., Farooqi, A., Al-Sultan, O., Alnasser, A.R.A., Bhuiyan, M.S.H., A strategic development of green manufacturing index (GMI) topology concerning the environmental impacts. *Procedia Eng.*, 184, 370–380, 2017.
6. Rajurkar, K.P., Zhu, D., McGeough, G.A. *et al.*, New developments in electrochemical machining. *CIRP Ann. Manuf. Technol.*, 48, 567, 1999.
7. Kozak, J. and Rajurkar, K.P., Hybrid machining process evaluation and development, in: *Proceedings of 2nd International Conference on Machining and Measurements of Sculptured Surfaces*, The Institute of Metal Cutting (IOS), Krakow, pp. 501–536, September 20–22, 2000.
8. Menzies, I. and Koshy, P., Assessment of abrasion-assisted material removal in wire EDM. *CIRP Ann. Manuf. Technol.*, 57, 195, 2008.
9. Shih, H.R. and Shu, K.M., A study of electrical discharge grinding using a rotary disk electrode. *Int. J. Adv. Manuf. Technol.*, 38, 59, 2008.
10. Yadav, R.N. and Yadava, V., Electric discharge grinding: A review. *Proceedings of the National Conference on Trends and Advances in Mechanical Engineering*, YMCA University of Science & Technology Faridabad, Haryana, pp. 590–597, October 19th-20th, 2012.

11. Kozak, J., Abrasive electro discharge grinding (AEDG) of advanced materials. *Arch. Civ. Mech. Eng.*, 2, 83, 2002.
12. Dabrowski, L. and Marciniak, M., Investigation into hybrid abrasive and electrodischarge machining. *Arch. Civ. Mech. Eng.*, 2, 5, 2005.
13. Jawalkar, C.S., Sharma, A.K., Kumar, P., Micromachining with ECDM: Research potentials and experimental investigations. *IJMAE*, 6, 7, 2012.
14. Brinksmeier, E. and Preuss, W., Micro-machining. *Philos. Trans. R. Soc. A Math. Phys. Eng. Sci.*, 370, 1973, 3973–3992, 2012.
15. Fung, Y.C. and Tong, P., *Classical and Computational Solid Mechanics*, World Scientific Publishing Company, Singapore, 2001.
16. Encyclopaedia Britannica, Engineering. *Definition of the Engineers' Council for Professional Development*, 2010.
17. Corbett, J., McKeown, P.A., Peggs, G.N., Whatmore, R., Nanotechnology: international development and emerging products. *CIRP Ann.*, 42, 523–567, 2000.
18. https://www.scopus.com/search/form.uri?display=basic#basic
19. Taniguchi, N., Current status in and future trends of ultraprecision machining and ultrafine materials processing. *CIRP Ann.*, 32, 2, 573–582, 1983.
20. Byrne, G., Dornfeld, D., Denkena, B., Advancing cutting technology. *CIRP Ann.*, 52, 2, 483–507, 2003.
21. Dornfeld, D., Min, S., Takeuchi, Y., Recent advances in mechanical micromachining. *CIRP Ann.*, 55, 2, 745–768, 2006.
22. Kibria, G., Bhattacharyya, B., Davim, J.P., *Non-Traditional Micromachining Processes*, Springer, Berlin, 2017.
23. Sarkar, B.R., Doloi, B., Bhattacharyya, B., Parametric analysis on electrochemical discharge machining of silicon nitride ceramics. *Int. J. Adv. Manuf. Technol.*, 28, 873, 2005.
24. Wei, C. and Ni, J., Electrochemical discharge machining using micro-drilling tools. *Trans. NAMRI/SME*, 38, 105, 2010.
25. Wei, C., Xu, K., Ni, J. *et al.*, A finite element-based model for electrochemical discharge machining in discharge regime. *Int. J. Adv. Manuf. Technol.*, 54, 987, 2011.
26. Zheng, Z.P., Wu, K.L., Hsu, Y.S. *et al.*, Feasibility of 3D surface machining on Pyrex glass by electrochemical discharge machining (ECDM), in: *Proceedings of AEMS07*, Nagoya, Japan, pp. 98–103, November 28–30, 2007.
27. Cao, X.D., Kim, B.H., Chu, C.N., Hybrid micromachining of glass using ECDM and micro grinding. *Int. J. Precis. Eng. Manuf.*, 5, 5, 2013.
28. Raghavan, S., Melkote, S., Hashimoto, F., Laser tempering based turning process for efficient machining of hardened AISI 52100 steel. *J. Manuf. Process.*, 15, 318, 2013.
29. Masood, S.H., Armitage, K., Brandt, M., An experimental study of laser-assisted machining of hard-to-wear white cast iron. *Int. J. Mach. Tools Manuf.*, 51, 450, 2011.

30. Gao, S. and Huang, H., Recent advances in micro-and nano-machining technologies. *Front. Mech. Eng.*, 12, 1, 18–32, 2017.
31. Bhattacharyya, B. and Munda, J., Experimental investigation on the influence of electrochemical machining parameters on machining rate and accuracy in micromachining domain. *Int. J. Mach. Tools Manuf.*, 43, 13, 1301–1310, 2003.
32. Parameswari, G., Jain, V.K., Ramkumar, J., Nagdeve, L., Experimental investigations into nanofinishing of Ti6Al4V flat disc using magnetorheological finishing process. *Int. J. Adv. Manuf. Technol.*, 100, 5, 1055–1065, 2019.
33. Kibria, G., Bhattacharyya, B., Davim, J.P., *Non-Traditional Micromachining Processes*, pp. 337–365, Springer, Berlin, 2017.
34. Debnath, S., Kunar, S., Anasane, S.S., Bhattacharyya, B., Non-traditional micromachining processes: Opportunities and challenges, in: *Non-traditional Micromachining Processes*, pp. 1–59, 2017.
35. Chavoshi, S.Z. and Luo, X., Hybrid micro-machining processes: A review. *Precis. Eng.*, 41, 1–23, 2015.
36. Sahu, A.K., Malhotra, J., Jha, S., Laser-based hybrid micromachining processes: A review. *Opt. Laser Technol.*, 146, 107554, 2022.
37. Roth, G.L., Haubner, J., Kefer, S., Esen, C., Hellmann, R., Fs-laser based hybrid micromachining for polymer micro-opto electrical systems. *Opt. Lasers Eng.*, 137, 106362, 2021.
38. Saxena, K.K., Qian, J., Reynaerts, D., A tool-based hybrid laser-electrochemical micromachining process: Experimental investigations and synergistic effects. *Int. J. Mach. Tools Manuf.*, 155, 103569, 2020.
39. Saxena, K.K., Qian, J., Reynaerts, D., Single-step meso/micro scale hierarchical surface-structuring using an innovative tool-based hybrid laser-electrochemical micromachining process. *Procedia CIRP*, 95, 821–826, 2020.
40. Jahan, M.P., Perveen, A., Rumsey, A.M., A review on the conventional, non-conventional, and hybrid micromachining of glass. *Mach. Sci. Technol.*, 23, 2, 264–338, 2019.
41. Zhumatay, N. and Perveen, A., Hybrid machining process for microfabrication of micro parts. *Mater. Today Proc.*, 18, 2209–2216, 2019.
42. Bhowmik, S. and Zindani, D., Combined variant of hybrid micromachining processes, in: *Hybrid Micro-Machining Processes*, pp. 61–70, Springer, Cham, 2019.
43. Bhowmik, S. and Zindani, D., Other assisted hybrid micromachining processes, in: *Hybrid Micro-Machining Processes*, pp. 49–59, Springer, Cham, 2019.
44. Feng, S., Huang, C., Wang, J., Zhu, H., Material removal of single crystal 4H-SiC wafers in hybrid laser-waterjet micromachining process. *Mater. Sci. Semicond. Process.*, 82, 112–125, 2018.
45. Kim, S., Kim, B.H., Shin, H.S., Chu, C.N., Hybrid micromachining using a nanosecond pulsed laser and micro EDM. *J. Micromech. Microeng.*, 20, 1, 015037, 2009.

13

Material Removal in Spark-Assisted Chemical Engraving for Micromachining

Sumanta Banerjee

Department of Mechanical Engineering, Heritage Institute of Technology, Kolkata, India

Abstract

The fundamental principles underlying the material removal mechanism in Spark-Assisted Chemical Engraving (SACE) are outlined in this chapter. In this process, energy required for machining is provided by electrochemical discharges. Till recent past, material removal in SACE μ-machining was attributed to melting of workpiece. However, recent state-of-art studies underline the important contribution of etching in electrochemical discharge on glass and other amorphous substrates. SACE-assisted micromachining uses either an active cathode (or anode) as tool, respectively in direct (reverse) polarity machining. As the discharge mechanism differs for these machining modes, the machining features and performance also vary significantly. This warrants fundamental understanding of the underlying mechanisms of the material removal processes. Experimental and analytical investigations indicate that several processes contribute to this micromachining mode, which includes electrochemical discharge-assisted melting and vaporization, etching at elevated temperatures, differential thermal expansion of components, random thermal stress cycles, as well as thermomechanical shocks attributed to expanding gases and movement of electrolyte.

***Keywords*:** SACE, micromachining, electrochemical discharge, etching, electrolyte

Email: sumanta.banerjee@heritageit.edu

Sandip Kunar, Golam Kibria, Prasenjit Chatterjee and Asma Perveen (eds.) Hybrid Micromachining and Microfabrication Technologies: Principles, Varieties and Applications, (283–312) © 2023 Scrivener Publishing LLC

13.1 Introduction

Humanity has put in significant efforts into mastering the art of material processing since the dawn of history and, possibly, even earlier. The fundamental technological significance of the machine as an instrument of human progress is well accepted. Any new technology calls for updating of machining infrastructure and abilities. In particular, increased use of specialized materials (silicon, ceramics, and composites) for customized applications has significantly expanded the vast know-how of machining technology.

The techniques for micromachining (μ-machining) on silicon substrates were developed in the last century. In particular, a wide range of economically affordable silicon μ-machining methods is currently accessible on a global scale. Similar scenarios apply to materials that conduct electricity, where two particularly potent methods are available: electrical discharge machining (EDM) and electrochemical machining (ECM). As a wide variety of electrically nonconducting materials (glass, composites) are useful for various technological applications, the machinability perspective of these materials evokes interest. More and more technically intricate applications of glass are being implemented in the fields of microfluidics and microscaled systems. The use of biocompatible materials in miniaturized medical equipment, or specially textured glass surfaces for car wind shields, are but two of countless allied examples.

The application areas of glass, in the realm of MEMS applications are on the rise. The acronym MEMS (μ-electromechanical) is used for a group of microsensors and actuators. The development of integrated circuit fabrication technologies in the 1990s led to emergence of MEMS applications. Applying electrostatic bonding or field-assisted thermal bonding to silicon makes Pyrex® glass a preferred choice for MEMS-fabrication. In these applications, Pyrex® glass is connected to the silicon matrix through anodic bonding, as in bulk micromachined accelerometers [1]. In addition, glass possesses a number of favourable attributes, which include biocompatibility and corrosion resistance. Due to its amorphous nature, it is amenable to chemical treatment. Being transparent, glass is frequently employed in situations where optical visualization of processes is necessary, or in (MEMS-based) optical applications.

Medical equipment (e.g. drug delivery devices, flow sensors), microreactors, micropumps, and microaccelerometers are some novel MEMS glass applications. In all of these applications, glass performs several tasks, which includes (a) providing a seal and requisite damping, (b) functioning as a capacitor when kept between electrically charged metal plates, and

(c) providing overload protection when required. Other than accelerometers, glass is frequently used in capacitive sensors [2].

13.2 Essentials of SACE

Various technologies are available for micromachining glass. However, its low machinability presents significant hindrance in fabrication of glass-based μ-devices. Other materials (e.g. ceramics and composites) are also difficult to machine. Spark-assisted chemical engraving (SACE), also known as electrochemical discharge machining (ECDM), presents a possible solution to these problems.

The underlying physical and electrochemical phenomena for machining glass by the SACE technique are outlined in [2, 3]. The workpiece is submerged in typically potassium hydroxide (KOH) or sodium hydroxide (NaOH) electrolytic solution. Between the *machine-tool* (also known as the *tool-electrode*) and the *counter-electrode*, a uniform DC voltage is applied. The tool-electrode is submerged in the electrolytic solution to a depth of a few millimetres. The counter-electrode is typically a sizable flat plate. The tool-electrode typically has a cathode polarization; however, it can be oppositely polarized as well.

Electrolysis takes place by conventional mode for low values of the cell terminal voltage, typically below the critical voltage (which normally ranges between). Depending on the electrolyte being employed as well as the polarization of the machine-tool and counter-electrodes, formation of H_2 and O_2 gas bubbles occur at the electrodes. More and more bubbles are created as the terminal voltage is raised. Around the electrodes, a bubble layer is gradually formed. Rising values of the current density increases the (volumetric) density of the bubbles and their average radius. When the terminal voltage is increased beyond a critical value, the bubbles eventually coalesce as a gas film, which surrounds the tool-electrode. Light emission can be observed (within this gas film), when electrical discharges occur between the tool and the electrolyte medium surrounding it. In the area adjoining the tool-electrode, the average temperature of the electrolytic solution rises to ~80 – 90°C. If the tool-electrode is placed in proximity to the glass sample, machining is feasible. When the separation distance between the tool-electrode and the workpiece gets below ~25 mm, glass can be machined [2, 3].

Nonetheless, the technology is more intricate than it apparently appears. The gas film that envelops the tool-electrode tends to become unstable in an environment of rapid rates of heat transfer. Microexplosions may result owing to film-collapse, which can locally destroy the machined surface.

The local temperature can rise so much during hole-drilling that thermal-induced cracking may result in (undesirable) heat-affected zones.

13.2.1 Instances of SACE Micromachining

As mentioned above, flexible μ-structuring of glass can be done using SACE technology. It is possible to create μ-holes and channel-like microstructures. Published literature [4] cites machining of a channel μ-structure at an applied voltage of 30 V, with a cylindrical 90 mm tool-electrode, where single-step μ-machining operation is completed with a tool speed of ~0,05 mms^{-1}. The depth and width of the machined channels are ~200 μm and ~100 μm respectively. The possibility of machining structures with sufficient depth (slender aspect ratios) is demonstrated by the machinability of μ-holes cited in [5].

The adaptability and flexibility of the SACE technology is a primary advantageous feature. The desired structure can be carved out without the need for a mask. There are typically two processing units in a four-axis SACE machining set-up. A prototype of the same is outlined in [2, 5]. Using wire electrical discharge grinding (WEDG) technology, the first unit (known as the WEDG unit) of this prototype enabled the production of tools in a variety of shapes [6]. The SACE-unit (or the second processing unit) is where glass machining is done. Glass wafers up to thickness can be machined in this SACE subsystem. This machining prototype annuls misalignment problems by offering flexibility in machining both glass as well as necessary tool-electrodes within the same facility.

13.3 Genesis of SACE Acronym: A Brief Historical Survey

Historically, the initial developments of SACE took place in late 1950s for die workshops applications in Japan [7]. Among the benchmark studies conducted on this novel technique, called *electrical discharge drilling* at that time, was the one conducted in [7]. Drilling tiny μ-holes in glass was demonstrated, and the authors investigated the influence of parameters like as chemical composition of the electrolyte and the material of the tool-electrode. However, the underlying mechanisms that govern the machining process were yet to be properly understood. Also, scientific concerns on similarity with prevalent ECM and/or EDM technology were raised, until the seminal publication by Cook *et al.* [8] about five years later. The authors of this study emphasized the distinction between the method (as described by [7]) and EDM (or ECM). A new name *discharge machining of non-conductors* was also proposed for the process. The authors not only probed the impact of the electrolyte in greater detail, but also demonstrated that the method can be used with a variety of non-conductive materials. Although the method for material removal still remained unclear, the drilling rates were quantified as a function of the depth of the μ-holes. The study also evaluated the machining rates of positively and negatively charged tool-electrodes.

By utilizing a wire as a tool-electrode, the experimental study [9] demonstrated a novel variation of the procedure outlined in [2]. The authors demonstrated that this procedure can be used to cut glass and various grades of ceramics, and named it *wire electrochemical discharge machining*. Jain *et al.* [10–12] presented a more refined version of this variant, and coined the process name as "electrochemical spark machining". The authors reckoned scientific resemblance (of this novel machining method) with a variant of ECM called the "electrochemical arc machining".

The initial studies on MEMS applications were published in the 1990s [13]. Several investigations on the fundamental principles that govern μ-scaled machining methods were reported as published literature concurrently [11–16]. For example, Ghosh *et al.* [17] established a direct connection between the electrochemical discharge phenomenon and the machining process, which countered the perception that melting of the workpiece was the only process for material removal. At the start of this century, a thorough investigation of the physicochemical aspects of material removal mechanism was conducted by Yang *et al.* [18, 19]. Their research findings proved as ground-breaking study in this field, as they identified the mechanism of material removal as an etching process associated with large temperature gradients. Fascio *et al.* [4, 6] also conducted research on this etching process induced by electrochemical effects. Based on their findings, and to distinguish this machining method from ECM or EDM, the acronym "SACE" was finally suggested. This abbreviation underlines the fundamental role that chemical etching plays in the machining process, and is used throughout the present book chapter. The same research group reported more studies on the underlying processes that govern the generation and dynamics of the gas film, with a view to optimize the machining parameters. Their research suggested the use of SACE technology in microfabrication [2], following which the first review paper on the procedure was released in 2005. Since 2006, literature review indicates a noticeable resurgence of interest in the technology, and substantial volume of research findings on the basics of the process has been published. Improvements to the process have also been suggested, based on systematic characterizations of various machining methods. The use of pulsed voltage-assisted machining proves to be promising technology, among much other advancement [20–22]. It has been shown how the SACE method can produce structures that are less than in size [23]. The incorporation of electrochemical discharges in nanofabrication is a more recent advancement in the field [2].

13.4 SACE: A Viable Micromachining Technology

The craft of microstructuring workpieces is possible using a variety of machining techniques. These various (machining) methods, classified either as mechanical, chemical, or thermal technologies, are seldom standalone in operation, and frequently complement each other. Occasionally, they are implemented in tandem. A concise overview of some of the most popular μ-machining methods can be found in the sections below.

13.4.1 Mechanical μ-Machining Techniques

- Glass can be machined by diamond cutting, which is essentially *mechanical μ-drilling* (with a diamond tool) with typical diameters ~400 μm. In diamond cutting, the tool size is a key limitation. As only drilling or cutting operations can be implemented (3D microstructuring not achievable as yet), this method is sometimes not included within the micromachining domain.
- In *water jet μ-machining*, abrasive materials are mixed with water to form a slurry, and projected on the workpiece at high pressure (usually ~0.7 MPa).
- *Abrasive jet μ-machining*, also known as *powder blasting*, is a process that uses a particle-laden jet to erode material from a target. The (particle) jet commonly contains Al_2O_3 particles that range in size from ~3 to 30 μm. For brittle materials (silicon, glass, and ceramics) powder blasting is a low-cost, quick (~25 μm min^{-1}) and directionally-precise etching process. A mask can be used to machine structures that are either intricate in shape and/or small in dimensions.
- The tool used in *ultrasonic μ-machining* operation oscillates at a high frequency (~20 kHz) and amplitude of ~25 μm. It is made of softer material than the workpiece. Ultrasonic machining removes material by grinding or eroding workpieces made of extremely hard ceramics. A liquid slurry

consisting of loose, abrasive particles surrounds the drill bit. The latter are smashed against the hard workpiece surface by the high-frequency vibrations, thereby producing more and more loose, abrasive particles and enabling material removal.

13.4.2 Chemical μ-Machining Methods

- Photoforming: *Photoforming* or *photofabrication* is an optical μ-forming technique, similar to stereolithography (technique of layering of photomasks, which enables the solidification of photo-sensitive resins through light exposure).
- Chemical etching: Dry and wet etching are two methods of *chemical μ-etching* for glass. Wet etching is typically carried out using hydrofluoric acid (etching rate ~6μmmin^{-1} using 50 wt% HF at 40°C) or potassium hydroxide (etching rate ~80 nmhr^{-1} using 35 wt% KOH at 40°C). Selective etching can also be accomplished (using masks).
- ECM: Through electrochemical dissolution of an electrically conductive workpiece, the *ECM* technique achieves material removal at the desired rate. This mode of chemical machining enables to achieve excellent surface qualities.

13.4.3 Thermal μ-Machining Methods

- Laser machining: Localized energy supply (to generate sufficient heat) that enables material removal is the foundation of the craft of *laser machining.* The coherent laser light dissipates as thermal energy at the material's surface, which can then be processed for machining. This heat-assisted machining process allows for sufficiently accurate μ-machining of glass.
- EDM: One of the popular non-conventional μ-machining techniques is *EDM*. Only materials that are electrically conductive can be processed. To remove material via melting and evaporation, controlled electrical discharges are generated between the tool and workpiece.

Chemical machining techniques often produce surfaces with great surface quality, but their limitation lies in sluggish machining rates, or machining high aspect-ratio structures. They are more suitable for batch processes. Although mechanical and thermal machining typically provide lower-quality surfaces, they are more flexible when it comes to processing high aspect-ratio components. SACE is an instance of hybrid technology, where removal of material accomplished by dual action of heat and chemical processes. As electrochemical dissolution of the workpiece does not take place, SACE warrants recognition as a novel machining mode instead of (incorrect) description as an amalgam of EDM and ECM.

13.5 Material Removal Mechanism in SACE μ-Machining

The preceding sections on electrochemical discharge μ-machining provided some glimpses of the material removal mechanism. Although electrochemical discharge was correctly reckoned as the energy source for machining, it was long held that melting the job (locally) accomplished the task of material removal (as in spark erosion-assisted EDM). Only recently has the important contribution of etching to SACE μ-processing of (mainly) glass substrates been recognized. This section addresses both aspects.

The following are some common terms associated with SACE. The cell terminal voltage is known as the *machining voltage*, the active electrode is known as the *tool electrode*, and the current flowing between the active electrode and the counter electrode is known as the *machining current* in connection to machining.

13.5.1 General Aspects

Since the benchmark work on glass μ-drilling by Kurafuji *et al.* [7], it is well known that electrochemical discharge can be used for machining. This innovation is not limited to glass substrates; some non-conductive materials (e.g. refractory brick, granite, alumina, quartz, and plexiglass) and some ceramics (e.g. Al_2O_3, Si_3 N, MgO, Y_2O_3) can also be machined [8]. Using a variety of electrolytes and salts ($NaNO_3$, NaF, NaOH, NaCl, etc.), holes (~1 mm in diameter) are commonly created in ceramic substrates with rather high processing voltages (about ~80 V as compared to ~30 V for glass and quartz). Compared to quartz and glass, the MRR (material removal rate) for ceramics is about an order of magnitude lower (~0.1 – 0.4 $mgmin^{-1}$); the quality of surface is also poor [24].

Active cathode (for direct polarity machining) or active anode (for reverse polarity machining) can also be used as tools [25], although the former is more common. The machining performance differs between the active cathode and the active anode due to differences in the discharge activity, particularly the discharge mechanism.

(i) Processes involved

Investigations into the material-removal mechanism by SACE technique is still an area of active research. A number of procedures can facilitate material removal, including (a) electrochemical discharge-induced melting and vaporization, (b) etching at high-temperatures, (c) thermal strains of components and weathering, (d) randomized distribution of thermal stress and µ-cracking, and (e) mechanical strains induced by expansion of evolved gases and electrolyte movement [2].

Review of published literature indicates, till date, only the thermal effects and corrosion processes have been systematically studied, mainly for active cathode-machined glass.

It is generally accepted that local heating by electrochemical discharge is at the origin of the µ-machining process. Depending on the substrate, melting (and eventual vaporization) or high-temperature corrosion facilitates removal of material.

The significance of thermal mechanism in µ-machining is supported by an increasing body of empirical research. For instance, during µ-drilling, the tool electrode can attain temperatures of up to ~500°C [16, 26]. Experimental demonstration by Allesu *et al.* [14, 15] showed that the thermal energy produced during electrochemical discharges is capable of etching glass. In their seminal work, the authors employed an electrolysis cell with two compartments divided by a glass wall. This partition has a small hole (1.5 mm ϕ) punched into it. A terminal potential difference (of 60 V) causes the electrolyte to evaporate locally, generating gas bubbles that quickly fill the perforation in the glass wall The bore diameter gradually enlarges from 1.5 mm to 2.5 mm owing to electrochemical discharges inside the gas.

By conducting various measurements, Kulkarni *et al.* [27] has shown that, after each discharge, localized heating elevates the surface temperature (of workpiece) above its melting (or even the vaporization) point. The study also estimates that only ~2 – 6% is used to heat the part; nearly ~77 – 96% of energy is used up to heat the electrolyte and the tool-electrode; It should be noted that the experiments in [27] were performed on metallic workpieces with very different thermophysical properties than those traditionally machined (e.g. ceramics or glass).

At high machining voltages, experimental evidence of thermal cracks formed within the machined material verifies the thermal mechanism in SACE µ-machining [28].

There is clear evidence of chemical contribution as well. For example, smooth machined surfaces are achievable in glass substrates even at low machining voltages. The bulk electrolyte also contains products from the etching reaction, as confirmed from chemical analysis. Based on the machining strategy, the temperature within the machining zone (for glass μ-machining) can be reduced to ~200°C, which is quite low as compared to usual values of ~500°C – 600°C. Within this temperature range, glass still remains highly viscous. Therefore, material removal is rendered possible only by etching [2, 25].

(ii) Modeling challenges

Various attempts in the form of analytical [29] and experimental [30] studies have been made to theoretically predict MRR as a function of SACE μ-process parameters. Two major technological challenges are posed when setting up working models.

Firstly, the *machining temperature* (= T_M) necessary for machining to take place must be known. A precise estimate of T_M, however, warrants knowledge of the exact mechanism of machining. For instance, T_M may be taken as the melting point of the job material if machining occurs by virtue of melting. However, amorphous materials (e.g. glass) that exhibit gradually decrease of viscosity with rising temperature and do not present a clear phase transition, this yardstick cannot be applied.

Where etching at high temperatures plays the dominant role, estimation of T_M is even more difficult. Although etching occurs at room temperature, the etching rate is very poor (at this temperature). Also, the process remains largely non-localized to sustain machining. The technical issue to be addressed is the estimation of local temperature that primarily determines the etch rate, and not just T_M. The rate of electrochemical corrosion is also influenced by mass transfer through (a) effective removal of already-etched material, and (b) introduction of fresh electrolyte to the machined area. To summarize, the estimation value of temperature required for μ-machining is still an open question.

The rate of energy release during discharge and the fraction (of energy) consumed for the machining process must both be considered for accurate prediction of MRR. Basak and Ghosh [29] estimated that each discharge has a duration of ~0.1 ms, and carries an average energy of ~2000 J/cm^2. Depending on the geometry of the tool electrode, the average energy release for each discharge varies between ~3 – 10 mJ, as estimated by Jiang *et al.* [30].

The estimation of the average heat released (= P_E) during discharge, in terms of the water decomposition potential (U_d), the machining voltage (U), inter-electrode resistance (R) of the electrolyte, and the average current ($\bar{I}$), is as follows [4, 11, 31]:

$$P_E = \underbrace{(U - U_d)\bar{I}}_{\text{Mean value of energy supplied}} - \underbrace{R\bar{I}^2}_{\text{Joule heating of electrolyte}} \quad (13.1)$$

The mean energy and the average duration of spark can also be estimated from statistical analysis of current signal [4, 6]. The latter approach provides an improved estimate of P_E (as compared to Eq. (13.1)). The experimental observations of MRR corroborates fairly well with the proposed models, except for voltages below ~30 V where actual machining rates exceed the predicted (machining) rate. This shows that the chemical contribution (to machining) at low voltages cannot be ignored [2]. These aspects are taken up in the following sections.

(iii) Glass μ-machining: material removal mechanism

In μ-machining of glass, the high-energy discharge sublimates material and forms μ-pits on striking the surface of glass [2, 32]. This process is diagrammatically illustrated and vividly explained in [33]. The major fraction of substrate surface is attacked OH radicals, with sodium silicate as probable etchant. In addition, local heating due to electrochemical discharge results in the formation of low-viscous vitreous region. The gap between heat source and job, the intensity of the source of power, the tool-electrode form, and the mass transport dynamics will all have an influence on the temperature of the machined region.

Presently, the influence of the first two factors on SACE μmachining performance is scientifically well-known. The feeding mechanism of tool-electrode determines the gap between the heat source and workpiece. The strength of the heat source can be regulated by the machining voltage. The influence of mass transport and geometry on metal erosion rates is still an open problem; only some known aspects are outlined here. Based on machining-zone temperature, the electrolyte exists either as molten salt or aqueous solution; the electrolyte's physical state, however, decides the machining mode. Therefore, depending on the machining conditions, there are multiple machining temperatures for glass.

Machining at low or high depths are both influenced by the local temperature and (temperature-dependent) corrosion rates; the latter, in turn, is influenced by mass transport. With increasing depth, the discharge activity and etching are observed to reduce, as the flow of electrolyte near the tip of tool-electrode is progressively hindered (when drilling μ-holes). Both effects lead to significant reduction in MRR. Thus, the depth of machining adversely affects the MRR. Two limiting situations are discussed. The first case is one of shallow-depth machining, where the rate of heat transfer inside the part delimits μmachining. The availability of electrolyte in the machining region and the technical challenges of removing material from the machining site delimits the machining rate in deep machining.

13.5.2 Micromachining at Shallow Depths

If the electrolyte can easily reach the machining zone during SACE μ-machining, the process is referred to as *low-depth machining*. In this instance, the electrochemical discharges happen when the tool is submerged in a non-confined electrolytic medium without a workpiece. Depending on the job material, material is removed by melting and/or chemical etching as soon as the machining temperature is reached. The models outlined in this section are applicable to situations where machining is constrained by heat transfer in the workpiece. Such circumstances are typical for a number of SACE μmachining techniques.

(i) Heat transfer modeling
In line with EDM, a simplified model is discussed here, where the thermal energy imparted by discharge is approximated by a constant-power source (= P_0) within a homogeneous material [34]. The temperature at far-off distance (infinity) is assumed constant (= T_0). The geometry of heat source depends on the tool geometry as well as the μ-machining mode [35]. Two cases of primary relevance are discussed here.

The first case employs a cylindrical tool electrode, as during gravity-feed drilling. The machining mode ensures that the workpiece is always pressed against the electrode. The heat source is modeled as a disk heat source with the following boundary conditions:

$$\lambda \frac{\partial T}{\partial z}(r, z = 0, t) = \begin{cases} 0 & r > b, \\ -\dfrac{P_0}{\pi b^2} & r \leq b \end{cases} \tag{13.2}$$

b: Radius of cylindrical tool-electrode
A second case discussed here is valid for cylindrical tool-electrode (as in 2D-machining or constant velocity-feed drilling), where electrochemical

discharges originate at its sharp edges. In this strategy, the (constant) speed of advancement of the tool-electrode is lower than the average value of MRR. This creates a gap between the tool and the job during μ-machining:

$$\lambda \frac{\partial T}{\partial z}(r, z=0, t) = \begin{cases} 0 & r > b, \\ -\dfrac{P_0}{\pi[b^2 - (b-e)^2]} & r \leq b \end{cases} \tag{13.3}$$

b: Radius of cylindrical tool-electrode; e: Thickness of cylindrical tool-electrode
In both cases (Eq. (13.2) and Eq. (13.3)), the spatial solution outside the domain of heat source satisfies the following temperature field T:

$$\frac{\rho c}{\lambda}\frac{\partial T}{\partial t} = \nabla^2 T \tag{13.4}$$

(ρ: Density, k: Thermal conductivity, c: Specific heat capacity
For the disk source, the solution of temperature distribution presents cylindrical symmetry. The spatial-temporal distribution of temperature [T = T(z, t)] is solved:

$$T(z,t) - T_0 = \frac{2P_0\sqrt{at}}{\lambda\pi b^2}\left[\operatorname{ierfc}\left(\frac{z}{\sqrt{4at}}\right) - \operatorname{ierfc}\left(\frac{\sqrt{z^2+b^2}}{\sqrt{4at}}\right)\right] \tag{13.5a}$$

$$\operatorname{ierfc}(x) = \frac{1}{\sqrt{\pi}}e^{-x^2} - x\operatorname{erfc}(x) \tag{13.5b}$$

a = λ/ρc: Thermal diffusivity; erfc (x): Complementary error function;
z: Distance measures along the axis of symmetry of the cylinder from the heat source

As t → ∞, the steady-state solution is given by:

$$\begin{aligned} T_\infty(z) - T_0 &= \frac{P_0}{\lambda\pi b^2}\left(\sqrt{z^2+b^2} - z\right) \\ T(z, t\to\infty) &= T_\infty(z) \end{aligned} \tag{13.6}$$

The point of maximum temperature lies on the axis of symmetry. Therefore, etching can only occur if the following conditions are met:

$$T_M - T_0 \leq \frac{P_0}{\lambda \pi b} (T_M\text{: Machining temperature}) \tag{13.7}$$

If uniform spark distribution is assumed, the heat source radius is same as the tool radius.

Equation (13.7) is only an approximation. The uniformity in spark distribution and the hypothesis of cylindrical shape is not always fulfilled. Also, the various thermo-physical properties of the job depends on temperature. In addition, the temperature required to perform machining (T_M) is difficult to estimate.

For estimation of P_0, the thermal power (= P_E) generated from discharge must be known. The slope of the average I-V characteristics in the linear portion of the curve (range ~5 – 15 V) measures the inter-electrode resistance (= R).

Out of the total heat (= P_E) generated, only a portion is imparted to the job ($P_0 = \epsilon P_E$). This fraction (= ϵ) depends on the ratio of the thermal conductivity of the electrolyte and the substrate as well as geometrical parameters. The thermal conductivity of the tool-electrode also influences ϵ, as most of the heat is dissipated through the tool electrode. In addition ϵ is influenced by the gap between the tool electrode and the job.

(ii) Estimation of MRR

The evolution of isotherms (in temperature field) T (z, t) = T_M and solution of the heat equation (Eq. (13.4)) are now analysed, in order to estimate the rate of material removal (or MRR) during μ-machining. Material removal is very rapid (occurs as soon as the value of T_M is reached), according to the technological standards of low-depth machining. In this case, the plate heat source (Eq. (13.2)) is considered.

Equation (13.4) is conveniently presented in non-dimensional form. The dimensionless temperature is defined by:

$$\overline{T} = \frac{T - T_0}{P_0 / (\lambda \pi b)} \tag{13.8}$$

$$\tau = b^2/4a \ (\tau\text{: Characteristic time scale}) \tag{13.9}$$

The dimensionless length $\bar{z} = z / b$ and the normalized time $\bar{t} = t / \tau$ are also defined. Making use of these variables, the solution for Eq. (13.4), with the set of boundary conditions specified by Eq. (13.2), is written as follows:

$$\bar{T}(\bar{z}, \bar{t}) = \sqrt{\bar{t}}\left[\operatorname{ierfc}\left(\frac{\bar{z}}{\sqrt{\bar{t}}}\right) - \operatorname{ierfc}\left(\frac{\sqrt{\bar{z}^2 + 1}}{\sqrt{\bar{t}}}\right)\right] \tag{13.10}$$

The distance reached by T_M (isotherm), or the evolution of z (t), is evaluated by solving Eq. (13.11):

$$\bar{T}(\bar{z}, \bar{t}) = \frac{T_M - T_0}{P_0 / (\lambda\pi b)} = \frac{1}{\kappa} \tag{13.11}$$

The number κ (in Eq. (13.11)) can have multiple interpretations, depending on the mode of material removal. The value of T_M is the same as the melting temperature of the substrate, if material removal is due to melting. The number κ, in this case, is expressed as in Eq. (13.12):

$$\kappa = \frac{P_0}{P_{min}} = \frac{P_0}{\lambda\pi b(T_M - T_0)} (\kappa\text{: Normalized thermal power}) \tag{13.12}$$

The value for T_M is no longer well-defined if machining is due to etching of the job (the etch-rate is determined by T_M. At sufficiently high temperatures, the etching rate is faster than the heat conduction rate within the workpiece, so that the heat transfer rate becomes the limiting factor. In this case, k is the reciprocal of the normalized temperature.

The starting time of machining $(= \bar{t_0})$ is indicated by the moment when the workpiece temperature attains T_M (machining temperature):

$$\bar{T}(0, \bar{t_0}) = \frac{1}{\kappa} \tag{13.13}$$

$$\bar{T}(\bar{z} = 0, \bar{t_0}) = \sqrt{\bar{t_0}}\left[\frac{1}{\sqrt{\pi}} - \frac{1}{\sqrt{\pi}} e^{-1/\bar{t_0}} + \frac{1}{\sqrt{\bar{t_0}}} \operatorname{erfc}\left(\frac{1}{\sqrt{\bar{t_0}}}\right)\right] = \frac{1}{\kappa} \tag{13.14}$$

As soon as T_M is reached, material removal occurs for low-depth machining. The MRR is governed by the temporal rate of propagation of T (z, t) = T_M through the material of the job. This propagation rate is numerically same as the slope of the curve $\left(\overline{T}\left(0,\overline{t_0}\right)=\frac{1}{\kappa}\right)$. Equation (13.15) is thus obtained, by applying the theorem of implicit functions:

$$\frac{d\overline{z}\left(\overline{t_0}\right)}{d\overline{t}}=\frac{1}{2\sqrt{\pi}}\frac{1-e^{-1/\overline{t_0}}}{\sqrt{\overline{t_0}}} \tag{13.15}$$

Equation (13.14) and Eq. (13.15) together yield the estimate of *rate of material removal* (or MRR). Two limiting situations are discussed here.

For large power dissipation $\left(\overline{t_0}\rightarrow 0,\kappa\gg 1\right)$ the MRR is approximated as:

$$\frac{d\overline{z}(\kappa)}{d\overline{t}}\cong\frac{\kappa}{2\pi}\left(1-e^{-\kappa^2/\pi}\right) \tag{13.16}$$

$$\overline{t_0}\cong\frac{\pi}{\kappa^2} \tag{13.17}$$

On the other hand, for small values of power dissipation $\left(\kappa\approx 1,\overline{t_0}\gg 1\right)$, the material removal rate (as quantified in Eq. (13.15)) is estimated by:

$$\frac{d\overline{z}(\kappa)}{d\overline{t}}\cong\frac{\kappa-1}{2\kappa}\left[1-\exp\left(\frac{-\pi(\kappa-1)^2}{\kappa^2}\right)\right] \tag{13.18}$$

$$\overline{to}\cong\frac{1}{\pi}\left(\frac{\kappa}{\kappa-1}\right)^2 \tag{13.19}$$

The value of κ must be estimated as a function of cell terminal voltage. A qualitative approximation can be obtained by assuming that each discharge imparts an identical thermal flux (= q_E) to the job. The power (P_0) is equated to the average number of discharges by means of the following equation:

$$P_0=\frac{d}{dt}[\langle N(t)\rangle\cdot q_E]=\lambda_d A q_E (\text{A: Surface on which discharges occur}) \tag{13.20}$$

Discharges usually occur on the lateral surface of the tool-electrode. It follows that:

$$\kappa = \frac{\lambda_d \pi b h_d q_E}{\lambda \pi b (T_M - T_0)} = \lambda_d \left\{ \frac{h_d q_E}{\lambda (T_M - T_0)} \right\}; (A \sim b h_d) \qquad (13.21)$$

h_d: Height at which discharges are visible
Equation (13.21) provides a rough estimate for κ. It shows that $\kappa \propto \lambda_d$ (λ_d: average number of discharges per unit time, a function of terminal voltage U).

(iii) SACE application in glass micromachining
Review of published studies presents the case of a circular heat source for glass μ-machining application [36]. This experimental study showcases machined glass parts at a specified machine voltage (29 V) in 50 wt% KOH, for different durations of machining. The cylindrical tool used has a diameter of 500 μm. The initial gap between the tool and the workpiece is a couple of microns, with the former held at a constant height. Cylindrical rings are machined for machining of short durations (less than 500 ms), which then develop as a perfect circular disc on the job surface. Comparison of ring diameters with different values of isotherms, as obtained from numerical solution of Eq. (13.4) (together with the boundary conditions (2)), indicates that the glass machining temperature (T_M) should be ≈ 185°C. The power (P_0) was then estimated from Eq. (13.1), with ring thickness taken as e = 10 μm.

According to estimate provided in Eq. (13.7), glass μ-machining requires low thermal power (less than ~1 W). Thus, Eq. (13.21) and Eq. (13.16) (written in dimensional form) can be used, respectively, to estimate the time required (= t_0) to attain T_M and to quantify the processing speed (= dz (κ)/dt):

$$t_0 = \frac{b^2}{4a} \left[\frac{P_0}{P_0 - \lambda \pi b (T - T_M)} \right]^2 \qquad (13.22)$$

$$\frac{dz(\kappa)}{dt} = \frac{4a}{b} \frac{\kappa - 1}{2\kappa} \left[1 - \exp\left(\frac{-\pi(\kappa - 1)^2}{\kappa^2} \right) \right] \qquad (13.23)$$

The discharge activity is observed to be lower with an active anode. Due to different discharge mechanisms, a wide range of machined forms and surface roughness (with sometimes even higher values of MRR) are observed for glass and quartz μ-machining [8, 25, 36]. In case of active anode, as pointed out earlier, the Hickling and Ingram mechanism produces a large number of OH radicals as a result of (electrochemical) discharges, which additionally contribute to etching of (glass or quartz) workpieces.

(iv) SACE application in ceramic μmachining
Ceramic micromachining requires much higher energy. As predicted by Eq. (13.7), alumina usually requires ~40 W. Although these energies can be reached at high values of voltages (typically ~50 V), large values of κ cannot be obtained (in fact $\kappa \cong 1$). In this case, Eq. (13.23) can be further approximated as follows:

$$\frac{dz(\kappa)}{dt} = \frac{2a}{b}\pi(\kappa - 1)^3 = \frac{2a}{b}\pi(\Delta\varepsilon)^3 \tag{13.24}$$

Δε: Ratio of excess energy supplied, as compared to the minimum heat output ($= P_{min}$) required for processing

As an example for aluminium oxide, assuming a few percent of Δε, one can expect a drilling speed of a few $\mu m s^{-1}$, and this estimate agrees well with experimental observations [37–39].

13.5.3 Micromachining at High Depths

When the electrolyte is not able to reach the machining zone as easily as at shallow depths, SACE μ-processing is termed as *high-depth machining*. This adversely affects material removal from the workpiece, either by melting or by chemical etching. The process is delimited by the ability to transport the electrolyte into the processing zone as well as flush out the machined material. Both of these processes are significantly geometrically-dependent.

As compared to machining at shallower depths, machining at greater depths yields significantly low values of MRR. Therefore, in order to develop an initial idea about the process, one can consider steady temperature distributions in the discharge zone.

Drilling of glass surfaces at high depths is considered as an example in [2]. The geometries of the tool and the (drilled) hole are adequately described. The steady-state heat equation is solved to obtain an initial estimate of the temperature distribution:

$$\nabla^2 T = 0 \tag{13.25}$$

Jalali *et al.* [31] have provided the numerical solution for this equation under various sets of boundary conditions.

The heat flux (= ϕ) of the gas film is approximated by Eq. (13.26) (P_E: electrical energy; A_g: cross-sectional area of gas film around the cylindrical tool electrode; A_e: cross-sectional area of cylindrical tool electrode):

$$\varphi = \frac{P_E}{A_e + A_g} \tag{13.26}$$

Electrochemical discharges can be modeled as constant-flux heat source at the bottom of the tool-electrode. The strength of the power source is estimated by Eq. (13.1). The model takes into account the thermal conductivities of the tool-electrode, the job and the gas film. As verified by calculations, a significant part of heat generated (by discharges) is removed through the tool-electrode. The gas film (homogeneous mixture of H_2 and steam) apparently plays a key role in thermal energy concentration in the vicinity of tool tip [2, 16, 26, 40].

This foregoing discussion suggests that thermal mechanism is not the only mode of material removal. In fact, the temperature at the bottom surface of the tool-electrode is insufficient to cause melting at greater depths. Possibly, machining is aided by etching due to small amounts of electrolyte that creeps into the machining zone.

As elaborated in [2], the temperature below the circumference of the tool exceeds that of the job beneath the bottom surface of the tool-electrode (the temperature difference exceeds 100°C). Consequently, the depth of SACE machining is more in vicinity of the edge of the tool-electrode.

At the edge of the tool-electrode, the heat near the gas film melts glass and sets up a crater. At large drilling depths, the molten electrolyte reaches this machining zone and assists chemical reactions to occur. This causes electrochemical erosion of the viscous glass material. The etching gradually proceeds towards the zone where the electrolyte progressively penetrates. The tool electrode move deeper into the workpiece as the underlying material gets removed. The contact force is a drag or shear force, and arises due to the penetration of tool into the viscous region [41].

13.5.4 Micromachining by Chemical Reaction

The effect of electrolyte on machining efficiency has been a focus of study, ever since the first set of published studies on μ-machining by electrochemical discharge. A benchmark study on the influence of parameters such as machining voltage, electrolytic properties, and others, on MRR has been conducted in [8]. In particular, the experimental study describes how the MRR enhances with the machining voltage, electrolyte concentration, as well as its temperature.

The effect of electrolyte on the machining process is complex, and the temperature and concentration of the electrolyte are not the sole parameters. In particular, the nature of the electrolyte influences machining behavior. Relative to other electrolytes (KOH, NaCl, NaF, $NaNO_3$, HCl, etc.), NaOH presents the most conducive properties [3, 7–9, 18]. Generally, acid electrolytes result in very low values of MRR, as compared to alkaline counterparts. The electrolyte influences the surface roughness of the machined job. In particular, electrolysis of molten salts can significantly improve the surface smoothness of machined parts.

In case of glass μ-machining with NaOH, the piece may be engraved by the following electrochemical reaction:

$$2NaOH + SiO_2 \rightarrow Na_2\,SiO_3 + H_2O \tag{13.27}$$

The $Na_2\,SiO_3$ formed is removed by the flow of electrolyte. The rate of chemical reaction is augmented with increasing temperature. Similar results have also been reported for ceramic materials [25]. Yang *et al.* [19] have defined the process as a high-temperature assisted corrosion process. The experimental study has concluded that SACE μ-machining is most probably a combined thermo-chemical attack on the surfaces of workpieces. Energy-dispersive X-ray analysis of the structures machined on soda lime glass substrates shows a decrease in Na^+ concentration, with an increase in H^+ concentration in the machined region [16, 36].

Didar *et al.* [42] has demonstrated that 2D μ-machining with electrochemical discharges can create (local) changes in surface topology of glass. Nano-indentation tests confirm that the density and hardness of a SACE-machined glass surface get increased. Even during machining, as the tool electrode moves over the glass surface, the surfaces (of machined grooves) are cooled by bulk electrolyte volume that reaches maximum temperatures of only ~80 – 90°C. Experimental tests also confirm that, irrespective of electrolyte concentrations, the density of machined glass increases. Studies confirm, for all electrolyte concentrations, the increase in density is ~20%. For more details, the interested reader is referred to [2].

13.6 SACE μ-Machining Process Control

An outline for controlling electrochemical discharge-assisted μ-machining process is outlined in this introductory treatise. For full information, the interested reader is referred to [2].

There has not been sizeable published literature on active control of micromachining with electrochemical discharges. There are two key problems that account for this trend. The primary challenge lies in the fact that the process cannot be vividly observed. It is still unclear how real machining states and/or parameters may be determined with limited amount of extractable data. Another challenge is to ascertain the optimum route for machining procedure. For instance, altering the machining voltage (= U) is observed to have limited impact on machining procedure. In addition, local parameters such as chemical composition or temperature are not solely dependent on U.

Given the challenges, a number of passive control techniques have been proposed and implemented. Some of these strategies, as described in this section, work well and produce outcomes that are quite promising. Using appropriate electrode shapes and motions, the first set of (control) strategies aims to promote electrolyte flow. The second group of control schemes aim to achieve localized heat generation. The most sophisticated methods use pulsed voltage processing, which combines optimization of heat generation and promotion of local sputtering through electrochemical discharges.

13.6.1 Analysis of Process

SACE assisted μ-machining involves mix of softening, melting, and etching of a substrate. Relative contributions of the respective processes depend on the workpiece material. Locally, the machining zone is heated up due to electrochemical discharges. Softening/melting and/or etching aids in material removal.

In order to optimize machining by electrochemical discharges, the aspects of stable discharge activity, localized heating, etching, and mass transport must be effectively monitored. On one hand, maximizing the etching yields the best surface quality; on the other, concurrent reduction of local heating prevents the creation of heat-affected zones. Satisfying both the above criteria requires optimal electrolyte supply to the machining zone.

The technological significance of the chemical contribution and, by natural extension, the implications of the local electrolyte flow, has only lately been realized. Only a few strategies have been tested and validated so far for SACE, as compared to other machining technologies such as electrical discharge machining (EDM) or electrochemical machining (ECM). Although the latter techniques meet similar concerns, a number of strategies have been developed to address these challenges. In addition, SACE machining presents a major difficulty in active control of separation distance between the tool-electrode and the workpiece (called the *machining gap*). This limitation makes it difficult to implement control mechanisms (developed in EDM and ECM) to modulate local flows. The influence of the discharges also presents a notable distinction. Each discharge in EDM locally melts the workpiece; keeping track on the current pulses thereby makes it possible to regulate machining rates (in EDM). In addition, as anodic erosion (of workpiece) serves as the material removal mechanism in ECM, μ-machining can be directly controlled by the current. Electrochemical discharges in SACE machining provide the heat required to enhance local etching; however, the impact of a single electrochemical discharge is yet to be clearly demonstrated. Significant obstacles to create effect control techniques for SACE machining stems out of this knowledge gap. The force applied to the tool-electrode during SACE drilling has recently been attempted to function as a control signal [2].

13.6.2 Etch Promotion

Etch promotion during machining is achieved by optimizing the local electrolyte supply to the machining area. This aspect is especially important during drilling, where the deeper the hole, the more difficult it is for the electrolyte to flow into the processing area. It can be noted that the quality of microcircuits deteriorates during drilling in hydrodynamic mode.

The supply of electrolytes to the processing area can be increased by using electrolytes of low viscosity (e.g. KOH). Selection of appropriate tool-electrode kinetics, tool-electrode shapes, or electrolyte flow also promote better etching rates. These strategies are briefly discussed below.

(i) Shape of tool-electrode
Published studies indicate that the electrode of the needle-shaped tool significantly increases the drilling speed at low values of voltage (typically less than ~30 V), compared to drilling with a cylindrical tool, [2]. However, for both the tools, drilling speeds increase at higher voltages. This follows from the fact that with the tool-electrode of needle-shaped geometry, the

discharges are concentrated at its tip. The discharge density is higher for smaller sections, resulting in higher values of MRR (compare Eq. (13.12) and Eq. (13.23)). This effect is significant in the discharge zone (up to a depth of ~200 – 300 mm) at low voltage. The drill speed becomes almost independent of voltage after reaching the hydrodynamic regime at high voltages. When compared with the tool-electrode of cylindrical shape, the average speed of drilling for a needle-shaped tool in the hydrodynamic mode is ~10 times higher.

Different shapes of tool electrodes have different effects. The concentrated discharge at the electrode tip for needle-shaped tool electrodes leads to an increase of the discharge regime. It is possible to enhance the discharge zone by choosing a tool electrode geometry that facilitates electrolyte flow within the microholes [43, 44]. Further improvements can be achieved by (a) increasing the rotation speed of the tool electrode, or (b) using pulsed voltage treatment. The effect of augmented flushing is observed in current signals. When drilling takes place at shallow depths, and voltages close to the transient voltage are applied, the current signal predominantly presents discharge spikes. As the drilling depth is increased, the signal indicates more gas film formation, which indicates that the discharge activity switches from the arc state to the unstable state. From ~300 mm depth onward, machining with a cylindrical tool initiates instability in the gas film (in 6 M KOH at 38 V).

Yang *et al.* [18, 19] have investigated analogous tool geometries in gravity-feed drilling of quartz work pieces. For this study, the research group used a tool with 150 mm diameter spherical end attached to a 10 – mm cylindrical body. For quartz drilling in 5M KOH at 40 V, as reviewed in [2], the problems associated with the hydrodynamic regime can mostly be resolved. Drilling time is significantly reduced compared to using pure cylindrical tools. This also reduces the inlet diameter. This effect is attributed to the shape of the tool. In spherical tools, the tool body is significantly thinner than the tool edges, and the μ-hole inlet walls are much less heated than in cylindrical tools. This is detailed in treatise [2], referred to the interested reader.

(ii) Vibration of tool electrode

Electrolyte flow inside the holes during μ-drilling can be induced by appropriate vibration modes of the tool. This processing strategy is sometimes called *tapping mode machining* [2].

An example of the effect of tool-electrode vibration is discussed in [2, 3]. The amplitude and frequency (of vibration) were respectively varied between 0-30 μm and 0-30 Hz.

For each mode of tool oscillation, 15 μ-holes were drilled at 33 V. Drilling is mainly done in discharge zone up to a depth of ~100 μm; the hydrodynamic modes trigger beyond ~200 μm.

In the frequency range (0 – 30Hz) and amplitude range (0 – 30 μm) of tool vibrations used in the above study, a clear dependence on the vibration amplitude can be observed; the effect of frequency is secondary.

The nature of variation of the average drilling time versus vibration amplitude, as reported in [2], is due to relative weightage between the two aforementioned effects. Increasing the amplitude of vibration favours enhancement in the drilling speed, promoting electrolyte-flow inside the μ-hole and also removal of material from machining site. However, the higher the amplitude of vibration, the farther the heat source is (locally) from the surface of the part. It should be noted that if the distance between the tool and the job exceeds 20 to 25 μm, the μ-machining of glass ceases to take place. This is why increasing the tool vibration beyond an amplitude of 10 μm makes no significant improvements in the drilling speed.

An alternative method to constant vibration of the tool is vertical oscillatory motion of the tool-electrode at specific moments (during μ-drilling) so that fresh electrolyte can flow into the hole. A possible algorithm could be designed on measuring the contact force between the job and the tool-electrode. The tool-electrode is pre-set to move up and flush the μ-hole, if this force is greater than some threshold level. However, this strategy does not enhance the drilling times. With larger gaps, the effect proves detrimental as the surface of the workpiece is no longer heated up effectively. A better strategy is creating a gap and vibrating the tool-electrode simultaneously (as the gap is created), to flush the machining zone. The vibration amplitude must slightly exceed the formed gap, so that they can periodically touch the workpiece and heat it effectively [2].

(iii) Rotation of tool electrode

Tool-electrode rotation also promotes flow of electrolyte. An example of tool-electrode rotation induced gravity-feed μ-drilling is presented in [45]. Two different regimes can be identified for a tool electrode of cylindrical symmetry. At low rotation speeds (not exceeding ~500 rpm), rotation of tool promotes μ-drilling. As rotational speeds increase, the effect proves detrimental. The machining zone is likely to cool down as the hot electrolyte is pushed away from the site. However, this effect is not observed with a flat sidewall tool-electrode fed with pulsed voltage; the interested reader is referred to [2] for greater details.

The MRR and cylindrical shape of the machined holes can also be improved by rotating the tool electrode. The effects of tool rotation speed is briefly touched upon. In the case of slow rotation (not exceeding ~25 rpm), machining efficiency increases. At rotational speeds exceeding 25 rpm, machining performance decreases and worsens compared to a non-rotary tool. This observation can be attributed to the instability of electrochemical discharge as the speed increases. Noticeable variations in efficiency are observed when eccentric rotation is employed for the tool; for a full discussion, the reader is referred to [2].

(iv) Motion of electrolyte

As viable alternatives to promote flushing of the machining site, appropriate kinetics can be imparted to the tool-electrode. To achieve this, different energy sources can be employed to trigger motion of the electrolyte. Two approaches have been proposed and investigated thus far: promotion of flow inside the μ-hole by means of MHD effects, and the use of acoustic waves.

By applying an external magnetic field near the tool electrode by using a permanent magnet, Cheng *et al.* [46–48] has demonstrated that the discharge regime over the substrate surface could be expanded.

Localized flow of electrolyte inside the μ-holes must be promoted by acoustic waves, and ultrasonic waves prove suitable due to their suitability to propagate through obstacles [49],

(v) Electrolyte with dispersed abrasives

The addition of an abrasive material to the bulk electrolyte does not in itself promote local etching. However, etching can be enhanced in conjunction with rotation or vibration of tool-electrode. In this way, the machining quality is improved by reducing the surface roughness [18, 19, 50]. As mentioned above, during the μ-machining process, addition of abrasives can help enhance heat transport rates.

13.7 Conclusion and Scope for Future Work

So far, μmachining aided by electrochemical discharges (or SACE) has been discussed as a novel technique for various materials that are hard to machine. Knowledge of glass μ-machining has reached sufficiently high levels of attainment to enable interesting engineering applications. Although some possible process-control strategies have recently been explored, there is still much to do in this field.

Also, there is much to investigate in exploiting the behavior of electrochemical discharges. Only the thermal energy produced by discharges has been effectively used. But more applications can be harnessed; for instance, electrochemical discharge transport electrons and reactions can be carried out at targeted sites for tailored applications, as in contact glow-discharge electrolysis. Localized surface modifications of machined surfaces offer just one of many possible applications.

The mechanism of particle growth is not yet clearly understood. However, the particles are thought to form at the interface between the gas film and the electrolyte. The size is possibly determined by the concentration of metal ions near the gas film and the amount of charge carried by each discharge. But, it is quite unlikely that formation of a large particle (~300 nm) happens in one step. Nucleation discharges, progressive over time, that reduce metal ions could be a probable growth mechanism. Some of these ions (born out of this discharge) form stable nuclei, and they continue to add atoms during successive discharges to grow in size. Another suggested mechanism of particle growth is reduction of metal ions in an atmosphere of hydrogen gas film. When the nanoparticles attain a certain critical mass, they stop growing any further and leave the reaction zone.

The potential of electrochemical discharges in μ-machining is immense. Full exploitation of this enormous potential is still quite far. Electrochemical discharges offer dual advantages of supplying thermal energy and highly energized free electrons over a localized zone. Synergistic combination of these effects and judicious use can open up novel and exciting avenues of research and applications.

References

1. Wolffenbuttel, R.F., *Silicon Sensors and Circuits: On-Chip Compatibility, Sensor Physics and Technology Series*, Chapman & Hall, London, 1995.
2. Wüthrich, R. and Abou Ziki, J.D., *Micromachining Using Electrochemical Discharge Phenomenon: Fundamentals and Application of Spark Assisted Chemical Engraving*, William Andrew, Elsevier, Norwich, NY, 2015.
3. Wüthrich, R. and Fascio, V., Machining of non-conductive materials using electrochemical discharge phenomenon-an overview. *Int. J. Mach. Tools Manuf.*, 45, 1095–1108, 2005.
4. Fascio, V., Wüthrich, R., Bleuler, H., Spark assisted chemical engraving in the light of electrochemistry. *Electrochim. Acta*, 49, 3997, 2004.
5. Masuzawa, T., Fujino, M., Kobayashi, K., Suzuki, T., Wire electro-discharge grinding for micro-machining. *CIRP Ann.*, 34, 431–434, 1985.
6. Fascio, V., Langen, H.H., Bleuler, H., Comninellis, C., Investigation of the spark assisted chemical engraving. *Electrochem. Commun.*, 5, 203–207, 2003.

7. Kurafuji, H. and Suda, K., Electrical discharge drilling of glass. *CIRP Ann.*, 16, 415–419, 1968.
8. Cook, N.H., Foote, G.B., Jordan, P., Kalyani, B.N., Experimental studies in electro-machining. *Trans. ASME, J. Eng. Ind.*, 95, 945–950, 1973.
9. Tsuchiya, H., Inoue, T., Miyazaiki, M., Wire electro-chemical discharge machining of glasses and ceramics. *Bull. Japan Soc Precis. Eng.*, 19, 73, 1985.
10. Jain, V.K., Rao, P.S., Choudhury, S.K., Rajurkar, K.P., Experimental investigations into travelling wire electrochemical spark machining (TW-ECSM) of composites. *Trans. ASME, J. Eng. Ind.*, 113, 75, 1991.
11. Jain, V.K., Dixit, P.M., Pandey, P.M., On the analysis of the electrochemical spark machining process. *Int. J. Mach. Tools Manuf.*, 39, 165, 1999.
12. Jain, V.K. and Chak, S.K., Electrochemical spark trepanning of alumina and quartz. *Mach. Sci. Technol.*, 4, 277–290, 2000.
13. Esashi, M., Matsumoto, Y., Shoji, S., Absolute pressure sensors by air-tight electrical feedthrough structure. *Sens. Actuators A Phys.*, A21–A23, 1048–1052, 1990.
14. Allesu, K., *Electrochemical Discharge Phenomenon in Manufacturing Processes*, Ph.D. Thesis Indian Institute of Technology, Kanpur, 1988.
15. Allesu, K., Ghosh, A., Muju, M.K., Preliminary qualitative approach of a proposed mechanism of material removal in electrical machining of glass. *Eur. J. Mech. Eng.*, 36, 202, 1992.
16. Basak, I. and Ghosh, A., Mechanism of spark generation during electrochemical discharge machining: A theoretical model and experimental verification. *J. Mater. Process. Technol.*, 62, 46–53, 1996.
17. Ghosh, A., Electrochemical discharge machining: principle and possibilities. *Sādhanā*, 22, 435–447, 1997.
18. Yang, C.T., Ho, S.S., Yan, B.H., Micro hole machining of borosilicate glass through electrochemical discharge machining (ECDM). *Key Eng. Mater.*, 196, 149–166, 2001.
19. Yang, C.T., Song, S.L., Yan, B.H., Huang, F.Y., Improving machining performance of wire electrochemical discharge machining by adding SiC abrasive to electrolyte. *Int. J. Mach. Tools Manuf.*, 46, 2044–2050, 2006.
20. Wüthrich, R. and Allagui, A., Building micro and nanosystems with electrochemical discharges. *Electrochim. Acta*, 55, 8189–8196, 2010.
21. Wüthrich, R., Comninellis, C., Bleuler, H., Bubble evolution on vertical electrodes under extreme current densities. *Electrochim. Acta*, 50, 5242, 2005.
22. Wüthrich, R., Hof, L.A., Lal, A., Fujisaki, K., Bleuler, H., Mandin, P., Picard, G., Physical principles and miniaturization of spark assisted chemical engraving (SACE). *J. Micromech. Microeng.*, 15, 268–275, 2005.
23. Wüthrich, R., Fujisaki, K., Couthy, P., Hof, L.A., Bleuler, H., Spark assisted chemical engraving (SACE) in microfactory. *J. Micromech. Microeng.*, 15, 276–280, 2005.

24. Bhattacharyya, B., Doloi, B.N., Sorkhel, S.K., Experimental investigations into electrochemical discharge machining (ECDM) of non-conductive ceramic materials. *J. Mater. Process. Technol.*, 95, 145, 1999.
25. Jain, V.K. and Adhikary, S., On the mechanism of material removal in electrochemical spark machining of quartz under different polarity conditions. *J. Mater. Process. Technol.*, 200, 460–470, 2008.
26. Basak, I., *Electrochemical Discharge Machining Mechanism and a Scheme for Enhancing Material Removal Capacity*, Ph.D. Thesis Indian Institute of Technology, Kanpur, 1991.
27. Kulkarni, A., Sharan, R., Lal, G.K., An experimental study of discharge mechanism in electrochemical discharge machining. *Int. J. Mach. Tools Manuf.*, 42, 1121–1127, 2002.
28. Gautam, N. and Jain, V.K., Experimental investigations into ECSD process using various tool kinematics. *Int. J. Mach. Tools Manuf.*, 38, 15, 1998.
29. Basak, I. and Ghosh, A., Mechanism of material removal in electrochemical discharge machining: A theoretical model and experimental verification. *J. Mater. Process. Technol.*, 71, 350, 1997.
30. Jiang, B., Lan, S., Ni, J., Zhang, Z., Experimental investigation of spark generation in electrochemical discharge machining of non-conducting materials. *J. Mater. Process. Technol.*, 214, 892–898, 2014.
31. Jalali, M., Maillard, P., Mishra, S.P., Kadem, L., Wüthrich, R., Modeling gravity feed micro hole drilling by electrochemical discharge, in: *2nd International Conference on Micro-Manufacturing*, Greenville, South Carolina, USA, September 10–13, 2007.
32. Abou Ziki, J.D., *Spark Assisted Chemical Engraving: A Novel Approach for Quantifying the Machining Zone's Parameters Using Drilling Forces*, Ph.D. Thesis, Concordia University, 2014.
33. Abou Ziki, J.D. and Wüthrich, R., Tool wear and tool thermal expansion during micro-machining by spark assisted chemical engraving. *Int. J. Adv. Manuf. Technol.*, 61, 481–486, 2012.
34. Abou Ziki, J.D., Didar, T.F., Wüthrich, R., Micro-texturing channel surfaces on glass with spark assisted chemical engraving. *Int. J. Mach. Tools Manuf.*, 57, 66–72, 2012.
35. Abou Ziki, J.D. and Wüthrich, R., Forces exerted on the tool-electrode during constant-feed glass micro-drilling by spark assisted chemical engraving. *Int. J. Mach. Tools Manuf.*, 73, 47–54, 2013.
36. West, J. and Jadhav, A., ECDM methods for fluidic interfacing through thin glass substrates and the formation of spherical microcavities. *J. Micromech. Microeng.*, 17, 403–409, 2007.
37. Chark, S.K. and Rao, P.V., Trepanning of A12O3 by electro-chemical discharge machining (ECDM) process using abrasive electrode with pulsed DC supply. *Int. J. Mach. Tools Manuf.*, 47, 2061–2070, 2007.

38. Sarkar, B.R., Doloi, B., Bhattacharyya, B., Parametric analysis on electrochemical discharge machining of silicon nitride ceramics. *Int. J. Adv. Manuf. Technol.*, 28, 873–881, 2006.
39. Tsutsumi, C., Okano, K., Suto, T., High quality machining of ceramics. *J. Mater. Process. Technol.*, 37, 639, 1993.
40. Kellogg, H.H., Anode effect in aqueous electrolysis. *J. Electrochem. Soc.*, 97, 133, 1950.
41. Tokura, H., Kondoh, I., Yoshikswa, M., Ceramic material processing by electrical discharge in electrolyte. *J. Mater. Sci.*, 24, 991, 1989.
42. Didar, T.F., Dolatabadi, A., Wüthrich, R., Characterization and modeling of 2D-glass micro-machining by spark-assisted chemical engraving (SACE) with constant velocity. *J. Micromech. Microeng.*, 18, 5016, 2008.
43. Zheng, Z.-P., Su, H.-C., Huang, F.-Y., Yan, B.-H., The tool geometrical shape and pulse-off time of pulse voltage effects in a Pyrex glass electrochemical discharge microdrilling process. *J. Micromech. Microeng.*, 17, 265, 2007.
44. Zheng, Z.-P., Cheng, W.-H., Huang, F.-Y., Yan, B.-H., 3D microstructuring of Pyrex glass using the electrochemical discharge machining process. *J. Micromech. Microeng.*, 17, 960, 2007.
45. Han, M.-S., Min, B.-K., Jo Lee, S., Geometric improvement of electrochemical discharge micro-drilling using an ultrasonic-vibrated electrolyte. *J. Micromech. Microeng.*, 19, 065004, 2009.
46. Cheng, C.-P., Wu, K.-L., Mai, C.-C., Yang, C.-K., Hsu, Y.-S., Yan, B.-H., Study of gas film quality in electrochemical discharge machining. *Int. J. Mach. Tools Manuf.*, 50, 689–697, 2010.
47. Cheng, C.P., Wu, K.L., Mai, C.C., Hsu, Y.S., Yan, B.H., Magnetic field-assisted electrochemical discharge machining. *J. Micromech. Microeng.*, 20, 075019, 2010.
48. Cheng, C.-P., Wu, K.-L., Mai, C.-C., Yang, C.-K., Hsu, Y.-S., Yan, B.-H., Study of gas film quality in electrochemical discharge machining. *Int. J. Mach. Tools Manuf.*, 50, 689–697, 2010.
49. Han, M.-S., Min, B.-K., Jo Lee, S., Micro-electrochemical discharge cutting of glass using a surface textured tool. *CIRP J. Manuf. Sci. Technol.*, 4, 362–369, 2011.
50. Yang, C.-K., Cheng, C.-P., Mai, C.-C., Wang, A.C., Hung, J.-C., Yan, B.-H., Effect of surface roughness of tool electrode materials in ECDM performance. *Int. J. Mach. Tools Manuf.*, 50, 1088–1096, 2010.

Index

Printed and bound by CPI Group (UK) Ltd, Croydon, CR0 4YY
25/06/2023
03229955-0003